ELOQUENT SCIENCE

A Practical Guide to Becoming a Better
Writer, Speaker, and Atmospheric Scientist

DAVID M. SCHULTZ

American Meteorological Society

Published by the American Meteorological Society
45 Beacon Street, Boston, Massachusetts 02108

For more AMS Books, see www.ametsoc.org/amsbookstore. Order online or call (617) 227-2426, extension 686.

Library of Congress Cataloging-in-Publication Data

Schultz, David M., 1965-
  Eloquent science : a practical guide to becoming a better writer, speaker, and atmospheric scientist / David M. Schultz.
    p. cm.
  ISBN 978-1-878220-91-2 (pbk.)
  1. Communication in science. 2. Technical writing. 3. Public speaking.
4. Scientists—Vocational guidance. I. Title.
  Q223.S23538 2009
  808'.0665—dc22

2009039865

⊕ Printed in the United States of America by Hamilton Printing Company. This book is printed on FSC-certified, recycled paper with a minimum of 10% post-consumer waste.

CONTENTS

PART IV : COMMUNICATING THROUGHOUT YOUR CAREER

PART V : APPENDICES

PREFACE

Mary Grace Soccio. My writing could not please this kindhearted woman, no matter how hard I tried.

Although Gifted and Talented seventh-grade math posed no problem for me, the same was not true for Mrs. Soccio's English class. I was frustrated that my first assignment only netted me a C. I worked harder, making revision after revision, a concept I had never really put much faith in before. At last, I produced an essay that seemed the apex of what I was capable of writing. Although the topic of that essay is now lost to my memory, the grade I received was not: a B–.

"The best I could do was a B–?" The realization sank in that maybe I was not such a good writer.

In those days, my youthful hubris did not understand about *capacity building*. In other words, being challenged would result in my intellectual growth—an academic restatement of Nietzsche's "What does not destroy me, makes me stronger." Consequently, I asked to be withdrawn from Gifted and Talented English in the eighth grade.

Another capacity-building experience happened when I was a postdoctoral research fellow. In writing the journal article that resulted from my Ph.D. thesis, one of my coadvisors, Dan Keyser, and I discussed revisions by phone while he lived in upstate New York and I in Oklahoma. My schooling was severe: fifteen one-hour-long phone calls where we would go through the draft together—one section at a time, sentence by sentence. Not all of Dan's lessons I embraced immediately, however. Sometimes we were frustrated by each others' stubbornness: me by his insistence to do things his way and he by my resistance to learning. Finally, something snapped inside and clarity came: I understood what he was trying to tell me about transition, coherence, and precision, and it made complete sense. Subsequent revisions went much more smoothly, and the manuscript made it easily through the review process and was published. Wherever that revelation came from, *Eloquent Science* would not have happened without that moment.

Throughout my career, mentoring by Dan, my other advisors, and my colleagues was essential to my development as a scientist and a writer. Unfortunately, not everyone has the benefit of such mentoring. The good news is that being a better writer, whether a student or a scientist with years of experience, does not require a revelation, merely an open mind. As I hope to convince you in this book, the essential skills can be taught. Moreover, it's not just the young dogs who can be taught new tricks. Everyone, no matter how experienced, can learn new skills to improve their writing.

Eloquent Science is an outgrowth of a scientific communication workshop I developed for the National Science Foundation–funded Research Experiences for Undergraduates program that the Oklahoma Weather Center (and its members the National Severe Storms Laboratory, the University of Oklahoma, the Cooperative Institute for Mesoscale Meteorological Studies, and Center for the Analysis and Prediction of Storms) hosted from 1998 to 2005, and has continued from 2007 to this writing. After seeing that we were not educating our students about how to write a scientific paper and make a scientific presentation, I created and led this workshop during 2000–2005. The workshop began as a collection of thoughts on a Web site, turned into an afternoon lecture, and became an eight-hour interactive workshop where students learned to critique their own and their peers' writing. I argued that these undergraduates would be my future colleagues, and I would likely be reviewing their papers and attending their seminars. Besides my desire to see them create excellent scientific work and present it effectively, I realized that if I could influence them not to write a bad paper or make a bad presentation in the future, I could be saving myself some subsequent heartaches.

As I developed the workshop from year to year, the organic approach took its toll. My slides, with new insertions each year, were characterized at best as verbose lecture notes rather than a clear and effective presentation. Also inadequate was the poorly organized collection of articles and handouts serving as a reference guide. Neither were even adequate examples of the instruction I was trying to give. The idea for turning the lectures into a book struck in summer 2005 while at a conference, frustrated by the pathetic presentations I was enduring. A book would solve both my problems, I thought. It would create a more effective vehicle to deliver the information on paper and free me to focus on improving the style of the presentations. An added benefit, I wishfully dreamt, might be to distribute this book to other atmospheric scientists to ease the kind of pain I experienced at that conference.

Writing a book about communicating effectively to a scientific audience is like speaking to an audience at a classical music concert about how to play a violin as a virtuoso would. Although some in the audience will learn quite a bit and benefit immediately, more experienced violinists need only specific advice to improve. Moreover, future performances by the speaker will be intensively scrutinized. As with that speaker, I fear that my words will come back to haunt me in the future. (I can already hear readers raising questions about my previous publications!) In my defense, few writers alive today believe that their previous work is impervious to revisions. And we should not expect perfection, either. In fact, many examples in *Eloquent Science* derive from my own writings and presentations: not only the best examples, but the imperfect, as well. For my future writing efforts, I can only plead forgiveness for a limited brain capacity to store and recall the abundant information contained within this book.

If you have any comments about the material in this book, I would appreciate hearing from you: eloquentscience@gmail.com.

ACKNOWLEDGMENTS

I am grateful to many people for their help in teaching me how to write. My parents supported me in my early development, bought me my first typewriter for college, and assisted me while I earned my Ph.D. My M.S. thesis advisor Cliff Mass is another who deserves praise. My 315-page M.S. thesis, although an unwieldy compilation of nearly everything I learned about occluding cyclones, was my first major lesson in managing a book-length manuscript. The result is that I improved somewhat with my Ph.D. dissertation, constraining the length to 198 pages. My Ph.D. advisors Lance Bosart and Dan Keyser were responsible for helping me further hone my writing and speaking skills. Lance and Dan would reign over rehearsals for presentations we students would give at national conferences, until we got it right. Dan was of particular help in the many hours he spent with me on the phone between Norman and Albany as we finalized the manuscript that arose from that Ph.D. dissertation. The writing process that Dan opened my eyes to was a turning point in my educational experience. Finally, in my National Research Council postdoctoral fellowship at the NOAA/National Severe Storms Laboratory, Chuck Doswell poked, prodded, and peeved me into further refining my writing style or defending why I chose to be different. The process of the two of us writing the *Guide for Authors, Reviewers, and Editors* for the *Electronic Journal of Severe Storms Meteorology*, which we helped cofound with other meteorologists, influenced several aspects of this book, as did the material from his Web pages. Furthermore, his extensive critiques of several chapters have made that material immensely stronger.

Because *Eloquent Science* is derived from the Research Experiences for Undergraduates program in Oklahoma, Director Daphne LaDue made the foundations of this book possible. Her insight into and support of undergraduate education and good communication skills makes her an extraordinary resource for our community. Stephan Nelson at the National Science Foundation provided the financial support to the program. Most importantly, I value the dozens of students who have endured the years I grew the workshop and provided feedback to improve the workshop. Much of the book was also test-driven in my Communication Skills for Scientists class at the University of Helsinki and at numerous workshops and conferences in North America and Europe. I appreciate those students' contributions to the material in this book, as well.

Colleagues that I have worked with have cajoled me to improve my scientific communication skills, and I thank them, especially Jim Steenburgh, Paul Roebber, John Knox, George Bryan and Fred Sanders. Some of the ideas about publishing papers (Section 2.1) and the review process come from C. David Whiteman. Some aspects of overcoming writer's block in Chapter 5 come from a writing workshop led by Norman author Darlene Graham and sponsored by the Norman Arts Council. Andy White has lectured in my classes previously about the principles of graph construction—some of his ideas were used in this book. Mary Golden, the chief editorial assistant extraordinaire of *Monthly Weather Review*, has been a great supporter of this book and teaching communication skills for English as a Second Language scientists. She provided numerous suggestions for the book, as well as extensive editorial support. I have been very proud of my association with *Monthly Weather Review*, which maintains rapid times for manuscript decisions (the best in the American Meteorological Society, as of this writing) without sacrificing high standards. David Jorgensen set the bar high and has served as a tremendous inspiration to me during his role as one of the chief editors. If I have inadvertently borrowed the thoughts of any of these people without attribution in this book, it is because I have integrated their lessons so well that I have lost the ability to distinguish their original ideas from my own.

Some people suggested specific material for the book. Paul Roebber offered "An Incoherent Truth" as the title of the book (which was later used for the Introduction, instead). Tracey Holloway suggested the Ask the Experts columns and the expression "manuscript-on-the-wall poster," and Peter Grünberg described to me his story of his Nobel Prize–winning discovery (page 249). Mary Golden interviewed and surveyed authors for the ESL chapter; I thank those individuals for their time and contributions: George Bryan, Huaqing Cai, Jielun Sun, Junhong Wang, and Yafei Wang.

Content for the book was graciously provided by Jelena Andrić, Svetlana Bachmann, Howie Bluestein, Lance Bosart, Chris Davis, Charles Doswell, Dale Durran, Kerry Emanuel, Robert Fovell, Alistair Fraser, Michael Friedman, Robert Marc Friedman, William Gallus, Mary Golden, Eve Gruntfest, Sabine Göke, Tom Hamill, Yvette Hancock, Vesa Hasu, Ken Heideman, Pamela

Heinselman, Bob Henson, Ron Holle, Robert Houze, Daniel Jacob, Jim Johnson, David Jorgensen, Stephanie Kenitzer, Dan Keyser, Petra Klein, Jaakko Kukkonen, Valliappa Lakshmanan, Gil Leppelmeier, Don MacGorman, Bob Maddox, Brian Mapes, Paul Markowski, Olivia Martius, Cliff Mass, Zhiyong Meng, Karen Mohr, Matthew Novak, Keli Pirtle Tarp, Petri Räisänen, Michael Richman, Paul Roebber Richard Rotunno, Elena Saltikoff, Roger Samelson, Chris Samsury, Joe Schall, Russ Schumacher, Alan Shapiro, Jim Steenburgh, David Stensrud, Mark Stoelinga, Neil Stuart, John Thuburn, Jari Tuovinen, Roger Wakimoto, David Whiteman, Johanna Whiteman, Dan Wilks, Warren Wiscombe, Fuqing Zhang, Ed Zipser, and Dusan Zrnić.

The following people made other specific contributions to the content: Harold Brooks, Adam Clark, Mark Fernau, David Gorski, Tracey Holloway, Jason Knievel, Ted Mansell, Michael McIntyre, Richard Orville, David Rust, Brad Smull, Jeff Trapp, Richard Tyson, Earle Williams, and Sandra Yuter.

Specific chapters were reviewed and edited by many qualified people: George Bryan, Li Dong, Charles Doswell, Evgeni Fedorovich, Mary Golden, Pamela Heinselman, Heikki Järvinen, David Jorgensen, Jaakko Kukkonen, Kaijun Liu, Cliff Mass, Zhiyong Meng, Heather Reeves, Miguel Roig, Chris Samsury, Joe Schall, and Warren Wiscombe.

I am honored that the American Meteorological Society gave this book their stamp of approval. I thank my editor Sarah Jane Shangraw for her efforts to allow my vision for this book to blossom with few compromises. I also thank Ken Heideman and Keith Seitter for their management and approval. The book was expertly copy edited by Ellen Goldstein. In addition to the editorial comments of Sarah Jane, Ken, and Ellen, the following AMS staff members also provided their input into the book through their content or comments: Beth Dayton, Mark Fernau, Michael Friedman, Lindsy Gamble, Jocelyn Humelsine, Stephanie Kenitzer, Jessica LaPointe, and Brian Papa.

My wife Yvette Hancock deserves recognition, as well. Her perspective from the theoretical physics community has brought a welcome breadth to the book, and our discussions on communicating science have opened my eyes to other approaches. I appreciate the time she gave from our relationship so that I could write this book.

Eloquent Science was largely written during my time at the University of Helsinki and Finnish Meteorological Institute starting in November 2006. I appreciate their patience with me during the writing process. I also thank the NOAA/National Severe Storms Laboratory and Cooperative Institute for Mesoscale Meteorology at the University of Oklahoma for their prior support (1996–2006) and Vaisala Oyj for their current support.

Finally, I thank Mrs. Mary Grace Soccio, who died while I was writing this book. I had the privilege to write to her a few years ago, letting her know of my journey since leaving seventh grade to becoming a better writer, including writing this book. Although I did not understand it in seventh grade, what she was teaching me was the value of repeated revision. I can assure her, as can my colleagues with whom I have written manuscripts, that I have learned that lesson.

—David M. Schultz
Helsinki, Finland
24 March 2009

FOREWORD

Professor Kerry Emanuel, MIT

Good communication is the lifeblood of science. Much of the thrill of discovery is wrapped up in the anticipation of sharing one's findings, and in this current age of highly collaborative science, discovery itself often involves intricate communication between colleagues. Among the most beautifully written documents in world history are scientific treatises, yet this history is littered with the refuse of virtually unreadable papers, some of which mask important discoveries now credited to other scientists who better knew how to present their findings.

In spite of the critical importance of communication to the scientific enterprise, few graduate students receive formal training in scientific communication. Almost all effort is devoted to developing the art of doing research; students are expected to pick up speaking and writing on their own. In a very real sense, students receive an excellent education in how to write bad papers and give boring presentations, simply because, in the course of their work, they must read dozens of papers many or most of which are badly written, and listen to poorly conceived and delivered talks. By this means, bad scientific writing and speaking perpetuate themselves.

Professional societies often contribute to the problem. The major one I belong to strongly encourages the use of the passive voice, and forbids the use of the active in abstracts. The idea, one supposes, is to convey an air of dispassionate professionalism . . . that dry sense of calculating logic so valued in Victorian doctors and Mr. Spock. We must never insert ourselves into our writing or speaking, lest we be suspected of having any passion for our work. This recipe for dull writing is honored in the breach by the best science writers—scientists like Richard Feynman and Carl Sagan, whose popular books and papers are eagerly read by a science-starved public, sometimes to the tut-tutting disapproval of their fellow scientists, steeped as they are in a culture of bland, dry, and passionless science writing.

Kerry Emanuel is a professor in the Department of Earth, Atmospheric and Planetary Sciences, Massachusetts Institute of Technology. He has written three books: Atmospheric Convection *(1994),* Divine Wind: The History and Science of Hurricanes *(2005), and* What We Know About Climate Change *(2007).*

Some enterprising graduate programs hire communications experts to coach their students in the arts of written and oral communication. While admirable, such efforts can be compromised by a lack of scientific training of the communications professionals, who may have degrees in literature or the arts, and may not understand the need for precision or the use of even rudimentary scientific terminology. Worse, their backgrounds in the humanities may have inculcated in them an active hostility to science, of the kind so well described by C. P. Snow in his "Two Cultures" lecture a half century ago. More than once have I seen such professionals turn moderately good student science essays into rubbish.

The challenge does not only rest with our writing and speaking skills. Even mature scientists well versed in the art of communication can have serious difficulties working with journalists, few of whom have a background in science. It is here, especially, that the clash of Snow's two cultures produces the most disturbing results. The scientist imagines that the reader/viewer shares his enthusiasm for nature, while the journalist assumes that his audience, like him, is bored by science and interested only in personal conflict, misconduct, and politics. Such orthogonal motives do not make for stellar journalism, and scientists are often caught off guard and may come across as wishy-washy, defensive, and/or petty, while the message they wanted to convey has been warped or omitted altogether.

Into this lamentable morass steps David Schultz, a working research scientist and editor of several professional journals, with a keen interest in scientific communication. Here before you is the complete guide to writing a good scientific paper, from the creation of an outline right through to the formalities of submission, review, and proofing. Just as important, Schultz provides invaluable guidance to the preparation and delivery of a scientific talk or poster, including techniques for soliciting and fielding questions, and fostering lively discussion. Finally, Schultz offers tips on the teaching of science, and on how to communicate effectively with the public and the media, avoiding those pitfalls that many have learned the hard way, often at a price to their careers. This book is also laced with advice from a wide spectrum of professional scientists, on subjects ranging from the use of scientific terminology to how to present at a conference. Although aimed specifically at atmospheric scientists, many of the important lessons you will find here are applicable throughout the sciences. So read on, and prepare to absorb what may prove the most valuable advice you will receive as a scientist.

HOW TO USE THIS BOOK

Eloquent Science is written so that students, early career scientists, and senior scientists can improve their communication skills. The book addresses the principal means by which we scientists communicate formally—we participate in the publication process by writing and reviewing scientific papers, and we attend conferences. In Part I, the focus is on writing a scientific document for a class project, conference extended abstract, thesis, or article in a scholarly journal. For brevity, I have not covered all the different types of documents that we might be called upon to write in our career, although the lessons herewithin are clearly relevant to them as well. Part II sheds light on the peer-review process and provides advice on how to participate as a reviewer and an author. Part III focuses on oral and poster presentations at conferences, although your hour-long seminars and speeches to lay audiences will also benefit from this material. Part IV discusses how to communicate outside of the scientific world, either to the public, particularly through the lens of the media, or in a professional setting. This part also contains the last chapter, which closes the book with suggestions on how to improve your skills. Two appendices help readers properly employ select punctuation and scientific terms. Each of the 31 chapters can be read largely independent of each other, so there is no need to read the book sequentially. Experienced scientific (and nonlinear) readers are unlikely to do so anyway.

This book contains four other features you may find useful:

- **Sidebars** highlight important information or discuss tangential topics.
- **Ask the Experts** include contributions from friends and colleagues to provide more than just my perspective.
- **Notes** provide specific citations and elaborate on items discussed in the text.

○ **For Further Reading** is an annotated list of sources of additional information culled from my many hours of research and featuring the best material of which I am aware outside of this book.

The figures, tables, and examples in *Eloquent Science* were derived from one of four sources. First, some of the examples come from American Meteorological Society (AMS) publications. Wherever possible, I tried to get the author's permission for these examples. Second, a few examples come from the public domain. Third, I created some of the other examples specifically for this book to illustrate certain points. Fourth, many examples come from my own writings or those of my coauthors. In some cases, the text or figure was revised to correct bad practices; in other cases, the bad practices were left in to illustrate a point. Although using my own material limits the breadth of the book and prohibits showcasing many other talented writers, it does mean that I can pick more effective material and be uncompromisingly critical of it.

HOW THIS BOOK COMPARES TO OTHERS

Although numerous books on communication skills for scientists have been written, *Eloquent Science* both distinguishes itself from and complements the others. With such a large topic, no single book can address all the issues in a manner appealing to everyone. My approach, therefore, is a practical one. I discuss what I see as the most relevant, topical, and important issues, which clearly may be different from others' opinions. More specifically, other books have not presented, or have done so only cursorily, certain topics that I wanted to emphasize, such as editing your writing, writing reviews for scientific journals, attending conferences, and presenting posters. In addition, because some aspects of formal communication are discipline specific, I draw nearly all of the examples from atmospheric science, even including a chapter on writing for the atmospheric sciences (Chapter 18).

DEFINITIONS

I use a few terms throughout this book that would be best to define here. A *document* refers generically to any number of types of writings that a scientist may produce: thesis, journal article, conference extended abstract, technical memo, etc. A *manuscript* is any unpublished document, whether completed or in draft form. An *article* is a published document in a scientific peer-reviewed journal. A *paper* is a document aimed at a scientific peer-reviewed journal, whether published as an article or not.

CAVEATS

The material in this book is a collection of good-use practices and tips that I have read, researched, or learned for myself. Many ways exist to write a journal article or make a presentation. Not every technique will work for every person or in every circumstance. Some people can deliver humor in their presentations flawlessly. Others should not even try.

Some readers might dispute my recommendations. I have tried to indicate topics where reasonable people can disagree. I would rather make a recommendation and let the reader make a conscious decision to disregard my advice than never to have considered the issue in the first place. Proceeding along the wrong path because "that's the way I was taught" is never an acceptable excuse.

INTRODUCTION: AN INCOHERENT TRUTH

> Too frequently, published papers contain fundamental errors. The presentation in many papers is careless. Some papers abound in unsupported claims stated as facts.

Was this an attack on global warming research by a climate skeptic? No. This quote comes from one of our own. Dr. Ronald Errico, then at the National Center for Atmospheric Research, published an essay in the *Bulletin of the American Meteorological Society* in 2000 that questioned whether we research scientists were being held accountable for our science. He continued, "the unnamed papers . . . are not obscure articles. . . . Both editors and authors have told me that some of these articles have sailed through the review process."

My own experience is similar to Dr. Errico's. Whether I am serving as a voracious reader of the scientific literature, as a reviewer for manuscripts submitted to scientific journals, or as an editor for one of four scientific journals, many papers I read lack sound scientific knowledge, properly constructed arguments, and basic language skills. As an editor, I rely on reviewers to provide recommendations about whether manuscripts should be published or not. Sometimes reviewers provide inadequate criticism of low-quality papers. If editors choose reviewers poorly or make hasty decisions, substandard manuscripts can slip through the review process and be published, officially blessed as The Scientific Truth.

The scourge of shoddy papers has also disturbed the respected fluid dynamicist, founder, and long-time editor of the *Journal of Fluid Mechanics*, G. K. Batchelor. On the 25th anniversary of the founding of his journal in 1981, he wrote a 25-page essay entitled "Preoccupations of a journal editor" in which he indicted such papers:

Papers of poor quality do more than waste printing and publishing resources; they mislead and confuse inexperienced readers, they waste and distract the attention of experienced scientists, and by their existence they lead future authors to be content with second-rate work.

I once saw a professor, someone for whom English is a second language, misspell a word in his presentation: *litterature*. I smiled to myself because he could not have known how often he was correct. Students may be shocked to learn that the quality of many published papers is less than ideal. The literature, or should I say *litter-ature*, does not meet even mediocre standards sometimes.

And the trend is getting worse. Geerts (1999) showed that the clarity of papers in 22 atmospheric science journals was either holding steady or declining. The reasons were the increasing number of words and figures, the increasing length and complexity of the abstract, and the increasing length of the conclusion section owing in part to tangential discussion topics. And these are the papers that survive peer review and get published. Most certainly an inconvenient—and an incoherent—truth!

Fortunately, most of the worst ones get rejected. Indeed, in 2006, the eight scientific journals published by the American Meteorological Society (AMS) rejected 685 manuscripts out of 2353 submissions, or 29%. Rejection rates for individual journals have been relatively constant over time and do not show much spread from this mean, ranging from 19% to 39%. These rejection rates are consistent with the rates from 46 atmospheric science journals, which range from 2% to 68% with a mean of 37%. Thus, more than a third of manuscripts submitted for publication were written by authors who have not demonstrated an ability to communicate effectively or perform high-quality science.

A CAREER COMMUNICATING

Why do we spend so much effort writing articles? Why do we pay as much as $2000 to attend scientific conferences around the world? We do this to communicate our ideas to, and learn from, others about the way nature works. Writing forces us to clarify our own thinking, leading to a much improved understanding. Conferences provide an opportunity for us to get direct feedback on our research and inform others of our results. Publications and conference presentations show funding agencies that their money was well spent, ensuring that they receive credit for their financial commitment. Science could not progress without communication. One of the most veracious statements I have heard is that *we write for our audience, not for ourselves*. This eight-word mantra reminds us *why* we communicate and the importance of doing it well.

Being a successful scientist means being an effective communicator. This may come as a surprise to those scientists with relatively low scores on the verbal components of standardized tests—the very same people who dread public speaking, who just want to be left alone in their offices to do their science. Suppose you had discovered the cure for cancer, but never communicated it to others before you died. Your discovery would be wasted, waiting for someone else to discover it again, perhaps not for decades. That is why senior scientists often write biographies or textbooks, summarizing their lifelong results and preserving their legacy for future generations of scientists to build upon. How unfulfilled the uncommunicated life must be!

Even those in nontraditional career paths need to write and speak well. Students may believe that, if they are not choosing teaching or research careers like their professors, they do not need communication skills. This is simply not true. As one example, forecasters need to convince their coworkers that their forecast scenario is the most probable one, and then they need to communicate their forecasts and warnings clearly to their customers or the public—people whose livelihoods, if not their lives, may depend on understanding the warning. A study conducted by the College Board's National Commission on Writing found that writing is part of the job of two-thirds of salaried employees in large U.S. companies, and writing is taken into consideration during hiring and promotions at half of those companies. Communication skills are not only needed in the workforce, but are in demand.

There are no boundaries, no walls, between the doing of science and the communication of it; communicating is the doing of science. —Scott L. Montgomery (2003, p. 1)

SCIENCE IS FUN

Scientists have one of the most exciting occupations I know. In general, we love our jobs. We get to learn new things every day, explore our own research interests, talk with other like-minded people, see our friends at conferences in exotic locations, and share the thrill of discovering the natural world with students. Yet, as I have shown in this introduction, scientists waste valuable and potentially enjoyable time by writing reviews rejecting poorly written papers and sitting through insipid conference presentations.

> Ah, there's nothing more exciting than science. You get all the fun of: sitting still, being quiet, writing down numbers, paying attention. Science has it all. —*Principal Seymour Skinner*, The Simpsons

How did we lose the fun? I believe part of the answer is that we are taught at an early age that science is impartial. Like Principal Skinner's vision of how science is done, we collect data and we report it, eliminating any evidence that science is done by real individuals. Yet, we scientists like a good mystery story.

The hunt for new knowledge excites us. We may even think something that no one has ever thought before. But, when we write or speak, we fail to convey our enthusiasm and to personalize our science within a proper context. Purging our personalities from our work sterilizes it. We scientists individually need to find our voices, our creativity, and our originality.

Improving our ability to communicate is a lifelong process. I hope this book excites you about your writing and presentations, encouraging you to make them better, interesting, and unique. How many manuscripts must be rejected before we say enough? How many boring presentations must we sit through until we demand better? I look forward to the day when all manuscripts I oversee as editor receive my recommendation to publish and all presentations I attend engage my scientific imagination.

WRITING AND PUBLISHING
SCIENTIFIC RESEARCH PAPERS

THE PROCESS OF PUBLISHING SCIENTIFIC PAPERS

Publishing a scientific paper involves interactions among authors, editors, reviewers, copy and technical editors, and the publisher, with the goal to publish the best-quality research as timely as possible. This chapter describes the publishing process, starting with how to submit a manuscript to a journal, what editors and reviewers do, how manuscripts navigate the peer-review process, and how an accepted manuscript undergoes layout and printing, finally becoming part of the scientific literature.

Scientific journals have been established since 1665 when *Journal des Sçavans* debuted on 5 January, followed by *Philosophical Transactions of the Royal Society of London* two months later (Fig. 1.1). Both are still published today. Despite scientific journals being around for over 300 years, many experienced scientists do not understand the publication process.

This chapter describes this process as it happens at many scientific journals. Although most articles have two or more coauthors, most of the time in this book I refer to a single author, specifically the corresponding author. The *corresponding author* is the person who represents all coauthors by being the one who submits the article to the journal, maintains correspondence with the journal, keeps coauthors informed about the status of the manuscript, and is responsible for revisions. The corresponding author may or may not be the first author listed on the manuscript.

1.1 SUBMISSION

Before the manuscript is written, the author usually has a vision for where it should be published, the *target journal*. Each journal has its own rules for submission. Some journals place few restrictions on submitted manuscripts,

LE

IOVRNAL

DES

SCAVANS

Du Lundy V. Ianvier M. DC. LXV.

Par le Sieur DE HEDOVVILLE.

A PARIS,

Chez IEAN CVSSON, ruë S. Iacques, à l'Image de S. Iean Baptiste.

M. DC. LXV.

AVEC PRIVILEGE DV ROY.

PHILOSOPHICAL

TRANSACTIONS:

GIVING SOME

ACCOMPT

OF THE PRESENT
Undertakings , Studies , and Labours

OF THE

INGENIOUS

IN MANY
CONSIDERABLE PARTS
OF THE

WORLD

Vol I.
For *Anno* 1665, and 1666.

In the *SAVOY*,
Printed by *T. N.* for *John Martyn* at the Bell, a little without *Temple-Bar* , and *James Allestry* in *Duck-Lane* ,
Printers to the *Royal Society.*

Fig. 1.1 The first scientific journals: *Journal des Sçavans* and *Philosophical Transactions of the Royal Society of London.*

as long as they have certain information on the cover page and are set in 12-point font, whereas other journals have strict rules about the format of their submissions.

When the manuscript is completely written and formatted as required by the target journal, the author submits the manuscript to the journal. Even as recently as the first few years of the millennium, the author would send four to six photocopies of the manuscript to the target journal by post, which cost paper resources and money for postage, as well as slowed down the review process. Today, nearly all journals have Web sites where authors can upload digital files. Typically, the manuscript, figures, and a cover letter are uploaded in their native format (e.g., Microsoft Word, LaTeX). Often, a PDF document is created from the uploaded files, and authors are required to approve the rendered PDF. Authors who fail to approve the rendered PDF document can delay the submission process, so pay special attention to the journal's requirements.

Other information that may be required at submission includes a complete list of coauthors, their contact information, and a list of suggested reviewers.

WHAT IS STYLE?

The word *style* means two things to an editor. The first meaning is that implied in the title *The Chicago Manual of Style*. Publishers refer to style in this sense as *house style* or *press style*—rules regarding the mechanics of written communication. . . . Authors more often think of style in its other sense, as a way of writing, of literary expression.
—*The Chicago Manual of Style* (1993, p. 65).

The American Meteorological Society has adopted *The Chicago Manual of Style* as its choice of press style, supplemented by its own online style manual (*Authors' Guide*; American Meteorological Society 2008). Authors are required to follow these guidelines; not doing so may hinder acceptance or prejudice reviewers and editors. If submitting a manuscript to a journal you may be unfamiliar with, read the Instructions to Authors on the journal Web site and look at papers that have already been published to get a sense of press style and literary styles that are acceptable to that journal. Although some authors view press style as oppressive, outdated, and sometimes nonsensical (e.g., why is punctuation placed inside the quotation marks in U.S. publications?), without widely accepted press style, the lack of consistent caption style, acronym expansions, and treatment of variables could be quite annoying, if not confusing.

In contrast, the second definition of style, literary style, depends on the individual author. Authors of scientific journal articles are usually given wide flexibility in determining their own tone and voice, with some dependence on the opinions and sensibilities of the editor and reviewers. Both of these definitions of style are used in this book.

Some journals even allow a declaration of people who should not act as reviewers because of potential biases or conflicts. Some journals may want the author to describe the manuscript's scientific contribution upon submission. A statement may be required that all coauthors agree to the submission of the current version of the manuscript. Authors may also have to state that the work has not been previously published and has not been submitted elsewhere. Authors (and sometimes all coauthors) often must sign a form that transfers copyright to the publisher. Some publishers may require a fax or electronic copy of this form before peer review can start.

The final step before peer review is an initial screening at the journal to ensure that the submitted manuscript meets basic standards of length, organization, and format for the journal. Following the format required of the target journal is essential for making the submission process go smoothly. Read about these requirements on the back pages of the journal or on the journal's Web site within the section for authors considering submissions.

1.2 EDITORS AND REVIEWERS

After the manuscript is approved to start the review process, notification is sent to the chief editor of the journal. The chief editor then decides which

COPYRIGHT

Ken Heideman, Director of Publications, American Meteorological Society

A number of publishers require that each author either transfer copyright for papers published in the publisher's journals or certify that the manuscript was prepared as a work of the government and in accordance with governmental regulations. By holding copyright, the publisher can act as a steward for the intellectual property of the authors, ensuring that authors always receive credit for their work and that their papers are preserved for the long term. Moreover, the publisher as copyright holder acts as a watchdog to preempt, identify, and respond to attempted plagiarism or improper use of the intellectual content contained in its suite of journals.

One additional advantage of the transfer of copyright is illustrated by the electronic legacy content composed of all articles published by the AMS prior to 1997, spanning well over 100 years. AMS makes these articles free and open to all, but, without the copyrights in hand, none of the articles could be posted without seeking out every single author to receive explicit permission to have their paper included in the legacy database. So, mandating copyright transfer is not an arbitrary policy. From a scientific, legal, and practical standpoint, the best interests of the author and the publisher are generally served.

For the author, the practical aspects of copyright policy depends on the publisher. The AMS copyright policy explicitly provides permission to authors to post their published articles on their own personal Web page. The policy, however, does not allow a copy of an AMS copyrighted work to be placed on a non-AMS server (e.g., a department Web site). However, authors are allowed to post a link to the article.

editor or editors will oversee the manuscript (or if the chief editor will handle it), and responsibility is transferred. The editor is typically someone who has a broad knowledge about the topic discussed in the manuscript. In some cases, the editor may recommend transferring a manuscript not appropriate for the target journal to another journal where the topic may be better received. In other cases, the editor may reject the manuscript before any peer review occurs because the manuscript is not written well, has questionable science, or both. This summary judgment by the editor spares potential reviewers the trouble of reading a poor-quality manuscript.

For most papers, the editor decides to start peer review of the manuscript and typically enlists two or three reviewers to provide comments. Reviewers are likely scientists who have done research on the topics in the manuscript. Sometimes reviewers may be outside of the discipline and thus provide a different perspective on a manuscript, especially for small, specialized research communities. The names of potential reviewers are obtained from the editor's friends and colleagues, the associate editors of the journal, the reference list of the manuscript, Web or publication searches, or the recommended reviewers provided by the author. Sometimes the most appropriate or most experienced scientist is unavailable to perform the review, so the reviewer may be someone

Journals may try new approaches to peer reviewing. Cartoon by Nick D. Kim.

Most scientists regarded the new streamlined peer-review process as 'quite an improvement.'

with less experience who is available. Reviewers typically remain anonymous so that criticisms can be made without fear of reprisal.

The reviewers read the manuscript and provide a written report on the suitability of the manuscript for publication. Reviews are merely recommendations that the editor uses to make a decision. As such, the term *reviewer* is preferred to *referee* (incorrectly implying the power to make decisions, like referees in a sporting match). Recommendations issued by the reviewers typically fall into one of five categories:

1. **Accept as is** occurs in less than 1% of papers submitted to AMS journals.
2. **Return for minor revisions** is a good outcome that portends eventual publication pending the author making small changes. This recommendation usually indicates the reviewer does not wish to see the manuscript again.
3. **Return for major revisions** usually indicates that the number and severity of the comments are such that the reviewer wishes to see a revised manuscript before recommending acceptance.
4. **Reject** means the reviewer recommends the manuscript not be published.
5. **Transfer to another journal** may be suggested because the subject matter is not appropriate for the target journal. (The author and editors of both journals must consent to the transfer.)

The editor weighs the reviewers' recommendations and makes a decision on the manuscript. How this decision is determined varies by the editor and the paper. Examples are provided in the Ask the Experts column "How editors make decisions" on page 226. The editor may follow the advice of trusted reviewers, let the majority rule, take the harshest (and unbiased) criticism, or weigh the likelihood of receiving an acceptable manuscript in a reasonable amount of time. Sometimes an associate editor may be asked for an opinion on the manuscript if the reviewers' recommendations contradict each other. The editor decides to continue the review process (return for major or minor revisions), end it (accept or reject), or transfer the manuscript elsewhere. If rejected, the editor may indicate in the letter accompanying the decision whether the author is encouraged to substantially revise and resubmit the manuscript.

If the editor continues the review process, the author has a chance to revise the manuscript and respond to the reviewers' concerns. After the revision is resubmitted to the journal, the editor reads the responses to the reviews. If the editor thinks the author has done an adequate job of responding to the reviews, then the manuscript is accepted. If the initial reviews were particularly critical, or if the editor wants the reviewers to see the revised manuscript and the author's responses, the reviewers may be asked to provide a second review. In this way, the peer-review process may iterate several times before the editor thinks the process has helped create a manuscript satisfactory to both the reviewers and author. Neither the authors nor reviewers may get their way all the time, but their interactions improve the manuscript.

If at any time the editor thinks the author failed to address the reviewers' concerns adequately, the manuscript may be rejected. Obviously, editors do not want to reject a manuscript if it had been progressing toward publication previously. In such cases, one of three reasons for rejection may be communicated to the author. First, the editor may see that the author failed to address one or more crucial concerns raised by the reviews, perhaps not even taking the revision process seriously. Second, the initial revisions may have uncovered more serious underlying flaws in the manuscript that favor rejection in this new light. Third, the editor may see that the rate of convergence between author and reviewers toward a manuscript acceptable for publication would take or is taking too long. Thus, authors should never view revising their manuscripts lightly. It is worth repeating that reviewers do not decide to publish the manuscript, editors do.

1.3 TECHNICAL EDITING, COPY EDITING, AND PAGE PROOFS
Once the manuscript is accepted, it is forwarded to the publisher who begins the process of copy editing and technical editing. Copy editors correct

COMMENT-REPLY EXCHANGES

If a reader of a journal article discovers an error, disagrees with the author's interpretation, or wishes to clarify or discuss certain issues publicly with the author, many journals have an option for such a public discourse: the comment–reply exchange. The exchange consists of a *comment* by the concerned reader presenting his or her side, published alongside a possible *reply* by the original author.

Given that science proceeds through such open discussion of ideas, you might think that comment–reply exchanges would be more common than they are. In fact, the decreasing number of comments, despite the increase in the number of published articles, leaves some wondering if this is healthy for science. Perhaps the dearth of comments may be because the process of writing and submitting a comment is mysterious to some. In fact, the process is quite simple.

Once a comment on an article is submitted to a journal, the editor, perhaps in consultation with the editorial board of the journal, assesses the comment to ensure that it is of sufficient scientific quality to eventually publish. (Personal attacks are not appropriate for comment–reply exchanges and are not published.) After the decision to proceed is made, the comment is forwarded to the corresponding author of the article in question to prepare a reply. Sometimes the corresponding author chooses not to write a reply, in which case the comment is published alone.

If the author writes a reply, it is forwarded to the author of the comments. Depending on the journal, both parties may have an opportunity to make revisions or withdraw their submission. Because all the parties are not anonymous, they may work out issues by themselves, presenting the editor with the finished exchange. Sometimes the editor may choose to adjudicate the process with additional peer review. Most of the time, the editor allows the comment and reply exchange to arrive at a resolution or "agree to disagree," leaving the ultimate disposition of the material published to the scientific community at large to resolve.

grammar and style of the text, whereas technical editors review the scientific meaning of sentences, abbreviations, symbols, and terminology, as well as the suitability of the abstract and technical aspects of the layout (e.g., equations, tables, figures). These two steps are a large part of the production stage of the journal article that forms the layout for how the paper will look. The editors also communicate technical aspects of layout, fonts, and symbols with the typesetter and printer.

The next step is when the publisher prepares page proofs, a draft layout of the way the manuscript will look once published. The copy editors and technical editors may also have queries for the author to answer, such as verifying references or checking that the meanings of sentences have not been changed after editing. Authors are expected to review the page proofs and make comments, identifying errors in transcription or layout within a few days. Despite this quick turnaround time, this is the last opportunity authors have to make minor changes to the manuscript (e.g., correcting typos and grammar, adding new references that have been published since acceptance). Publication follows successful return of the manuscript within a few weeks to a few months, depending on the journal.

SHOULD YOU PUBLISH YOUR PAPER? QUESTIONS TO ASK BEFORE YOU BEGIN WRITING

2

How do you decide whether a research project is worthy of publication? How do you identify and attract an audience? How do you select a journal for your manuscript? This chapter answers these questions.

Scientific research is often portrayed in the noblest terms, as in the poem on the next page. One result of scientific discovery should be a scientific publication, communicating that discovery to others in our profession. In science today, many researchers labor under publish or perish. Not publishing is looked upon unfavorably by funding agencies, laboratory management, and university administrators. Thus, an author's list of publications is one measure of success.

Most of Part I of *Eloquent Science* discusses the mechanics of assembling a scientific manuscript: the organization, language, and conventions of effective scientific writing. Mastering these skills of presentation, however, does not portend that your papers are destined for awards. All the best mechanics cannot save a manuscript that fails in the science. Many research projects are attempted, some are completed, but even fewer deserve publication.

This point can be illustrated on a graph of the quality of presentation on one axis and quality of science on the other (Fig. 2.1). Presentation includes such aspects as the organization of the manuscript, neatness, effectiveness of the figures, grammar, spelling, and format consistent with the style guide. Scientific content comprises the idea, execution, choice of data and methods, results, and interpretation. To create a publishable manuscript, its contents must possess both a high quality of presentation and a high quality of science. The first section of this chapter asks questions to help frame your manuscript, putting it on the path to being high-quality science. Subsequent sections help target your manuscript to the right audience and the right journal.

MOTIVATING THE PUBLICATION OF SCIENCE

Y. Hancock, Lecturer, Department of Physics, The University of York, United Kingdom

The beauty of nature
Lies in her hidden secrets
And mysteries not given
Away
But by the act of discovery
And efforts to perceive

New knowledge
Using the creativity of the mind
And going beyond what was previously
Explained
And painstakingly understood
In classrooms and laboratories
Not found in answers at
The back of the book
But by persistence
To deliver the complexities of
The scientific truth

2.1 IS THE SCIENCE PUBLISHABLE?

The germ of a scientific publication, in fact its essence, is the scientific question. What is the question that you want to answer with your research? That question is followed by a well-designed research study that develops testable hypotheses to answer the question. From the results of this study, what new scientific knowledge is gained? This approach from question to research to results evokes a list of questions that prospective authors should ask *before* writing and submitting a manuscript. Successfully answering these questions in advance can minimize unnecessary effort and heartache later.

Have you asked a good scientific question? Knowing the relevant scientific issues of your specific topic through an up-to-date knowledge of the scientific literature is essential to being a good scientist. Some of the most powerful scientific articles are those that contradict commonly held beliefs within the science. As Mark Twain said, "Sacred cows make the best hamburger."

Is the science original in your paper? The most valuable publications are those that deliver something novel, through either the development and application of a new technique or the creation of a new explanation for how the atmosphere works. All papers should state their purpose and original contributions.

Are your conclusions supported by evidence in your paper? Evidence supporting your conclusions is required. Rarely can all questions be addressed within the confines of a scientific paper, so some speculation is generally accepted. Fred Sanders, long-time editor of *Monthly Weather Review*, used to say that speculation was like dessert. If you eat all your dinner, then you are entitled to a little dessert, but you cannot rely on dessert for the entire meal.

Fig. 2.1 Where does your manuscript lie on this graph?

Is this information substantial enough to warrant publication? Manuscripts can be rejected because the research is not of sufficient depth. Whether your research has reached the criteria to be considered a least publishable unit (LPU) or publon may not be easy to assess. Comparing your research to other published research in your target journal can help determine if you have succeeded.

Does your work have current relevance or allow for future impact? Research articles that present results, then ask more questions than they answer, can be effective vehicles for advancing a field or opening up new avenues of research for others. What is the potential impact to your field by publishing this research? Communicate these opportunities within the paper.

Not every paper you write has to be groundbreaking. In fact, most papers only incrementally advance the field in small steps, as Kuhn (1970) has pointed out. Some people have lost career opportunities because they did not want to add to the glut of scientific literature. This is the wrong attitude. Some research may take years to appreciate. For example, although the basic tenets of a stochastic–dynamic approach to weather forecasting had been formulated by the late 1960s, almost 20 years would pass before its operational implementation could be achieved, in what we know now as ensemble weather forecasts (Lewis 2005). Other research may energize a different audience than you intended. For example, Prof. Lance Bosart of The University at Albany/State University of New York received numerous requests for reprints from his study of the Catalina eddy (Bosart 1983)—a circulation pattern in the Los Angeles basin and "a nonevent on the meteorological Richter scale"—not from

The formulation of a problem is often more essential than its solution, which may be merely a matter of mathematical or experimental skill. To raise new questions, new possibilities, to regard old problems from a new angle, requires creative imagination and marks real advance in science. —Albert Einstein and Leopold Infeld (1938, p. 95)

meteorologists, but from those studying, monitoring, or regulating air-quality in southern California.

Furthermore, scientific research is rarely 100% bulletproof. Observational holes, computational limitations, and theoretical walls prevent us from being certain. Allowing yourself to be defeated by every limitation is not healthy. If you are honest about the limitations and make assumptions that are standard for your discipline, your paper will have a better chance of surviving the review process. If we expected every published article to be unassailable, then the published literature would be quite small indeed.

There is no secret recipe to creating the perfect paper. There are simply too many ways to write a paper to offer a single strategy to ensure the work reaches the right audience, gets a fair hearing in the review process, and gets published. If you doubt the quality of your research, get feedback through internal peer review of your manuscript or give a seminar at your home institution. Ask for support from your colleagues. When publishing, you are not only representing yourself, but also your institution. Remember, their reputation is also at stake!

2.2 WHO IS THE AUDIENCE, AND WHAT ATTRACTS THEM TO YOUR PAPER?

When an audience reads your writing, there is no way for them to interact with you and to ask questions to clarify or expand upon what you wrote. You have to write assuming that you know exactly who your audience is, as well as what is their education level, their level of scientific knowledge about your topic, and the whole wealth of their personal experiences. Your audience might be quite clear to you and be tightly focused—a manuscript about forecasting tornadoes in Idaho, for example, would likely appeal to a small group of people with specific goals in mind. Your audience might be broad and diverse—a new way of thinking about atmospheric convection could appeal to theoreticians, observationalists, and forecasters. Or, your audience might be people you never even anticipated—as we just read, a manuscript on the meteorology of the Catalina eddy might be of great interest to air-quality regulators.

Consequently, take some time to imagine who your readers might be. Your audience may be diverse: students, professors, research scientists, broadcast meteorologists, and forecasters. Or, your manuscript might only appeal to a specialized audience, say numerical modelers. Your manuscript cannot be written to speak to everyone all the time, but you can pick a level that captures most of your member constituencies and at least touches upon topics that appeal to each part of your audience while remaining focused.

The next step is to attract that audience to your paper. Is the topic something that the audience is wondering about? One way to increase readership is to write about their interests in a way that appeals to them. For example, your paper might answer questions that they may have (or not know they have) about their topic of interest. Even people who disagree with you may be extremely interested in reading your paper. But those who do not know about your paper are more of a challenge.

How do others discover your published papers? If scientists receive paper copies of the journals, they may scan the table of contents for familiar authors or topics of interest to them. They may also perform an author or keyword search on databases or look for the published version of a conference presentation that you once gave. Others may forward your name or paper to their colleagues, perhaps hearing about your article through the news media. Many people may discover your article serendipitously, browsing through journals (online or in paper format) looking for an entirely different article. Sadly, of those who discover your paper, some may never be motivated to read it, even despite its importance to their own research.

2.3   WHAT IS THE TARGET JOURNAL?

Reading a paper is a voluntary and demanding task, and a reader needs to be enticed and helped and stimulated by the author. —G. K. Batchelor (1981, p. 8)

Usually one of the first decisions an author makes when beginning to write a paper, or perhaps even upon beginning the research, is to decide on the target journal. Selecting the target journal early in the writing process based on the intended audience, and formatting the article to be consistent with the style expected for that journal, will garner a more favorable outcome from the review process.

The topic of the target journal should be compatible with the topic of the manuscript. Many journals allow submissions from related disciplines, which, as long as the manuscript addresses the foci of the journal, can be a magnificent way to communicate with a different audience than you may be accustomed to.

Choosing a target journal by considering the audience, however, is a deceptively simple problem. You need to consider *actual readership* and *potential readership*. Whereas actual readers are those that subscribe to or follow the journal because of its subject matter, potential readers are those who may use other methods to find your paper (e.g., through a press release, colleagues, keyword search). Determining your audience is essential to meet your goals for your paper. Consider the following example. Publishing meteorological studies in geophysical journals has the potential to attract a different audience to your paper than the actual readership you will get by publishing in a meteorological journal. Do you wish to introduce yourself to a potentially new audience with your latest research, or do you wish to communicate with the colleagues in your particular community (including the big players in your field) who are most likely to find your paper interesting? Although exposure to a larger or different audience is worth considering, you run the risk of sending the manuscript to a less-than-ideal audience because their target journals may be different. One unfortunate result is that your paper may not get recognized by either audience.

The prestige of the journal may affect where you submit your manuscript. Typically, *Science* and *Nature* are perceived to be two of the most prestigious journals, publishing groundbreaking, cutting-edge research and newsworthy scientific discoveries having relevance to a larger scientific audience. But, *Science* and *Nature* have high rejection rates (over 90%) compared to what is typical in the atmospheric sciences (30%–40%). Many submissions are not even sent out for peer review, not because they are bad science, but because they do not meet the criteria of the journal.

Despite the difficulty in getting their work published in a prestigious journal, people may choose this route for a couple of reasons. One reason is that high-impact journals offer a potentially broader readership, and possibly even send news releases to the world's media. Another reason is self-preservation of an academic department or an individual, which is a result of the unfortunate overemphasis placed upon these journals by administrators.

On the other hand, not everyone you are targeting with your research may read such high-prestige journals or even have access to them through their libraries or online subscriptions. A new generation of *open-access journals* allows free online access to their published articles to everyone. For example, if you are trying to reach a wide audience of international forecasters, then publishing in an online open-access journal aimed at forecasters (e.g., *Electronic Journal of Operational Meteorology*, *Electronic Journal of Severe Storms Meteorology*) may be the best choice.

FACTORS TO CONSIDER IN CHOOSING A TARGET JOURNAL FOR YOUR MANUSCRIPT

- Topic
- Audience
- Prestige or impact factor
- Format and length of manuscript
- The similarity between your manuscript and the types of articles the journal publishes
- How the public can read your published article (e.g., open access, subscription required)
- Online access to published articles
- Previous experience getting manuscripts published
- Rejection rate
- Language and geographical location
- Urgency to get the manuscript published
- Editorial board members and network of reviewers
- Page charges
- Ability to use color figures, animations, and electronic supplements

THE JOURNAL IMPACT FACTOR

Eugene Garfield fathered a new field by proposing to calculate statistics of the number of citations to the published literature (Garfield 1955). Some scientists now curse these citation indices, whereas others examine their own statistics weekly. Some probably do both. Despite the limitations and assumptions of citation indices (e.g., Seglen 1997; Garfield 2006; Campbell 2008; Todd and Ladle 2008; Archambault and Larivière 2009), bibliometrics, which is the science of measuring and analyzing texts and information, and scientometrics (the bibliometrics of science) have become important disciplines in information science.

One of the more popular statistics from bibliometrics is the *journal impact factor* (Garfield 1972). The impact factor is the ratio of the number of citations to the journal across the whole field divided by the total number of citable items across the whole field and is calculated over the previous two-year period. Thus, if every article in atmospheric science published over the previous two years cited a particular journal exactly once, the journal would have an impact factor of 1.0. The journals with the highest impact factors are mostly in hot fields such as medicine, biology, and biochemistry. For example, *Annual Review of Immunology* had the highest impact factor in 2004: 52.4 (Garfield 2005). *Nature* and *Science* have impact factors near 31–32. By comparison, in 2006, the journal with the highest impact factor in meteorology and atmospheric sciences was *Atmospheric Chemistry and Physics* at 4.4, compared to the average impact factor across all ranked meteorology and atmospheric science journals of 1.4.

Some journals have predetermined categories of submissions of various lengths and for various purposes. For example, some journals publish *letters*, short contributions with a more direct message that only amount to a few pages each when published. Other journals may publish longer review articles, which are syntheses of previously published research.

Geographical region and language can play a role in your choice of journal. You may have a better chance of reaching your target audience by publishing in your native language in a regional or national journal. For example, if your research is relevant to those in Romania, then *Romanian Journal of Meteorology*, a journal distributed primarily inside of Romania, may be the most appropriate journal for your research results.

Is there an urgency in the field to receive your work? Some journals target submissions where the research requires urgent dissemination, so-called rapid-communication articles. In addition, examine the time between submission and acceptance of papers in your target journal via the *submission* and *in final form* dates listed in the articles. If you are hoping for rapid publication of results, you need to consider the length of two time periods: the period the paper is in review and the period from acceptance to publication. The first is a measure of the efficiency of the editors and peer-review process. Some journals require their reviewers to provide reviews within two weeks. Many

THE ADVANTAGE OF OPEN ACCESS

How often do you find the title or abstract of a potentially interesting article on the Internet only to find it unavailable unless you pay?

Open access is a movement to make published scientific articles freely and permanently available online. There are two main approaches to open access: self-archiving and publishing. In self-archiving, authors may publish in a subscription journal, but make their articles freely available via some online depository (e.g., their own Web site, university repository, arXiv, PubMed Central). In contrast, open-access journals make all their articles freely available online.

There are three ways to fund the publication of scientific research:

1. pay to publish: page charges paid by the authors or their institutions;
2. external funding: advertising, subsidies by professional societies, and grants; and
3. pay to subscribe: journal subscriptions to individuals and libraries, pay access to archives, pay to download individual articles.

Open-access proponents argue against the pay-to-subscribe model because:

1. some users cannot afford subscriptions;
2. many researchers pay for publications out of their research grants;
3. most research is funded by the government, so taxpayers have already paid for the research—they shouldn't have to pay to see the results; and
4. freely available articles are more likely to be downloaded, read, and cited.

Articles that have been self-archived have two to six times more citations than non-self-archived articles because prospective readers can obtain the document more easily. *Atmospheric Chemistry and Physics* is an open-access online-only journal founded in 2001 by the European Geosciences Union. That this young journal has become the atmospheric science journal with the highest impact factor since 2005 and is open access cannot be discounted.

Libraries favor open access because the increasing number of journals, the volume of published literature, and the costs of print journals stress their limited staff, storage space, and budgets. Therefore, open access ensures the largest possible access to your publications and is a sustainable form of publishing.

journals allow a month, although some allow even more time. Once your manuscript is accepted, the efficiency of the publisher becomes a primary determinant in how rapidly the paper will be published. Some journals make accepted papers available online almost immediately, before the final published version is ready.

Some journals make quick decisions on whether to publish your manuscript (days or weeks). Others may take several months or longer. For example, *Science* and *Nature* will usually send you notification about whether they will send your manuscript out for peer review within a few days. In contrast, many other journals, including those published by the AMS, may take 1–3 months to get the initial decision on the manuscript and, pending the author's attentiveness to performing revisions in a timely manner, a final decision in 4–12 months.

The composition of the editorial board may be worth investigating before submitting to an unfamiliar journal because the people on the board may determine the likely disposition of your manuscript. Will the board be sympathetic to your research topic? Is the board knowledgeable enough to choose appropriate reviewers, recognizing that reviewers are potentially drawn from the readership of the journal?

If animations or lots of color figures would improve your ability to communicate your science, you may consider publishing in an online journal. There are no extra fees for color graphics in online journals, which raises the next issue with selecting a journal: the cost of publishing. Although some journals may have no publication fees, others can cost up to several thousand dollars per article. Students, unaffiliated or retired scientists, and scientists at some foreign institutions whose governments do not pay page charges may have to publish in journals without page charges. These journals, however, tend to have high subscription rates, which may limit readership and access to your article once published.

A final point to consider is that journals can only publish from among the manuscripts received as submissions. Editors may sometimes solicit papers directly from authors, but most papers arrive unannounced and unheralded, as discussed by Batchelor (1981, p. 3) for the *Journal of Fluid Mechanics*. Despite explicitly stating a desire to publish on all aspects of fluid mechanics (theoretical, mathematical, and experimental), rumors persisted about the journal favoring certain types of papers. Thus began the cycle of people saying that *Journal of Fluid Mechanics* did not publish particular types of papers, further limiting the scope of the journal. I have seen similar behavior at other journals, including ones for which I have served as editor. Occasionally, an author will approach us with a submission that is not typically what we have published in the past but are willing to publish in the future. If you are writing a paper and your target journal is not clear, ask an editor if they would welcome your submission. Doing so may direct your manuscript to a more appropriate journal sooner or ensure a smoother peer review later.

WRITING AN EFFECTIVE TITLE

3

The title is your first opportunity to attract an audience to your paper. A well-worded and catchy title can lure reluctant readers to take a closer look at your paper. This chapter discusses the characteristics of effective titles and provides examples of how to write accurate, concise, and attention-commanding titles.

A catchy headline in a newspaper often entices peoples to read a newspaper article that would not have interested them otherwise. Similarly, a well-written title in a journal can entice scientists to look at a journal article that they might otherwise have bypassed. Unfortunately, a poorly written title may even scare readers away, regardless of the manuscript's relevance to the readers' interests and the quality of the science inside.

Because the title is likely the first exposure of your paper to a potential audience, the title should be constructed with care and with purpose. Do not just quickly throw it together! Begin with a working (or draft) title to give your writing scope and perspective. Never underestimate the warm feelings from seeing a titled document on your word processor to motivate further work. When the manuscript is completed, reevaluate the working title to ensure that it still represents the work contained within the manuscript.

3.1 CHARACTERISTICS OF AN EFFECTIVE TITLE

The five characteristics of a desirable title (Lipton 1998) are:

1. **Informative.** Identify one or two main points in the paper to communicate to the audience; a good title is capable of conveying those points. Be as specific as possible without adding unnecessary details. Titles that are too vague or too general do not help the reader distinguish your work from

EXAMPLES OF TITLES THAT ARE INFORMATIVE, ACCURATE, CLEAR, AND CONCISE

- Life history of mobile troughs in the upper westerlies
- Numerical instability resulting from infrequent calculation of radiative heating
- The sensitivity of the radiation budget in a climate simulation to neglecting the effect of small ice particles
- Vertical structure of midlatitude analysis and forecast errors
- What are the sources of mechanical damping in Matsuno–Gill-type models?
- Columbia Gorge gap winds: Their climatological influence and synoptic evolution
- Potential predictability of long-term drought and pluvial conditions in the U.S. Great Plains
- Giant and ultragiant aerosol particle variability over the eastern Great Lakes region

other work. Choose words carefully, being cognizant that prospective readers will often find your article through electronic searches.

2. **Accurate.** The title should be truthful about the contents of the paper. Do not overpromise the results of the paper in the title.

3. **Clear.** The audience should not have to think about what the title means. Different people may interpret the title differently, so ask a number of people to critique your title and tell you what they think the paper is about before they even read it.

4. **Concise.** Short titles are instantly recognizable and jump off the page. Every word should have a reason for being present, and each word should contribute to the message of the title.

5. **Attention commanding.** Not all research projects can produce an attention-commanding title, nor do all projects need them. But, if you can meet the other four criteria and have a choice between a pedestrian title and one that is a bit provocative, consider the provocative one.

Ideally, titles should strive to adhere to these five characteristics. However, not all may be met or can be met in one title. For example, to write an attention-commanding title, often you have to sacrifice being more clear or informative. How much concision are you willing to give up in order to be accurate? These are decisions for the author to make.

3.2 STRUCTURING THE TITLE

Begin with the principal one or two points addressed in the paper and construct a draft title. Include words and phrases in the title that identify your work as unique relative to previously published papers, but at the same time being recognizable to others working on similar research. Construct a phrase that includes these elements. Focus on putting the most important information in the title first or last. These are the positions that are likely to catch the reader's eye. Then, examine the five characteristics on the previous page and Day and Gastel's (2006) definition (in the right margin) to see how the draft can be improved.

[A good title is] the fewest possible words that adequately describe the contents of the paper. —Robert Day and Barbara Gastel (2006, p. 39)

Including keywords. Electronic searches have nearly superseded the use of hand-searching paper copies of journals in the library. Thus, selecting the right keywords is crucial to getting your article found. People reading through lists of titles online will want to know immediately, based just on the title, whether your paper is of interest to them. If your manuscript discusses a famous flood, but the date and location is not listed within the title or abstract, individuals searching for "Johnstown flood 1977" may not find your manuscript. Using common word order will also help your article be found more easily. "Potential vorticity inversion" would be more commonly used (as well as shorter and more clear) than "the inversion of potential vorticity."

First words of the title. The first words in the title should be bold and alluring. Avoid having a weak and often unnecessary word such as "the" or "an" occupy such important real estate. Words such as "study" and "investigation" are generally unnecessary and bury important information deeper into the title. For example, "An observational study of . . ." could become "Observations of . . ." or be eliminated entirely if "observations" is unimportant.

Word choice and acronyms. Avoid words that can have multiple meanings or are vague. Also, take care with jargon, acronyms, and abbreviations in titles. For example, CSI has two common meanings in atmospheric science (conditional symmetric instability and critical success index), let alone the more popular meaning from the TV show (crime scene investigation). Not only may acronyms be unfamiliar to many readers, but people searching on the full words may not find your paper. For the same reason, care must be taken with chemical formulas (e.g., consider spelling out CO_2 as "carbon dioxide"), as the formulas can be problematic for online searches.

Word order and "using." Be careful about the order in which you place words and phrases. Often, authors who try to include as much information as possible in the title create misplaced modifiers (Section 9.7). When including the word "using" in titles, be careful of misplacing modifying phrases. Consider a typical example: "Reexamination of the 1979 Presidents' Day Storm using current numerical weather prediction models." Is the storm actually employing the numerical weather prediction models? To eliminate the misplaced modifier, this title could be reworded as "Using current numerical weather prediction models to reexamine the 1979 Presidents' Day Storm." A quick search through recent articles published in the AMS journals having the word "using" in the titles indicated that about 80% are misplaced modifiers. Moreover, poorly considered word order may make understanding the title difficult for those for whom English is not a primary language.

Titles starting with "on." A manuscript titled "The formation of tropical cyclones" sounds like it reaches more definitive conclusions than one titled "On the formation of tropical cyclones." Some scientists think titles starting with "on" sound pretentious, whereas other scientists appreciate the distinction implied by the "on." If you choose to title your article with the "on" beginning, beware that you may raise the ire of some of your readers.

Assertive sentence titles. Sometimes titles can be sentences, termed *assertive sentence titles* (Rosner 1990). Titles such as "Antarctic ice is melting two times faster than prior measurements indicate" sound more like newspaper headlines than titles of scientific articles. These types of titles seem to be popular in high-profile journals such as *Science* and *Nature*. If you choose a title such as this, your results had better be solid. For example, an article with the above title and a statement in the conclusion that the error bars are three times as large as the effect will lessen your credibility. Or, what if later research shows that the ice was not melting at all?

Assertive sentence titles annoy some scientists for the following reasons:

- Such a declarative statement implies some "eternal truth" that a traditional title does not.
- If the principal conclusion of the paper is proven wrong, the (incorrect) title remains a part of the literature.
- Useful aspects of the paper (methods, data, other results) could be overshadowed by the title, especially if the work is later proven to be incorrect.
- Such title statements are often unprovable, as not all possible counterexamples can be tested.
- The title may overstate conclusions that have many caveats and lack generality.
- Assertive sentence titles "trivialize a scientific report by reducing it to a one-liner" (Rosner 1990, p. 108).

Thus, authors who wish to employ assertive sentence titles should wade carefully into the waters.

Colons. The advantage of a colon in titles, called *colonic titles* by Thrower (2007), is that important or attention-commanding text is moved to the front of the title where the audience is more likely to see it, whereas more specific information or detail follows the colon. Consider the example: "Overturning circulation in an eddy-resolving model: The effect of the pole-to-pole temperature gradient." The authors could have titled their article "The effect of the pole-to-pole temperature gradient on the overturning circulation in an eddy-resolving model," which would also have been a reasonable title. By putting "overturning circulation" first, the authors emphasize the circulation rather than the temperature gradient. Authors choosing to use colons in their titles should be alert to the likelihood of making the title longer than is necessary.

3.3 MULTIPART PAPERS

Multipart papers (titled "Part I," "Part II," etc.) allow readers to identify common papers on a single theme, usually from a single research group. Any advantages to multipart papers (e.g., linking two separate papers) are usually outweighed by many disadvantages:

- Multipart papers are more difficult to write than stand-alone papers. Often, the multipart manuscripts are too long and contain too much redundant text. As a result, tangential information that otherwise might be trimmed in a single manuscript remains. Finding the right balance between connectivity and separateness between the two papers without excessive cross references is an enormous challenge.
- Multipart papers submitted together almost always face difficult reviews.
- Multipart papers submitted sequentially usually receive *even worse* reviews. Decent Part I papers go down in flames because the reviewers wanted information that they have not seen in the still-to-be-submitted Part II papers.
- If multipart papers pass through the review process at different rates, how is Part III being published before Part II to be interpreted? What happens if Part I gets rejected and Part II gets accepted?
- Even if multipart papers get published, readers may avoid them, thinking the task of reading them too onerous. For example, to read Part III, does the reader have to be familiar with the material in Parts I and II, even if Part III was written to stand alone?
- Too many examples exist in the literature of Part I papers awaiting their unpublished companions. Consequently, some readers may be on a long, possibly fruitless, quest searching for the never-published Part II.

Thus, publishing papers that stand alone is the best strategy. With appropriate trimming, many two-part papers often are better one-part papers, anyway. Thus, if you have related papers on a single theme and you intend to submit them sequentially, write them independently and do not include

Part I, II, . . . , N in the title. If you wish to submit multipart papers, take special care in writing them and submit the manuscripts together to ensure that they progress through the review process at the same time.

3.4 EXAMPLES

To discuss some of the lessons from this chapter, here is a list of titles that have been, or could have been, published. Some readers may disagree with my assessments. What do you think?

"The use and misuse of conditional symmetric instability." I wrote this title to grab your attention. It is clear, concise, and attention commanding, albeit at the expense of failing to be somewhat informative (the use and misuse in what contexts?). Some critics think it trite. The original title was "The use and abuse of . . ." and was intended to appear in the Forecasters Forum department in *Weather and Forecasting* where opinion pieces appear, often with provocative titles. Informal reviews by colleagues suggested that the paper might better be served as a review article in *Monthly Weather Review* instead. Thus, the purpose of the paper and its title were changed. Another issue is that although this is a review article, potential readers might not recognize it from the title. A better title in that regard, although less attention commanding, would be "A review of conditional symmetric instability."

"Is the tropical atmosphere conditionally unstable?" An appealing and provocative title, it makes me want to read the paper even if I am uninterested in the tropical atmosphere. I like titles with questions, but authors should not make a habit out of using them. Use them once or twice in your career. The good thing about this title is that the authors do not even have to answer the question in their paper. Just asking the question is intriguing enough. A downside of this title is that someone searching titles for "tropical convection" would not find it. Another downside is that some view these titles as too quaint or nonscientific. As always, use attention-commanding titles with full knowledge of the ramifications.

"Diagnostic verification of wind forecasts." More information is needed, unless this is a book chapter or review article. What is "diagnostic verification"? Is "diagnostic" even needed? What types of wind forecasts? 500 hPa? The surface? Forecasts for where? Salt Lake City? Bhutan?

"Snowbands during the cold-air outbreak of 23 January 2003." Although this title is accurate, clear, and concise, it is not particularly informative. "Snowbands" and "cold-air outbreak" are the only searchable words or phrases in the title. How the snowbands occurred or the uniqueness of the snowbands is not apparent from the title. What made these bands unique was that they were associated with a boundary layer circulation called horizontal convective rolls, which had not been identified previously as being deep enough to produce precipitation over the continent. A good title should contain all this information without being unwieldy. Starting the title with "snowbands" emphasizes this feature, so leave that in place. A better title would be "Snowbands associated with horizontal convective rolls during the continental cold-air outbreak of 23 January 2003." If the date is not interesting or relevant to readers, it could be eliminated from the title as well.

4

THE STRUCTURE OF A SCIENTIFIC PAPER

Organization is essential for a well-written scientific document. The readers must know where to quickly find the information they seek, from the cover page to the reference list. This chapter explains the parts of a typical scientific document, how to structure these parts into a well-organized document, and how to write each part to effectively communicate the science.

All scientific documents generally have the same underlying structure, but they may be organized in many ways, depending on the intent of the author and the nature of the research. Some scientific writing books take a more conservative stance, saying that scientific documents should conform to a strict organizational structure. They argue that scientific documents are not literature, so the author's individual literary style should be suppressed. Indeed, some types of studies, such as laboratory experiments where consistent methodology facilitates experimental duplication, may require such rigidity.

In atmospheric science, however, an author must develop a more flexible style with slightly different reporting strategies and organization to cover a larger variety of topics and scientific methods such as case studies, climatologies, field program summaries, and theoretical studies. As an introduction to writing scientific documents, this chapter will introduce you to this underlying structure, presenting the components of this structure sequentially and explaining how best to write them. After this background, the last section of this chapter presents a few alternative organizations, to hint at just a bit of the variety that is possible.

4.1 PARTS OF A SCIENTIFIC DOCUMENT

Although Table 4.1 lists the parts of a generic scientific document, not all manuscripts may contain all these sections. For example, a theoretical derivation may not have a data section. Furthermore, as your career develops, you may be asked to write literature reviews, articles for laypeople in popular science magazines, or field program summaries. These documents may require a different organization with different parts, and their organizations will depend on the material to be covered, article length, audience, and format requirements of the publication, among other factors.

4.2 NONLINEAR READING

Given this structure in Table 4.1, you might think that people would read scientific documents as if they were reading a novel, front to back. In fact, after the title and abstract, the introduction is often *not* the next section that people read. Perhaps, they look at the figures or the conclusion, or even the discussion. Only if the paper is of supreme interest to the reader is the whole paper likely to be read from beginning to end. Thus, of the potentially large number of people who may read the title of your paper, only a small fraction may commit to reading the entire manuscript.

This jumping around among the different sections, or *nonlinear reading*, occurs because busy scientists want to avoid committing the 30–90 minutes, or even more, to reading a paper that may have only limited relevance to them. And, scientists are not the only ones protective of their time in this way, as

---

Table 4.1 Parts of a generic scientific paper, including reference to other tables with more specifics. The * represents sections that are unnumbered.

Cover page* (Table 4.2)
Abstract* (Table 4.3)
Keywords*
Introduction (Table 4.4)
Literature synthesis/background/previous literature (Table 4.5)
Data and methods
Results
Discussion (Table 4.6)
Conclusion/conclusions/summary
Acknowledgments* (Table 4.7)
Appendices
References*
Tables and figures*

---

the results from a study of managers at the Westinghouse Corporation shows. Although every manager had read the executive summary of a report (i.e., a long abstract), only 60% read the introduction, 50% read the conclusion, 15% read the body of the report, and 10% read the appendix. Thus, nine reports sat on their shelves largely unread for every one that was read in its entirety. Those managers who read the entire report said that their interest in the complete report was due to their deep interest in the topic, involvement in the project, skepticism of the conclusions, or concern that the urgency of the project demanded their attention. Scientists would probably express similar views about the articles they choose to read. What is *your* ratio between the number of full articles that you read to the number of abstracts that you read?

Nonlinear reading has a profound influence on how we write papers to best attract an audience. By putting our strongest, most effective, and most convincing writing in the sections that people are most likely to read, we increase our chances that our paper will be read.

4.3 COVER PAGE

The first page of a scientific manuscript should be a cover page containing information about the manuscript. At a minimum, the cover page should contain the items listed in Table 4.2.

The author list including accurate affiliations is necessary for readers to contact the authors if they have questions and for the authors to receive credit for the publication on abstracting services and scientific search engines. Choose a form of your name that you wish to use throughout most of your career and try to stick with it (marriage, obviously, may change that). Use your complete first name and middle initial, if you have one or more. Otherwise, retrieving meaningful results when searching for "J. Menendez" or even "Donna Franklin" may be difficult.

The affiliation should be the location where the author performed the bulk of the research or the most recent location. If an author has changed affiliation after the research was completed, most journals can identify the current affiliation as a footnote, so include this information on the cover page, as well.

Table 4.2 Parts of the cover page

Title of the manuscript
List of authors and affiliations
Type of document, target journal, and status of the manuscript
Date of last revisions
Corresponding author name, mailing address, phone, fax, and e-mail address

---

Include a statement on the cover page about the type of manuscript you are submitting, the target journal, and current status. Examples include:

- Submitted as an Article to *Quarterly Journal of the Royal Meteorological Society*;
- Revised as a Picture of the Month for *Monthly Weather Review*;
- In preparation for submission to *Journal of the Atmospheric Sciences* for the Special Collection on "Spontaneous Imbalance";
- ATM 495 Research Project; and
- Research Experience for Undergraduates Final Report.

Always include the date of last revisions on the title page to avoid confusion about the most recent version. Having an accurate date on the manuscript also helps journal editors and reviewers who may be juggling multiple versions of your manuscript ensure that they are looking at the most recent version. Finally, include the name, mailing address, and e-mail address of the corresponding author.

A well-prepared abstract enables readers to identify the basic content of a document quickly and accurately, to determine its relevance to their interests, and thus to decide whether they need to read the document in its entirety. —ANSI (1979), cited by Robert Day and Barbara Gastel (2006, p. 52)

4.4 ABSTRACT

The first section of the manuscript is the abstract (or summary as it is called in some journals). Because the abstract is a synopsis of the manuscript, the abstract is often the last part of the manuscript written. Only when authors have an overview of their entire manuscript do many of them write the abstract. Some authors draft the abstract early in the writing process, for many of the same reasons that they may write the title first. By the time the manuscript nears completion, check the content of the abstract against the rest of the manuscript for consistency.

Effective abstracts describe the contents of the manuscript and help potential readers know whether the manuscript is of interest to them or not. As discussed in Section 4.2, the abstract is the first part of the text that most readers read, and sometimes the only part of the text that gets read beyond the title. Therefore, a compelling abstract attracts the audience to your manuscript and should contain the basic information in Table 4.3.

Table 4.3 Information contained within the abstract (Day and Gastel 2006, p. 53)

1. Principal objectives and scope of the investigation
2. Methods employed
3. Summary of the results
4. Principal conclusions

Many journals have limits on the length of the abstract, so authors should always read the Instructions to Authors within the journal's end pages or on the Web page. Most abstracts should not exceed about 250 words. Abstracts for dissertations, of course, may be longer. Because of this short length, the abstract should be dense with content. Avoid sentences that are so vague as to be worthless (e.g., "Differences between two numerical forecast models are examined, and the cause of these differences is discussed."). Be specific. As in the title, avoid abbreviations and unnecessary jargon in the abstract. Too much introductory material can burden the abstract; instead focus on the research results.

Because abstracts of published papers often appear alone on Web pages and abstracting services, abstracts should not have any referential material in them: no undefined abbreviations, figures, tables, or external references. Citations to specific papers should be avoided.

Even if your target journal does not require an abstract for your manuscript, consider writing one anyway. Not doing so will limit potential readers who may not find your article when doing literature searches or may not know the article is of interest to them solely from the title.

4.5 KEYWORDS

Although not required for all journals, keywords are used to organize by topic the articles in the journal's year-end index, for abstracting services such as *Meteorological and Geophysical Abstracts*, and to aid those performing electronic searches. If authors do not choose their own keywords, the editor will. Prof. John Thuburn, editor of the *Quarterly Journal of the Royal Meteorological Society*, says, "[M]y advice to the person choosing keywords for their own article would be to put themselves in the position of someone doing the search and try to imagine what keywords someone would search for in the hope of finding the material in the article." In this way, the keywords should be specific information about the manuscript not already in the title, but not too general either (e.g., picking "meteorology" as a keyword for your paper being submitted to *Meteorology and Atmospheric Physics*). Avoid unnecessary prepositions and articles. Commonly recognized acronyms (e.g., CAPE, NWP) are allowed at some journals. List the keywords alphabetically or in the manner expected by the journal.

4.6 INTRODUCTION

The first numbered section of the manuscript is the introduction. After the title and abstract, the introduction is one of the most frequently read parts of a paper, so the importance of a good introduction cannot be overstated. A

Table 4.4 Three components of the introduction (Booth et al. 2003, pp. 222–234)

1. Contextualizing background information
2. Problem statement
3. Response to the problem

good introduction is your chance to show the audience why the content of the manuscript is important to them, even if they are not specialists on your topic. An audience unimpressed by your efforts to convince them that you are working on interesting problems may not venture any further than the introduction.

A successful introduction usually has three components: contextualizing background information, the problem statement, and a response to the problem (Table 4.4). The *contextualizing background information* helps ground the reader in familiar material, and how it is presented will depend on the audience. No one wants to pick up a paper and immediately have unfamiliar information thrown at them. Once common ground is established, the *problem statement* is the hook to gain the reader's attention and draw them into your paper. Just as movies engage the audience by conflict, so, too, should a scientific paper focus around a conflict. This conflict may entail some kind of paradox, error, or inconsistency in the previous literature; the lack of knowledge on the subject; or a general misunderstanding of the problem. If your paper does not have a hook, then ask what is unique about the research and why does it need to be communicated to others. Why do they need to pay attention?

Consider the following introduction:

The classical conceptual model of a cold front typically is manifested as a baroclinic zone that monotonically tilts rearward with height over the cold postfrontal air. At the leading edge of the cold front, a narrow band of ascent occurs that sometimes produces a rope cloud, and, if precipitating, a narrow cold-frontal rainband. The passage of a classical cold front at the surface typically is marked by a relative minimum in sea level pressure (pressure trough), cyclonic wind shift, and temperature decrease. In some cases, however, cold fronts do not possess these characteristics: they can be tilted forward with height, possess prefrontal features (e.g., troughs, cloud bands), or both.

Notice how the author first establishes the common ground of what a cold front is in the first three sentences, then, in the last sentence, hits the audience hard with several contradictions: cold fronts can tilt forward, possess prefrontal features, or, do both.

Once common ground is established, then disrupted, the readers anticipate a *response*. At this point in the introduction, promise them the solution. You can be vague or cagey about it, but give them some expectation of what the purpose of the paper is. Then, say how you are going to address the problem and resolve the conflict. Avoid the tendency to include too much information about the results in the introduction. Some papers give away the punchline too early, so few surprises are left for the reader.

Returning to the previous example, after presenting examples of prefrontal troughs and forward-tilting cold fronts, the author subsequently addresses what the response to the problem will be:

> The purpose of this paper is to address these nonclassical aspects of the cold front associated with [the March 1993 Superstorm] (the forward tilt of the cold front and its associated cloud bands) as it moved equatorward along the eastern slopes of the Sierra Madre in Mexico. The observed data presented previously, although suggestive, were often inadequate to provide additional details about the evolution of this case and, therefore, to ascertain more confidently its structure and dynamics. Consequently, a mesoscale model simulation is used to provide a high-resolution four-dimensional dataset for analysis and diagnosis.

Recall published introductions that engaged you as a reader, especially those on topics that you may have been marginally interested in. You will likely see these three components in them, but used in different ways. Although the formula is successful, do not make your introductions formulaic. Each writing project will require a different introduction.

Going a bit further into this three-component model of an introduction, the contextualizing background information of a good introduction may be the reason that attracted you to study the problem. Explaining your personal motivation is one of the most underappreciated ways of attracting an audience. Perhaps because we are trained to be impartial in our reporting, we eliminate a potentially interesting motivation to the introduction. Scientists enjoy reading other scientists' success stories about how a problem was discovered, addressed, and resolved. Having your paper resonate with that desire in your audience can be quite acceptable.

Some people establish the background information with a bland opening statement, what Nobel laureate Peter Medawar would describe as a "resounding banality": "Tornadoes are frequent occurrences across the Plains, causing much death and destruction." Try to avoid making such statements that nearly everyone knows are true, unless you are going to contradict that statement with your research. Williams (2004, p. 31) likens these slow openings to an orchestra tuning up before the concert. They are necessary for the author to warm up, "but the audience does not necessarily want to listen in."

The purpose of the paper should be clearly stated in the introduction. A statement of purpose is one place where a bit of formulaic prose can go a long way. Specifically, I recommend that each introduction have a statement saying something similar to this: "The purpose of this paper is to. . . ." Writing such a purpose statement also forces you to condense your goals for the paper in one or two sentences.

The scope of the document, what it will and will not be able to address, is also appropriate for the introduction. Because introductions can include broad overviews of particular fields, some readers might infer a much larger purpose to your paper than you intended. A clear purpose statement, possibly supplemented by what topics lie outside the scope of the paper, helps ground readers and prevents any disappointment at the end of the paper when you instead only bite off a small piece of the pie.

The last paragraph of the introduction is often a statement of the organization of the rest of the paper. This text provides readers with the expectation of what they will find out by reading your paper. Some journals require this paragraph, as in the following example, although some authors think that this paragraph is unnecessary: "Section 2 is a review of the previous literature, Section 3 is the data and methods, and Section 4 contains the results." I have heard one author argue, "Does Stephen King lay out the outline to his story in the introduction to his novel?" Good point, but scientific articles are not Stephen King novels. (They should most certainly not be as long as one!) To avoid the repetitiveness of the "section X is . . ." structure, try to provide more context for why the layout of the paper is the way it is, as in the example below.

> In Section 2 of this paper, previous literature attributing observed cyclone/frontal structure and evolution to the large-scale flow is reviewed. Also, two well-known conceptual models of cyclone/frontal structure and evolution are discussed: the Norwegian and Shapiro–Keyser (1990) models. These two models exhibit characteristic differences from each other and, as such, may be thought of as representing two realizations on a spectrum of possible cyclone evolutions. In Section 3, two observed cyclone cases are presented and compared, each representing one of the conceptual models discussed previously. The case resembling the Norwegian cyclone model developed in large-scale diffluence, whereas the case resembling the Shapiro–Keyser model developed in large-scale confluence. In Section 4, the observed cyclones are abstracted to a nondivergent barotropic framework by placing an idealized vortex in various background flows with potential temperature treated as a passive tracer. The evolution of an initially zonally oriented frontal zone is examined first for an isolated circular vortex in the absence of background flow, and then for an initially circular vortex placed in diffluent and confluent background flows.

The resulting frontal evolutions in these simulations are compared to those of the respective observed cyclone cases and their associated conceptual models. Finally, Section 5 concludes this paper.

For many, but not all, of my own papers, I do not write the introduction first. I have a general outline of what I am going to do with the body of the paper, and I start writing that material, but the compelling introduction often has not come to me yet. This is especially true if I start writing before the research is completed. As the research progresses, my perspective on the problem, and perhaps even the main point of the paper, may change, and I may have to discard a perfectly good introduction. Better to lay out some thoughts and wait until the paper has more substance before committing a lot of effort toward the introduction. For other manuscripts, writing the introduction first helps me write a better-organized manuscript.

4.7 LITERATURE SYNTHESIS

The less you know about a topic, the more authoritative the sources sound.
—The Tongue and Quill
(U.S. Air Force 2004, p. 30)

The literature synthesis is potentially one of the most important sections of the manuscript as it can motivate the manuscript by showing the historical and scholarly context of the problem and can justify the manuscript by showing that good research is needed to solve existing problems. Thus, the literature synthesis can demonstrate that an author's manuscript is a meaningful contribution to a meaningful problem.

The best literature syntheses are the equivalent of a box set of music. They are more than just a greatest hits collection of the chronological recitations of the most popular, previously published papers. Instead, the best box sets contain demo tracks to show the evolution of the music, unreleased and lost songs, underappreciated album tracks and B-sides, and live performances. Similarly, the best literature syntheses contain critical evaluations of previous literature, dead-end research directions, forgotten gems from the literature, unpublished conference preprints, and questions still to be addressed, pointing toward potential avenues for further investigation. Synthesizing the previous literature also holds you up to the scrutiny of all the previous authors who you cite—authors who are depending on you to understand and accurately cite their literature. Scholarly literature syntheses can become the equivalent of textbooks for students and scientists who are surveying the field. Thus, every effort should be made to be complete, accurate, and fair.

When and how to cite a paper is discussed in Chapter 12. Instead, this section focuses on how to organize the literature synthesis. I named this section *literature synthesis* (instead of one of the traditional names: literature review, background, or previous literature) because I want to emphasize that describ-

Table 4.5 Rubric for determining the quality of a literature synthesis (excerpted from Table 1 in Boote and Beile 2005)

Category	Criterion
Coverage	Justified criteria for inclusion and exclusion from review
Synthesis	Distinguished what has been done in the field from what needs to be done
	Placed the topic or problem in the broader scholarly literature
	Placed the research in the historical context of the field
	Acquired and enhanced the subject vocabulary
	Articulated important variables and phenomena relevant to the topic
	Synthesized and gained a new perspective on the literature
Methodology	Identified the main methodologies and research techniques that have been used in the field, and their advantages and disadvantages
	Related ideas and theories in the field to research methodologies
Significance	Rationalized the practical significance of the research problem
	Rationalized the scholarly significance of the research problem
Rhetoric	Was written with a coherent, clear structure that supported the review

ing the previously published literature should be more than just a review. Authors should synthesize and critique the past literature, showing explicitly how their manuscript compares and contrasts to previously published work.

The author has a great deal of flexibility to discuss the previous literature in the manuscript. The literature synthesis does not have to appear as a separate section or in a certain place in the manuscript. Nor does it have a required length. For example, the literature synthesis for a review article might be most of the manuscript. In other manuscripts, a literature synthesis that is long enough might be its own section. Alternatively, the previous literature may be discussed only as part of the introduction.

Although citing previous literature should be a component of every research manuscript, discussing all the cited literature at the start of the manuscript may not be universally wise. The author may know that discussion of certain literature to be discussed in Section 6 is needed in the literature synthesis, but the audience may not be prepared to receive that information until more of the manuscript is revealed. Weave the discussion of the literature throughout the narrative of the paper where the literature needs to be discussed (e.g., comparing the author's results to that of the previous literature in the discussion section).

To measure the quality of a literature review, Boote and Beile (2005) constructed a rubric consisting of twelve criteria grouped into five categories (Table 4.5). These criteria reveal four common weaknesses in literature syntheses that are often identified by reviewers.

1. Coverage not well defined—Too little or too much. The literature synthesis should be clearly focused on a specific theme. For example, if an author were writing a manuscript on applying a specific data-assimilation technique to radar data, the topic would be too broad to cite every published paper on radar data assimilation. Yet, a more thorough investigation of the literature might be needed to find out how other research groups have assimilated radar data for the benefit of the author's knowledge, education, and context. Rather, the scope of the literature synthesis should be bounded, clearly defined, and stated upfront. There are two ways in which the coverage category is lacking: too little and too much.

- *Too little.* Literature reviews can be hard work and time consuming. As such, some authors may only cite a few articles that they deem most important or relevant. They skimp on this section, perhaps only listing those articles that they or their close colleagues have coauthored. In addition, some literature syntheses, especially those from early career scientists, may suffer from what severe-storm scientist Charles Doswell calls "temporal myopia," the tendency to cite only papers published within the last ten years. Another situation is when authors fail to balance their review by picking only those references that support their argument. Disagreeing with a source is no reason to exclude it from your reference list. Discuss the source, say why you disagree with it, then present the evidence that supports your argument versus the evidence that does not support it.
- *Too much.* In contrast, the literature synthesis section is not a place where every single paper that has influenced, inspired, or infuriated the author is listed. In such cases, the literature synthesis may be lengthy and dominate the early part of the manuscript, a sign of insufficient focus and restraint. An unfortunate side effect of having a bloated literature synthesis at the start of the paper is that the reader becomes exhausted wading through the previous literature instead of being energized by reading about the present work.

2. No synthesis or discussion of methodologies—"Just the facts." To be effective, the previous literature must tell a story that is relevant to the audience. Unfortunately, many authors treat the background literature lightly, providing a list of references with little explanation or interpretation. This approach of delivering "just the facts" robs the reader of any context for why the information is being presented.

Such reviews can be improved by first shedding unnecessary sources, leaving only those that contribute to understanding the manuscript, then integrating the remaining ones into a cohesive narrative. Remember to describe the topic, not the papers themselves. The organization for such narratives could be thematic or chronologic. More specific advice is found under the two categories "synthesis" and "methodology" in Table 4.5, which list eight ways by which authors can avoid "just the facts" reviews.

3. Significance not discussed. Not describing for the readers the practical or scholarly significance of the research goes beyond the weaknesses in the literature synthesis and indicates a failure in the manuscript as a whole.

4. Poor structure–The grocery list. "Dunn (1983) showed this. Carpenter (1993) did that. Onton et al. (2001) demonstrated the following." I call such a literature "synthesis" *the grocery list* because this listing of articles is devoid of context and often organization. Perhaps authors feel compelled to give the articles lip service, but assume the audience knows why they were cited or believe that such a list is sufficient to convey insight. Or, worse yet, the authors may not even have read them. Grocery lists are often "just the facts" reviews, too, requiring further elaboration on why the articles are cited and their relevance to the present manuscript.

When writing about the literature, avoid sentences lacking quantifiables or specificity. An example is "Very little research has investigated bow echoes in Europe." In contrast, the following statement is more easily defended: "Compared to the United States, less research on the climatology of bow echoes has been done in Europe in the last 50 years."

Finally, the literature synthesis, even if a specific section of the manuscript, should not be isolated from the rest of the manuscript. For instance, connections are often made between the discussion section and the literature.

4.8 DATA AND METHODS

Scientists are naturally skeptics. In fact, science advances because of a healthy dose of skepticism: the advances trumpeted by one research group are tested for their veracity independently by other groups. Only through verifiable testing can claims be shown to be valid. Consequently, for science to proceed, the data and methods used in the study must be clearly described in the manuscript, as these often contain the critical distinctions between experimental success or failure. Therefore, one of the tenets of successful scientific paper

Dan Keyser once asked Chester Newton why certain ideas in atmospheric science were repeated at different times in history. Dr. Newton responded, "That's why they call it research instead of search."

writing is whether a reader has enough information from the paper to duplicate the study.

The data and methods section needs to be complete. Only by describing the data and methods with sufficient detail and precision, or by providing references to other papers that do, can reviewers and the audience evaluate and potentially duplicate your study. Incomplete, incorrect, or inappropriate methods, once in the literature, are hard to eliminate, with later researchers often citing past bad work. You do not need to cite Excel, MATLAB, or other software applications as an analysis technique, unless the specific way in which the calculations were performed is necessary to know.

As with the literature synthesis, the data and methods may not be separate sections. For example, if you are presenting a case where the North American Model (NAM) failed, there is probably little reason to make a separate section to describe the NAM from the observational data. Simply describe the essential aspects of the model for readers who may be unfamiliar with the model and provide a reference or two at the time you first introduce the model. An exception may be where the failures of the NAM are due to a complex interaction between physical parameterizations in the model. In that case, the author may wish to have a separate section describing the details of the model to prepare the reader for the upcoming discussion about the model output.

As a final point, Day (1995, p. 128) advocates using "method" instead of "methodology," arguing that "methodology" strictly means "the study of methods," which is a legitimate scientific discipline. The *Oxford English Dictionary* offers only "the study of methods," whereas *Webster's American Dictionary* has one of the three definitions of methodology as "a series of related methods or techniques," which appears to be consistent with its use in the scientific context. Ultimately, the decision will lie with the journal and the author, but if you can use "methods," why not choose the more precise and concise word?

4.9 RESULTS

The results section is the meat of the paper. Although the section does not need to be called "Results," it should nevertheless provide some indication that this is where your results sit in the paper. Most of the material in this section should be firmly based on the available data, and most, if not all, of the figures and tables will likely be found here. As with the other sections, this material may be broken up into different sections by the different tasks that were performed or by different datasets.

Begin the results section with an overview or big picture of your results. If presenting a tornado climatology of Australia, the results section could begin by stating that you will show the spatial, diurnal, annual, and synoptic distributions of the tornadoes. The results in the form of figures and text should then follow. Readers expect the results to be presented from the most obvious to the least obvious, although developing your argument may require you to deviate from this generality.

A focused presentation of the results needs to be presented. Not every method you tried needs to be presented in the paper. Such a focused presentation may require not presenting even good results that are not relevant to the story being told. If so, these results can be saved for a later paper, or never presented formally at all. Every component of the narrative should advance the story in a logical progression, as transparent to the reader as possible.

In this day of electronically generated plots, it is tempting to make your manuscript a comprehensive examination of the data. Turns out that John Wesley Powell, president of the American Association for the Advancement of Science in 1888, and also the leader of the first expedition down the Colorado River through the Grand Canyon, had seen this before. He said, "the fool collects facts; the wise man selects them." Prepare the minimum of plots that it takes to tell your story and convince an audience. Audiences have limited patience for repetitious plots and unnecessary tangents. Be selective and have some sympathy for your audience.

We have a habit in writing articles published in scientific journals to make the work as finished as possible, to cover all the tracks, to not worry about the blind alleys or to describe how you had the wrong idea first, and so on. So there isn't any place to publish, in a dignified manner, what you actually did in order to get to do the work. —Richard Feynman, physicist (Nobel Lecture, 11 December 1965)

Some people argue that negative results should not be included in the manuscript. I disagree. Description of such negative results can be quite short: "The relationship between wind speed and growth rate of aerosol particles was not significant at the 95% level." Furthermore, negative results can serve some important purposes. First, the atmosphere does not always produce destructive storms, and those forecast decisions for when nothing happens are just as worthy to investigate. Second, presenting negative results shows that not all problems are solved with certain methods. Finally, your negative results published now may save some future graduate student years of work down a dead-end path.

4.10 DISCUSSION

The discussion is the section in the manuscript to explore alternative interpretations, discuss unresolved issues, introduce speculative material, and present overarching themes to integrate, extend, and extrapolate your results for the audience (Table 4.6). A discussion section is optional in many papers, especially shorter ones, but if the material you wish to put into a results section falls into one of the categories in Table 4.6 and exceeds a few paragraphs, consider creating a new discussion section for this material.

Remember that some speculation and inferences are acceptable in a paper, but do not expect the paper to stand on this material. Separating this material

Table 4.6 Material to put into the discussion section (some material adapted from Perelman et al. 1998, p. 196, and Day and Gastel 2006, p. 70)

- Present the theory, relationships, and generalizations revealed by the results that go beyond the results, offering explanations for them. Do not restate or summarize the results, discuss and interpret them.
- Discuss any exceptions to or outliers in your results.
- Discuss alternative interpretations of your results.
- Debate substantial issues that are left unresolved by your results.
- Compare or contrast your results and interpretations with the previous literature.
- Hearken back to the questions you raised in the introduction of the manuscript. Were you able to address these questions satisfactorily? If not, why not? What needs to be done to answer these questions?
- Provide explanations for your disagreements with previous work or discrepancies with expected results.
- Expound upon the theoretical implications and practical applications of your research. What is the significance of the research?
- Elaborate on speculative material that is generally inappropriate for the results section.
- Identify the limitations of your results and explanations. How do the assumptions and scope of your study affect your results?
- Discuss overarching themes that extend beyond the scope of the paper.

into a separate section, with explicit statements such as "I speculate that . . ." or "If our results are correct, then the hypothesis offered by Smith (1996) is invalid for the following reasons . . ." is the best way to avoid scolding reviews that say you misunderstand the difference between results and speculations.

4.11 CONCLUSION, CONCLUSIONS, OR SUMMARY

The last numbered section of the paper is traditionally called the conclusions or summary section. For linear readers, this section will be the last thing they read, so leaving the most important parts of the paper in the readers' minds is imperative. Nonlinear readers commonly read this section before many other sections of the paper. Therefore, regardless of the reader, as with the title, abstract, and introduction, the conclusion section deserves extra special care.

Speaking about the conclusion section, Gil Leppelmeier of the Finnish Meteorological Institute says, "I want to know where this work *leaves us* (i.e., the summary) and where it *leads us* (i.e., what are the questions raised by this work, the conclusions)." Unfortunately, this ideal is rarely met in practice.

Many authors run out of enthusiasm for writing by the time they get to this section, so the conclusion section is often an afterthought, duplicating text (sometimes even verbatim) from other parts of the manuscript. Although a good way to draft the conclusions (or the abstract) is to go through the manuscript and grab all the important sentences, those sentences must be reworded into a well-written narrative.

I like making a distinction between a summary/conclusions section and a conclusion. Sometimes, the author may choose to conclude a paper in a slightly different manner than a retelling of the principal results. In this case, the last section should be titled the singular *conclusion* because it indicates to the reader that the summary of the paper may not be contained within this section. This is especially true if the paper is a shorter contribution, where an explicit retelling of the principal conclusions of the study may be wasting space.

In contrast, a *conclusions/summary* section should be brief, listing the principal conclusions of the paper in a bulleted or numbered list or as short paragraphs. Text is usually best as it allows the author to integrate the conclusions properly into a coherent story, summarizing briefly the evidence for each conclusion. The conclusions section should not contain new material that was not in the text previously (Geerts 1999).

Many authors are confused about the differences between the results, discussion, and conclusion sections. If your manuscript is to contain a discussion section (not all do), then the results section should focus on presenting the experimental or theoretical results of the paper. Inferences from the data should be reserved for the discussion. Finally, the conclusions section should describe the principal conclusions, summarizing the research.

Some authors feel compelled to include a sentence, a paragraph, or more on future work that needs to be done. However, most authors write hastily or think little about this material. They may claim that studying more cases will unlock additional secrets. Other times they simply want to do more research using the same or slightly modified methods. Papers should conclude forcefully, not on a whimper like this. If you feel compelled to address what the next steps are, why not argue for specific objectives to advance the science based on the unanswered questions in your manuscript? Can you offer *testable* hypotheses for future researchers? Were there questions you would have liked to answer, but did not or could not? What approaches would you have chosen knowing what you now know? These are much stronger ways to end the paper if you want to include a discussion of future research directions.

Finally, be careful of saying something like, "Further numerical experiments, currently ongoing and to be reported in future research, will address . . . ," because you may never finish that manuscript or it may not get published. I believe it is best to avoid such statements in most situations.

4.12 ACKNOWLEDGMENTS

The acknowledgments (acknowledgements, in British English) is often not a separate section, but a statement at the end of the text of the manuscript about who has provided help or support in creating the paper but does not warrant authorship (to be discussed more in Chapter 14). Thanking everyone involved, such as those listed in Table 4.7, is gracious. Also, thank colleagues who have provided reviews of the manuscript, including the editors who provided substantive comments and anonymous reviewers. If you had contributed to someone else's paper, but not at the author level, would you not want to receive an acknowledgment in the paper? As an author, if you have a doubt about whether to include people in the acknowledgments, better to include them, or ask them if they want to be included. There is no sense to risk hurting someone's feelings.

Acknowledgments often include the funding agency and grant number responsible for the funding. For example, the National Science Foundation requires this statement in all publications: "This material is based upon work supported by the National Science Foundation under Grant No. X." Citing any sources of funding (either direct or indirect) may be required in some journals to address potential conflicts of interest. For example, people evaluating the effectiveness of a new instrument who received support from the manufacturer should disclose this information to the journal and acknowledge that support in the manuscript. More information on conflicts of interest can be found in the Ask the Experts column by David Jorgensen (page 186).

Sometimes authors must accept full responsibility for the paper with a statement: "Any opinions, findings, and conclusions or recommendations expressed in this material are those of the author(s) and do not necessarily reflect the views of the National Science Foundation." As of this writing (July 2009), this disclaimer should appear in all work sponsored by the National Science Foundation except scientific, technical, and professional journals (e.g., conference extended abstracts, Web pages, press releases).

Table 4.7   Who/what to list in the acknowledgments

Internal, informal, formal, and anonymous reviewers
Editors who have made substantive comments
People who provided specific suggestions on methods or techniques
Funding agencies
Data providers
Software providers (Do not include commercial providers, in general.)
Disclaimers

When writing the acknowledgments, do not write "I wish to thank . . . ," "I would like to thank . . . ," or "I want to thank. . . ." Instead, consider the simpler and more literal, "I thank. . . ."

4.13 APPENDICES

Appendices are not part of the body of the paper, but are self-contained sections of the paper where explanations, theory, or derivations too complicated, tangential, or unsuitable for the main text lie. Appendices can also consist of tables, lists, or the questions on a survey. Most appendices should be given titles and should be referred to within the body of the manuscript. A single appendix is titled "Appendix," but if more than one exists, they are to be lettered sequentially: Appendix A, Appendix B, etc. Because dumping unnecessary content into an appendix can be a convenient way to not integrate text into the body of the manuscript, question whether any appendix is really needed.

4.14 REFERENCES

After any appendices, include the references. Follow the specific instructions for the target journal. Because of the variety in formats for references, always refer to the authors' guide of the target journal, and follow recently published examples for guidance. Chapter 12 has more information on references.

Following the references, some journals (including AMS journals) require the figures and tables on single pages with captions. More will be said specifically about figures and tables in Chapter 11.

4.15 ALTERNATIVE ORGANIZATIONS TO YOUR MANUSCRIPT

Although many scientific papers follow this typical organization, many others do not follow this format. Not every paper has to have the same sequencing of sections. For example, a basic paper may have the following organization:

1. Introduction
2. Previous literature
3. Data and methods
4. Results
5. Discussion
6. Conclusion

Alternatively, the discussion of the previous literature could be folded into the introduction (or results or discussion section). Consider a different structure for a manuscript where two different tasks were accomplished (say, a climatology and case study):

1. Introduction (includes previous literature)
2. Task 1
   a. Data and methods
   b. Results (compare results to previous literature)
3. Task 2
   a. Data and methods
   b. Results (compare results to previous literature)
4. Discussion
5. Conclusion

Or yet a different structure:

1. Introduction (includes previous literature)
2. Data and methods
3. Task 1: Results and discussion
4. Task 2: Results and discussion
5. Conclusions

How the paper is organized will depend on what the major components of the research are that need to be described to the audience and a recognition of a logical way to tell the story. Although a paper can be written in many ways, some ways are clearly better than others. Carefully consider how to organize your paper to achieve the best presentation. One indication that a paper needs reorganizing is when you frequently reference figures or text either well forward or well backward from other parts of the paper. Experiment with different organizations if you think that your draft is just not working.

THE MOTIVATION TO WRITE

Writing can be a struggle, or it can be fun. Most likely it is both. The attitude with which we approach the writing project can determine its success. Lack of motivation and writer's block can prevent us from beginning, continuing, completing, and enjoying writing projects. This chapter provides strategies for overcoming these obstacles to our writing.

Many scientists hate to write. And it shows.

Some very smart people do not like to write, describing the process as too hard or not worth the effort. This issue is common, even among professional writers. American author and poet Dorothy Parker said, "I hate to write, but I love having written." What frightens horror author Stephen King the most? "The scariest moment is always just before you start [writing]."

Much of this intense dislike of writing may stem from childhood when many of us started losing interest in writing. Learning vocabulary and diagramming sentences can dampen a young mind's enthusiasm for creative expression. In addition, most writing assignments force students to write about topics they have little interest in.

We as scientists should be immune from those burdens. We write grant proposals about research we are excited to perform. We write papers about research results we are excited to communicate to others. Ideally, we, of all career-oriented people, should love to write, but some of us do not.

Often you will hear someone say that a particular person is a natural-born writer. Such trite sayings embed themselves into our consciousness, implying that writing is a skill that you either have or you do not. But, writing is not a quick process. Even if the initial draft flows easily from the brain through the fingers into the word processor, editing will take a substantial amount of time.

COMBATTING WRITER'S BLOCK

- Clearly define and focus the topic.
- Clearly define the audience.
- Write throughout the research process.
- Develop a plan for writing.
- Set an external or internal deadline.
- Motivate yourself by submitting your work to a conference.
- Make appointments with yourself to write.
- Create a writing ritual that puts you in the mood to write (e.g., favorite writing spot, certain time of the day).
- Break the writing project up into smaller components.
- Do not let "the editor" dominate during composition.
- Try stream-of-consciousness writing.
- Leave unfinished work for the next day.
- Meditate.
- Change your mode of writing. If you usually use a computer, try writing longhand.
- Do something different or creative for stimulation (e.g., knit a scarf, play your flute).
- Talk with others about your project.
- Get feedback from others on the draft manuscript.
- Do not procrastinate—it creates more stress to produce.
- Reward yourself for small accomplishments.

Sometimes we provide the excuse of "writer's block" as if it were some kind of disease external to us, but the problem lies entirely within. One cause of writer's block is having so much to say we do not know how to say it or how to start. In this case, the writer needs to focus the topic of the manuscript by limiting the content. Another cause is not knowing what to say, perhaps because of a lack of knowledge or a lack of understanding the assignment. Further research may be necessary to develop the theme of the paper.

Any writing project requires four things: something to write about, a means to communicate it, someone who will read it, and the desire to write it. We have already discussed the first three in Chapter 2. This chapter primarily addresses motivation and how to get it. Once you have the motivation, the mechanics of outlining, composing, writing, and editing will come.

5.1 THE IMPORTANCE OF ATTITUDE

A positive attitude facilitates the best writing. If you lack that positive attitude, ask yourself why. Many potential excuses arise out of fear: fear of not saying the right things, fear of the time taken away from other responsibilities, fear of missing the grant deadline, or fear of being judged on what you have written.

Remind yourself of why it is important to write this document. Remind yourself that writing records your methods and observations. Remind yourself that writing helps flesh out your arguments and makes your science better.

Remind yourself that you have other deadlines and the more quickly you can finish this writing assignment and do it well, the more time you will have to do something else. Do whatever it takes. For some, the impending deadline is the only motivation. (As we will discuss later, good ideas may arise under deadlines, but often the execution of those ideas is less than desirable because the attention to detail in writing needs time.)

Author and writing workshop instructor Darlene Graham recommends developing a sense of immediacy to your writing. Carry around a notebook or scrap paper to take notes on. Keep a pad of paper next to your bed if you wake up in the middle of the night with a great thought or phrase. Given opportunities to write all the time, we will.

Fairbairn and Fairbairn (2005) say, "[T]he truth is that writing is just a job, like any other—like washing the dishes, or mowing the lawn, or digging a hole in the ground. None of these would get done if you waited for the ideal time to do them." Begin writing projects now! Do not wait until your children are out of college to write the Great American Journal Article.

5.2 REDUCING THE HEIGHT OF THE HURDLE

One way to avoid the pressure to produce is to write a little bit at a time. Begin writing before the research is done. Often research projects start with the author having performed a review of the literature and developing the data and research methods. Why not write them, or at least drafts of them, first while the ideas are fresh? Because these sections are more factual and descriptive, they may ease you into the manuscript more gently. In fact, most technical writers do not write linearly (introduction, data, methods, results, conclusion), just as most technical readers do not read linearly (Section 4.2).

Writing these sections early forces you to begin writing before the research is finished. Writing should strengthen your arguments. Allow the development of the paper to flesh out weaknesses in your argument, suggesting further sections needing to be written or further figures needing to be created.

Another strategy is to develop a plan to write the manuscript in pieces. The plan keeps you from being overwhelmed by a large writing assignment and allows you to focus on short-term goals. This advice can be helpful for people who only respond to deadlines or cannot see how to tackle a big project such as a thesis. When I wrote my Ph.D. dissertation, my advisors and I decided the best way to proceed was for me to write a draft of each chapter, and, when I finished with it to the best of my ability, to submit the draft to them. While I waited for their comments, I began writing the next chapter. In this way, we were able to make efficient use of our time. Because I anticipated seven chapters of my thesis and it was the end of the summer when I began to write, I budgeted the seven months from September through March to

I find that the creative process is continued into the writing-up stage of the more theoretical type of scientific paper. Clear writing is possible only on a foundation of clear thinking, and my attempts to draft a paper usually lead to considerable clarification of my thinking about the problem and often to further useful developments. —G. K. Batchelor (1981, p. 9)

complete my dissertation—one chapter a month. The first part of April was for final revisions and submitting it to my committee. I would defend in late April and graduate in May. The seven months and seven chapters provided a natural deadline for each chapter. I stuck to the plan and graduated on time. Such a system, however, implies that you accept responsibility for executing this plan and sticking to the schedule.

Do not overpromise your writing within too short a time, especially if it needs to be a quality product such as a published paper or a grant proposal. You may struggle writing some parts of the document, need to do some more literature research, or even rerun some simulations to refine your argument. Always be generous in your estimates, especially if you are working on a deadline. Start early.

Most of us are busy as it is. How do we find time in our schedule to write? Easy. Make the time. If your life is overrun with appointments, make an appointment with yourself to write. Set aside that time (at least several hours), close your office door, work at home or the library, and do it. Unplug your Internet connection, and turn off your e-mail. Focus. Pick an ideal time during the day when you are most focused. Is it in the morning? In the evening? After going for a run or playing tennis? Avoid writing after meals when your body slows down a bit. Clearing your schedule and your brain will allow you to focus better.

Furthermore, write when you have the urge to write. Take advantage of windows of opportunity when thoughts flow easily onto the paper. Such times are precious—rearrange your schedule if you find yourself in one of these moods. Do not let the editor side of your brain dominate. Do not lose momentum by fact-checking, looking up words in the dictionary, spell-checking, and surfing the Web. Ride the wave when it comes.

Set up a daily writing schedule. That is the best advice I can offer any aspiring writer. . . . After a few months of sticking to your schedule, you should be rewarded with an astounding improvement in your writing. If not, there's always computer programming. —Patrick McManus (2000, p. 14)

Keep your writing lively by thinking of it like music. It is important to be grounded in the traditions of a particular form; but just as a great musician knows almost reflexively when to deviate from the form, so should a writer. —Paul Roebber, University of Wisconsin–Milwaukee

5.3 PREPARING THE WRITING ENVIRONMENT

Discover what style works for you. Do you like to compose in front of a computer or on paper? Do you like to write a detailed outline first or do you have more of a free spirit? Your personal style will greatly influence how you best like to tackle your writing. Try different approaches.

The environment you write in can make a big difference in your productivity. Some people can write anywhere. Others need a specific place designated as a writing space. Try different locations for writing to see where you can be most productive. Make the environment as inviting, focused, and efficient as possible. Some writers have an old computer stripped of all other applications except for word processing software. Sitting down to this computer means they are taking writing seriously. Make the room temperature and your clothes comfortable. Prepare your favorite beverage.

Prepare the resources you will need to write, and have them in front of you at a spacious desk or table. These items include all the papers you will cite (hardcopy preferred) and other reference material such as a dictionary, thesaurus, style guides, *Eloquent Science* (of course!), templates for manuscript formating, and English–Finnish dictionaries (if your native language is Finnish). Not having these resources readily available will be an unnecessary distraction.

5.4 OPENING THE FLOODGATES

Let's say that we have set aside a whole week to start writing—how do we make the thoughts flow? To open up the creative writing process, we need to understand a bit about how the brain works. Both hemispheres of your brain—the left hemisphere (rules, science) and the right hemisphere (creative side)—are stimulated during the writing process, and both are needed to write well. While the left side is committing attention to details such as correct grammar and punctuation, obsessing about these details at the composition stage can prevent the creative expression from the right side. By excessively focusing on the left side, the connection to the creativity essential for good writing can be lost. The result is that we may think of ourselves as *bad* writers, losing the self-confidence we need to be *uninhibited* writers.

If fears from the left hemisphere are inhibiting your ability to write, turn it off. Just commit fingers to keyboard or pen to paper and forget about grammar and spelling. Do not even think about writing in complete sentences—write in a stream of conciousness. Beginning writing will open you up. Simply put, stop making excuses for why you cannot write and begin to write. Do not be afraid to put first drafts on paper or in the computer. Revisions can always be performed later. Often, this process of putting anything down accomplishes two things.

First, stream-of-consciousness writing can start the creative juices flowing. Even when impending deadlines and writer's block prevent you from writing, sit down and do it. Even a trickle of vapid thoughts about your topic may help open the floodgates eventually. Of course, do not flagellate yourself unnecessarily for not producing. Sometimes some of my best writing periods happened when I did not initially feel in the mood to write. As with a thunderstorm, a vast reservoir of convective available potential energy may be waiting to be released, if the cap can be breached.

Second, your initial draft, if flawed, suggests one way to approach the problem that may not work. At least you got it out of your system! A common aphorism goes, "It's easy to edit stuff—it's hard to create." Getting material, *something*, *anything*, out of your head into a computer file or onto paper is an essential, initial step to any writing project.

Many writers depend heavily on inspiration because it produces their best, most efficient, and most satisfying writing. Many believe inspiration comes from the outside and must simply be waited upon; most have no effective recourse when it fails. Unfortunately, many writing problems are thinking problems which inspiration is ill-adapted to solve. —Linda Flower and John Hayes (1977, p. 451)

If writing the introduction is challenging you at this moment, try writing sections of the paper that are ready to flow more easily. Work on the reference list or figures if you cannot get excited about writing the text. Waste no time thrashing about for the perfect start to your manuscript when other sections could be written instead. Alternatively, you can blow through the stuck material, writing "BLAH BLAH BLAH" to alert yourself to fill in this material later when your mind is functioning better. If stuck between two words or phrases, place both in parentheses, allowing you to pick the better choice later. Anything that can keep the brain focused on writing is fruitful.

Are you still looking for inspiration? If you find yourself in a deadlock, have coffee with your friends, and talk about your topic. Often just talking about the inability to write opens the floodgates. You may even wish to record conversations you have about your topic in hopes of capturing some spoken moments of brilliance that could be harnessed in your writing. In a similar vein, pretend you are writing a letter to a friend about your work in plain language.

Look for inspiration from other authors whom you admire (or least admire). Reading well-written journal articles could inspire you to similar levels of greatness. Or, pick up a manuscript that you dislike either because you disagree with it or because it is poorly written. Knowing that you can do better is often one way to motivate yourself. You may even try reading one of your own favorite works from the past. Reminding yourself that you once had written something really good can be a tremendous inspiration to achieve similar heights again.

Or you might do something out of the ordinary for inspiration. Go to a museum and be inspired by the art. Take a walk in the forest. Visit a historic place.

If your day is over, you might try writing a note to yourself about the topic you want to write next or even writing the first few paragraphs of the next section, then walking away. "Leaving water in the well" was what American writer Ernest Hemingway called it. That way, the next time you sit down to write, your mind, either consciously or subconsciously, has been preparing for that topic.

Finally, when you reach those milestones you set for yourself—the first chapter is written, the draft is in the hands of the Ph.D. committee, figures are done—celebrate a bit. Go to a movie or have dinner at an expensive restaurant. Take a day trip that you have been dreaming about for years. Reward yourself with something enjoyable for the accomplishment. Remember that carrots generally work better than sticks.

BRAINSTORM, OUTLINE, AND FIRST DRAFT

6

The prewriting process consists of two components: brainstorming and outlining. Brainstorming allows the author to think randomly, covering the topic both deeply and widely; whereas outlining organizes and focuses those thoughts into a framework that can be explained clearly to others. This chapter describes the prewriting stage, leading up to, and including, the production of the first draft.

The truth is that badly written papers are most often written by people who are not clear in their own minds what they want to say. —John Maddox (1990)

Looking back on sections of papers I have written that I have never felt completely happy with, I find a lot of truth in the above statement. I wrote them during stream-of-consciousness sessions, but the text never seemed to reach a level where I could justify the science inside. Sections of such papers may have been inconsistent with other parts of the text or may have included vague statements lacking substance. The text may have served a purpose at one time, but not in the final vision of where I wanted to go.

In middle and high school, we are taught the route to a successful paper is to brainstorm ideas on paper and write an outline, before starting on any draft. How many times do we do that now before we start a writing project? I suspect very few. Perhaps we think we are sufficiently well organized in our minds that we can skip this process without regret. Or we may be in a hurry, thinking that brainstorming would be a waste of time. I wonder sometimes if a little bit of brainstorming and outlining would not benefit more papers I read.

6.1 BRAINSTORMING

Brainstorming is a core dump of all the ideas that are rolling around in your head before starting to write. If the brainstorming session is a particularly fruitful one, ideas from your deeper consciousness may be realized for the first time and new connections between aspects of the research may be made. Before brainstorming, write down the topic sentence or purpose of your document (also called *nutshelling*). This statement forces you to confront the topic and make explicit your writing goals. Doing so will also keep you focused during the brainstorming and outlining processes.

Because brainstorming is a very personal process, whatever way works for you is the way to brainstorm. Set aside a block of uninterrupted time of at least 90 minutes. One approach is to start by writing down all the issues you want to address in the paper. What do you know? What do you *not* know? What questions need to be considered? Write down everything that comes into your head about your topic. You may do so graphically, to show the relationships between your ideas, or you may just create a list as the ideas come to you. Follow your intuition. Do not censor bad ideas—try to find the nugget of insight hidden in those bad ideas. Not every concept necessarily needs to get incorporated into the final document, but at least identifying potential ideas is valuable, even if they are incomplete or inappropriate for the present paper (perhaps for another paper in the future!). Despite this free association, remain focused on the problem at hand. Be creative.

Brainstorm well beyond the point where you feel you have written down everything you can think of. Extending yourself often produces extraordinary insight.

After you have exhausted yourself, look at the result of your brainstorming. Group your thoughts into common themes, especially those that will provide organization to the paper, making sure you have made the important points you wish to make. Cut out the themes from the paper, and lay the pieces out on a table to try possible groupings and arrangements.

A very different approach to brainstorming can occur on a longer time scale as a much less organized activity. Some people brainstorm by writing down thoughts as they have them during their daily activities. Documentation consists of a file (either on paper or computer) of ideas. I carry around a notebook, making notes as I think of them. Periodically, I transfer thoughts from my notebook into these files.

6.2 OUTLINING

As with brainstorming, outlining should be done in any manner you prefer. Some choose to develop their outlines in stages, progressing toward greater complexity, eventually with entries written as complete sentences, until the

WORD PROCESSORS AND BACK UPS

A most upsetting occurrence is to have written the most beautiful paragraph, only to lose it in a computer crash. If this has happened to you, even when writing an e-mail, then you know how frustrating losing even a small bit of your work can be. To avoid such misfortunes when writing, you can do three things. First, set the preferences on your word processor to save automatically and frequently. To ensure versions are saved, remember to save manually, too. Second, maintain multiple versions of your document, if ever you need to go back to an earlier version of the text.

For example, save the results in a new filename containing the date (e.g., article-080323.tex) or by version number (e.g., article-v28.doc). Third, back up everything and store the copies in more than one location, in case of loss, theft, or fire. If the worst happens and the file is accidently deleted, data recovery software may be able to retrieve your lost file from the hard drive. For *Eloquent Science*, I kept three copies of the book at all times: one on my desktop computer, one on my laptop, and one on a memory stick. At the end of the day, I would synchronize all versions to be the most up-to-date across all three.

complete paper nearly exists, albeit in its outlined form. Others use the outline as a skeleton of the paper, then quickly start writing the first draft. In either case, make the outline as thorough as you need it. Section headings are a good start, but more detail is usually needed. Outlines do not have to be neat and well structured, but they should be useful to the author.

Around the time you start outlining, draft a list of figures that you think you might use in the paper. Preliminary figures, especially if laid out on a table in front of you, can be an excellent way to test possible organizations for the manuscript. This approach has several additional benefits. First, unnecessary figures can be identified because you can see material that does not easily fit into the flow of the paper. Second, your writing becomes more focused on telling the story through the figures. Tangential text is more clearly identified when you stray off the trail laid by the figures. Finally, gaps in the story may indicate additional figures you may need but have not created yet.

6.3 WRITING THE FIRST DRAFT

The time has begun to write. Which type of writer are you? Turtle or rabbit? Are you the type of writer that carefully constructs the manuscript piece by piece until it is complete? If so, then you are a turtle, taking the slow and clean approach. Writing a manuscript this carefully is best accomplished with a thorough, well-considered outline that undergoes minimal major changes during the writing process. If writing goes smoothly for a turtle, the first draft needs only a few editing sessions before becoming the completed manuscript.

Are you the type of writer that dashes off with only the roughest of outlines? If so, then you are a rabbit. Because a first draft is reached relatively quickly, rabbits spend (or should spend) much more time revising their drafts. The revision process requires a lot of patience though, something that rabbits sometimes lack.

Naturally, these two extremes are rarely observed in their pure form in any given person. Most authors probably adopt a strategy somewhere in between. Moreover, the same author may use different strategies, depending on the writing project. Sometimes the path to the completed manuscript is very clear and the turtle approach is more feasible; other times the path is less clear and the rabbit may be employed instead.

While writing and organizing the manuscript, I may have some false starts, half-completed ideas, and wonderful pieces of prose that do not belong anywhere in the present document, but I cannot stomach throwing them away. Removing this material from the document focuses the draft, but deleting such material may be difficult emotionally for us pack rats. For such text, I maintain a document called outtakes.tex or outtakes.doc. Larger sections of text (whole chapters, germs of separate documents) may even become a whole new document. In this way, moving large chunks of text that I am not likely to use again from the current article to the outtakes file can be quite cathartic.

As your writing progresses, avoid the tendency to "fall in love with your own text." Nothing that is written down in your manuscript is sacrosanct; everything could possibly be written better.

7

ACCESSIBLE SCIENTIFIC WRITING

Scientific writing does not need to be turgid, dense text written for a handful of specialists. Indeed, authors should strive to present well-reasoned arguments using clear, accessible language for the audience. This chapter challenges us to write so that others will be able to grasp what we say through the approaches we take to writing the manuscript and the way we organize our writing.

Stereotypical scientists are not known for their communication skills. Perhaps for good reason. Maybe that is why scientists who are talented at explaining to nonscientists the complexities of the universe (Carl Sagan), physics (Stephen Hawking), or evolution (Stephen Jay Gould) are so highly regarded by the public. These people have taken their skills at science and scientific writing and have crossed over into the realm of literary writing.

Many of us are voracious readers, whether it be novels for relaxation, newspapers for current events, or nonfiction for learning. How do we use this experience in our scientific day job? Why do we not translate some of that enjoyable experience to our writing? Is there a fundamental difference between literature writing and scientific writing? Yes, and no.

7.1 THE DIFFERENCES BETWEEN LITERARY AND SCIENTIFIC WRITING

Poetry and prose convey different facts and emotions than scientific writing. Perelman et al. (1998, Section 1.1) define the characteristics of effective technical communication as accuracy, clarity, conciseness, coherence, and appropriateness. In some ways, literary writing violates many of these. Literary writing does not need to be factually accurate if it is fiction. Good literary writing sometimes relies on ambiguity to develop the story. Literary writing does not need to possess clarity if the author wishes to engage the reader's imagination. And, concision is certainly not a hallmark of literary writing—fans of James Michener testify!

Ideally, our goal as scientific writers is much the same as for literary writers. We want to convey information to our audience, and we want to invoke a response, whether it be informational, emotional, persuasive, or a call to action. Sure, we have more jargon and terms with complicated definitions than literary writing, and our work is found in the nonfiction section of the bookstore (we hope!). Nevertheless, if we visualize our scientific writing being more accessible to the public (imagine writing for your parents or your friends), then we will have gone a long way toward making it more accessible to *scientists*.

I make it a point to read papers or books by authors whose writing style I have a high regard for. This can be anything—classical fiction, scientific papers written during the Victorian period, etc.—to erase the unfortunate memory of the numerous dry, badly written papers one inevitably has to read as background to the research one is presenting. —Kerry Emanuel, Massachusetts Institute of Technology

7.2 MAKING WRITING MORE ACCESSIBLE

Recognizing that we write for our audience, not for ourselves, we need to become considerate of the group for whom we write. Here are some tips for making writing more accessible to the audience.

Demonstrate your points to the audience with clear, specific examples. Every statement should contribute positively toward the paper by presenting evidence, citing a reference, indicating speculation, or offering a hypothesis, for example. Readers are puzzled by statements like, "Nor'easters cause extensive damage to beaches along the East Coast of the United States." Yes, such statements are obvious, but what does the audience do with this information? More specifics on the area of beach lost, the volume of sand washed away, what period of time, and how many houses and buildings have fallen into the ocean give the reader much more context.

Assume your audience is not as knowledgeable about the topic as you are. Explain nuances, jargon, and assumptions. Given the choices of stating or eliminating information that much of your audience may know, err on the side of backtracking a bit and providing your audience with a little more information than you think they may need. The audience wants to feel comfortable reading your article. Starting slowly—but not too slowly—will ease them into the article.

Justify your assumptions. Each study, no matter how carefully designed and executed, makes assumptions. As such, the strongest papers are those that anticipate the rebuttals and address them up front without apology. Even if your assumptions are relatively commonplace for specialists such as yourself, future specialists reading your paper or nonspecialists today might not know what those assumptions were. Not justifying your assumptions leaves you open to reviewer criticisms, annoying for author and reviewer alike, and lengthening the time to publication. Describe your assumptions from the most plausible to the least plausible or the most general to the least general.

Explain the limitations and alternative explanations of your research. Whether the lack of potentially important measurements, limited grid spacing in your model, issues with the way the data were collected, or instrument calibration problems, being forthright in the paper will enhance, not reduce, your credibility. Do not pretend that you are being smart by not stating the limitations of your work. Astute readers will recognize the limitations anyway and may wonder about your intentions. Reviewers who identify the limitations and alternative explanations for your results will ask you to address them (or reject your manuscript), so you might as well declare them and discuss them on your own terms. Acknowledging and stating limitations also keeps you from overgeneralizing your research results. One way to evaluate your proposed explanations is to take the opposite point of view and try to shoot holes in your arguments. If you were to play devil's advocate (or your arch-enemy) to your paper, what issues would you raise that would be most damaging? Unfortunately, more papers should address the limitations to their work and evaluate alternative explanations for their results. Instead of a sign of weakness, it should be a sign of an honest author.

The right to search for the truth implies also a duty; one must not conceal any part of what one has recognized to be the truth. —Albert Einstein

Consider how your audience will receive your argument. Will they be skeptical or hostile to your conclusion? If so, then develop the text to provide all the evidence first. Do not jump right in with your controversial ideas before they have seen the evidence, and expect them to go along with you. Let them arrive with you to the conclusion that perhaps previously they were not ready to embrace.

Create a document that is accessible to the audience. Everything from the organization of the manuscript to the paragraphs, sentences, words, and figures should be explained to your audience. Be concise without omitting substance. Write so the words sound natural, but professional.

If it's boring to you, it's boring to your reader. —from the poster "The Only 12 ½ Writing Rules You'll Ever Need"

7.3 STRUCTURING LOGICAL ARGUMENTS

Much as the organization of the paper has a certain order that should be followed (e.g., data, methods, results, discussion), arguments also need a certain presentation to maximize reader comprehension. Remember that you are presenting new results to your readers, and you expect them to follow your logic. Therefore, present it in a manner that will make sense to them, as follows:

data → results → interpretation → inference → speculation

The whole chain begins with the *data*. Present the data in the text, and present figures that will support your later argument. After showing the data, the reader is ready to hear the *results*, or what the data are saying. Especially when describing a figure, authors often present results first, to precondition the reader to interpret the figure the way they want.

Next, the reader is open to *interpretation* of that data, a slightly more in-depth analysis of what the data mean. After interpretation, the reader is primed for *inference*, or an extension of the data outside the limited focus of the present argument. Finally, *speculation*, or an educated guess of what the data might imply in an entirely different context, should be presented last. When injecting opinion or speculation, be clear to your audience that it is not fact, but it follows from your data and reasoning.

Rearranging the links in this chain is possible to some extent, and not all steps in this chain will occur in all situations, but too much rearrangement could confuse and frustrate your audience. Consider the following example.

> **DRAFT:** We speculate that buoyant convection caused by the release of conditional instability above a region of low-level frontogenesis was organized into bands by the midtropospheric inertial instability. (*speculation*) Negative absolute vorticity in the Northern Hemisphere implies the presence of inertial instability. (*interpretation*) Calculations are performed on the output from the Rapid Update Cycle from 0000 UTC 20 July. (*data*) The 500-hPa absolute vorticity is negative in the area where the bands form (Fig. 5). (*results*) The occurrence of the bands in the region of negative absolute vorticity indicates inertial instability could have been released. (*inference*) The elimination of the negative absolute vorticity after 0600 UTC shown previously (Fig. 3) suggests that the inertial instability was released, returning the atmosphere to an inertially stable state. (*inference*)

If you felt slightly offended that the author offered the speculation first, and supported the argument later, then you are not alone. Although the author may have felt that the audience was informed of all the necessary information, in fact, the audience was conditioned to be skeptical of this argument by having the speculation presented before the evidence upon which that speculation rested. The author did not allow the structure of the text to carry the reader to the conclusion.

> **IMPROVED:** Calculations are performed on the output from the Rapid Update Cycle from 0000 UTC 20 July. The 500-hPa absolute vorticity is negative in the area where the bands form (Fig. 5). Negative absolute vorticity in the Northern Hemisphere implies the presence of inertial instability. The occurrence of the bands in the region of negative absolute vorticity indicates inertial instability

could have been released. The elimination of the negative absolute vorticity after 0600 UTC shown previously (Fig. 3) suggests that the inertial instability was released, returning the atmosphere to an inertially stable state. We speculate that buoyant convection caused by the release of conditional instability above a region of low-level frontogenesis was organized into bands by the midtropospheric inertial instability.

Similarly, how you organize the text may improve the audience's ability to understand your argument. Organize your text from general to specific, or specific to general, or case study to climatology, or vice versa. Avoid jumping around among topics.

7.4 WRITING IS LIKE FORECASTING

Writing a scientific document is a little like making a weather forecast. Snellman (1982) described the process of making a forecast using the forecast funnel analogy (Fig. 7.1a). The forecast funnel provides a framework for forecasters to visualize the analysis and forecast process sequentially through the different scales of motion in the atmosphere from the planetary scale to the

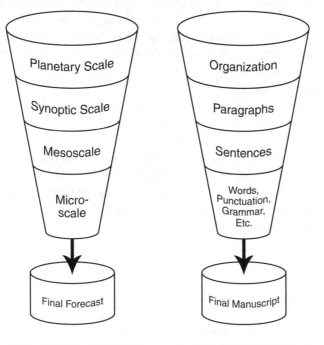

Fig. 7.1 (a) The forecast funnel (after Snellman 1982) and (b) the writing/editing funnel.

(a) THE FORECAST FUNNEL
(after Snellman 1982)

(b) THE WRITING/EDITING
FUNNEL

microscale. By focusing on the largest scales first, forecasters understand the environment that may favor or inhibit certain types of smaller-scale weather phenomena. As the forecaster progresses down the funnel, greater attention is paid to how the mesoscale and microscale details evolve within that specific synoptic pattern.

Similarly, writing and editing a manuscript can be considered like the forecast funnel (Fig. 7.1b) in that it requires a focus, first on the largest scales (the organization of the manuscript: chapters in a thesis or sections in an article) before a consideration of the paragraphs, sentences, and words. The chapters (in a thesis) or sections (in an article) are analogous to the planetary-scale flow, organizing and shaping the writing. Paragraphs serve as the synoptic-scale flow, the regular flow of pressure rises and falls that deliver the sensible weather. Sentences are the mesoscale components of the flow, and words, punctuation, grammar, spelling, etc., are the microscale components. High-quality scientific writing requires all scales of the writing/editing funnel to be high quality, in the same way as all meteorological scales need to be properly understood to make a high-quality forecast.

In planning, writing, and especially editing a manuscript, remembering the writing/editing funnel will produce a better organized paper and make the most effective use of your time. For instance, jumping right in at the microscale and spending a lot of time revising word choice and fixing misspellings on a stream-of-consciousness idea when the organization of the paper has not even solidified (planetary scale) may result in an extremely well-written paragraph, but no place for it within the eventual manuscript. Smart authors consider the organization of the paper first before starting on much of the smaller-scale work.

The components of the writing/editing funnel are described in this book: Chapter 4 presented the parts of a scientific paper, the planetary-scale organization to the manuscript. The next three chapters explore the rest, starting with the synoptic scale and working down to the microscale. Chapter 8 focuses on writing effective paragraphs, Chapter 9 focuses on effective sentences, and Chapter 10 focuses on effective words and phrases. Although this book generally does not cover grammar and spelling, some punctuation is discussed in Appendix A.

CONSTRUCTING EFFECTIVE PARAGRAPHS

8

A well-written scientific manuscript demands strong, effective paragraphs for support. An effective paragraph is characterized by unity of theme, and those themes from all paragraphs together provide the constituents of the manuscript. This chapter describes how to construct potent paragraphs, focusing on the coherence internal to a paragraph centering about the unitary theme, as well as the coherence between paragraphs that make the manuscript fluid.

A s atoms are to matter, paragraphs are the fundamental organizational unit of a paper. Paragraphs serve this role because each one contains only one theme, which is explored within the bounds of the paragraph. Subsequent paragraphs deliver different themes, and the accumulation of themes with each paragraph builds the content of the manuscript. Thus, effective paragraphs bind the manuscript together.

Effective paragraphs possess two primary characteristics: unity and coherence. Unity means a paragraph consists of one theme only. Everything within that paragraph should be related to that one theme. The focal point of the paragraph, the *topic sentence*, defines the theme of the paragraph. Although typically the first sentence of the paragraph, the topic sentence may sometimes appear at the end of the paragraph for additional emphasis. Should more than one theme be in a paragraph, three options exist—break up the paragraph, one new paragraph for each theme; revise the topic sentence and, hence, the scope of the paragraph to encompass multiple themes; or delete one or more themes. The importance of the topic sentence should not be underestimated. As part of their outlining, some authors write the topic sentences for each paragraph, ensuring a logical flow between topics early in the writing process.

Coherence within a paragraph derives from the ordering and relationship between sentences. Sentences within each paragraph should proceed in

The writer has much control over the paragraph. The main sections of the paper are largely determined by convention, and the structure of sentences is determined by the syntax of the language. The paragraph however has no such formal constraints; the chief constraint is content. —Antoinette M. Wilkinson (1991, p. 437)

65

a logical order, introducing new concepts sequentially. (An example of how improper ordering can affect coherence is presented in Section 7.3.)

8.1 COHERENCE WITHIN PARAGRAPHS

When I was living in Norman, Oklahoma, colleagues at Iowa State University in Ames, Iowa, invited me to visit. I had never been to Iowa State before and I wanted to see some other places on my way back home, so I drove the 600 miles. From the Web, I determined the following directions to the building that housed the Meteorology Program:

1. From Norman, take Interstate 35 north to Ames.
2. Leave the highway at Exit 111.
3. Drive west on Highway 30.
4. Turn right on University Boulevard.
5. Turn left on Lincoln Way.
6. Turn right on Union Drive.
7. Turn right on Wallace Road.
8. Turn left into the parking lot of Agronomy Hall.

Imagine if I misread the directions, rearranged the order of the instructions, forgot one of the eight steps, or made a wrong turn. With a little concerted effort, I probably could still get to Agronomy Hall. The more the directions were altered, the more effort (and gasoline and time) would be wasted. For travelers familiar with Ames, these directions would probably suffice, even with a few transcription errors. But, for me, making a mistake, being confused, and getting lost were possibilities.

To supplement these directions from the Web, I asked one of my colleagues, Prof. Bill Gallus, to send me directions. Here is what he sent:

1. From Norman, take Interstate 35 north to Ames.
2. Take the first exit for Ames, which is exit 111 (Highway 30), with signs mentioning Iowa State University.
3. Drive west on Highway 30 until the third exit, which is University Boulevard.
4. Take a right on University Boulevard and drive past the big football stadium and the large coliseum.
5. Just beyond the coliseum will be Lincoln Way. Turn left at this light.
6. Be sure to get in the right lane, because you'll be making a right onto Union Drive in only two blocks. This road takes you past the president's mansion.

7. Turn right onto Wallace Road after a block or so. This intersection is at the bottom of the hill.

8. Stay on Wallace for about two blocks until you see the Agronomy Building on your left. It is the large, red-brick building on the southeast corner of the intersection of Wallace and Osborne Drive. Turn left into the parking lot.

Had I misread Bill's directions, rearranged their order, forgot some steps, or made a wrong turn, the additional detail would have been incredibly helpful in returning me to my desired route. Bill's directions are more informative and longer, but the turns, where I potentially could have made an error, are more descriptive. Sometimes his directions repeat elements from the previous step. For example, the coliseum was mentioned at the end of the fourth step and the beginning of the fifth step. Had I omitted step 4 inadvertently, I might have still found my way knowing that I was to pass the coliseum. As a result of the additional detail and repetition, Bill's directions gave me additional confidence during my drive.

Writing is like providing directions to the reader. You could provide directions such as the terse first set and wish the audience luck on their journey through your manuscript, hoping that they fully understand what you wrote and make no mistakes. Or, you could provide clear, detailed directions, describing how each turn relates to the next, as with Bill's directions. Readers, like travelers, appreciate being led through all the steps. The transitions may be clear in the author's mind, but the author needs to inform the readers of those transitions, especially if the audience is unfamiliar with the topic, just like the traveler unfamiliar with Iowa will want detailed directions. Anticipating how the audience will interpret your writing is one challenge of coherent writing.

The secret to creating a coherent paragraph lies in recognizing the structural expectations that the audience places on the text they read (Gopen and Swan 1990). As the audience reads text, they have "old information," material that they have already been exposed to, and "new information," material that they are just being exposed to. Just as the beginning of a paragraph has a topic sentence, the beginning of the sentence has a *topic position* (Fig. 8.1a). Placing old information in the topic position comforts the reader, providing links backward and context forward. The topic position connects the material previously introduced in the text (e.g., the prior paragraph) and the new material to be introduced in the present paragraph. In this way, writing is linking up information in a logical, flowing manner (Fig. 8.1b), just like steps 4 and 5 in Bill's directions were linked through his repetition of "coliseum."

In the same way that the beginning of a sentence or paragraph is important, the end also has special significance. New information to be emphasized

I aim for the happy medium between too much and too little information. I don't know of any formula that directs one toward the optimal amount of information. Inasmuch as the optimal amount depends on the receiver as well as the transmitter—I try to be sensitive to audience response to see what works and what doesn't and adjust accordingly. —Richard Rotunno, National Center for Atmospheric Research

Fig. 8.1 (a) A single link in the chain: a sentence with a topic position at the beginning and a stress position at the end. (b) Creating a chain of links: linking the sentences together by connecting the stress position of one sentence to the topic position of the next sentence.

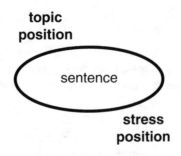

(a) Structure of a sentence

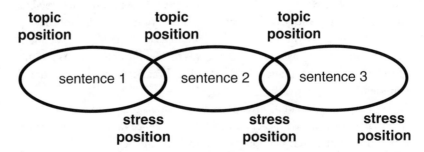

(b) Maintaining transition in a paragraph by linking the stress position of one sentence to the topic position of the next

should appear at the end, in the *stress position* (Fig. 8.1a). Readers naturally emphasize the material at the end, whether it be at the end of a sentence, the end of a paragraph, or at the end of a novel. Secondary stress positions within a sentence may also occur before colons or semicolons.

Read this paragraph out loud. Notice how you naturally place the emphasis in your voice at the end of each sentence? Material improperly occupying the stress position might receive undue attention from the reader, and, therefore, the author would fail to communicate the most important point. Furthermore, the material in the stress position typically links forward. Such linkages help the reader infer the relationship between one sentence and the next, thus helping to keep that link in the chain intact.

8.2 EXAMPLES OF COHERENCE

There are many ways to maintain coherence within a paragraph. Here are three examples: repetition, enumeration, and transition.

8.2.1 Repetition

Repeating key words and phrases (what Michael McIntyre of the University of Cambridge calls lucid repetition) is one of the easiest ways to maintain coherence. The words or phrases do not have to be identical, but the linkage should be clear. In the paragraph below, the topic, the life cycle of a cyclone, appears in the first sentence. Each subsequent sentence is linked to the previous one by the italicized words.

> The life cycle of a Bjerknes and Solberg (1922) cyclone, hereafter the Norwegian cyclone model, begins with a small-amplitude *disturbance* on the polar front. This *disturbance* consists of a cyclonic circulation that advects cold air equatorward west of the cyclone center and warm air poleward east of the cyclone center, forming *cold and warm fronts*, respectively. Because the *cold front* is observed to rotate around the system faster than the *warm front*, the cold front eventually *catches up* to the warm front, forming an occluded front. Originally, Bjerknes and Solberg (1922) believed that this *catch up* initially would occur away from the low center.

Pronouns can also be used to link sentences, if the pronoun has a clear noun to which it refers.

> *Galway (1975)* developed an outbreak definition that included three classifications of family outbreaks: small (6–9 tornadoes), moderate (10–19 tornadoes), and large (20 tornadoes). *He* found that 73% of the tornado deaths from 1952 to 1973 were attributed to outbreaks with 10 or more tornadoes.

In the two examples above, despite being excerpted from journal articles and devoid of the surrounding text, the text makes sense because the grouping of sentences exhibits coherence.

8.2.2 Enumeration

Organizing a list of items through enumeration helps readers follow your argument. If more than a few sentences for each item are needed, start a new paragraph for each item. Make this enumerated list within the text painfully clear to the audience. Use "first," "second," etc., as the extra "-ly" in the adverbs "firstly," "secondly," etc., is not needed. Alternatively, for longer enumerations, the topics could be listed as a numbered list, as a bulleted list, or as a table.

> These jet-streak winds could play *three* roles in the resulting convection. *First*, the jet streak provides upper-level synoptic-scale ascent leading to development of cirrus, reducing insolation and slowing the removal of the low-level

capping inversion. *Second*, the low-level synoptic-scale ascent associated with the jet streak favors the removal of the cap through adiabatic cooling, which would counter the cloud-radiative effects and promote the development of deep, moist convection. *Third*, the strength of the incoming winds affects the magnitude of the deep-layer shear and storm organization, favoring long-lived, isolated, rotating storms.

8.2.3 Transition

Read the following paragraph.

> **DRAFT:** Whether or not the center of a mammatus lobe is warmer or colder than ambient depends on the individual lobes and the height at which the temperature is examined. Lobes simulated in experiment M2 have both warmer and colder cores than ambient (Fig. 4a). Lobe 1 has a warmer-than-ambient core near the bottom of the lobe. Lobes 2 and 3 have colder-than-ambient and near-ambient cores (Fig. 4a). Lobe 1 is warmer than ambient at lower heights near the base of the lobe; at higher heights, the perturbation is colder than ambient. Lobes simulated in experiment M3 have core temperatures near ambient for most of the depth of the mammatus lobe (Fig. 4b).

The paragraph seems to read as a list of observations about lobes 1, 2, and 3 and two experiments, M2 and M3. Why are these observations important and how do they relate to one another? Although repetition of "lobe" and "experiment" provides some comfort, meaning may still elude the reader.

Transitional devices are words or phrases that are used to maintain coherency by indicating relationships between sentences and sentence fragments. Transitional devices can indicate similarity, contrast, sequence, emphasis, causality, or summary (see the sidebar). By inserting just a few transitional devices into the paragraph (seven italicized words in a 119-word paragraph), the relationship between these observations becomes much clearer.

> **IMPROVED:** Whether or not the center of a mammatus lobe is warmer or colder than ambient depends on the individual lobes and the height at which the temperature is examined. *For example*, lobes simulated in experiment M2 have both warmer and colder cores than ambient (Fig. 4a). *Specifically*, lobe 1 has a warmer-than-ambient core near the bottom of the lobe, *whereas* lobes 2 and 3 have colder-than-ambient and near-ambient cores (Fig. 4a). *Furthermore*, lobe 1 is warmer than ambient at lower heights near the base of the lobe; at higher heights, the perturbation is colder than ambient. *In contrast*, lobes simulated in experiment M3 have core temperatures near ambient for most of the depth of the mammatus lobe (Fig. 4b).

COMMON TRANSITIONAL DEVICES FOR SCIENTIFIC WRITING

Sequence

again, and, besides, then, further, furthermore, next, moreover, in addition, first, second, third, etc.; (a), (b), (c), etc.; 1), 2), 3), etc.; following this, subsequently, to enumerate, also, another, last, plus

Comparison and contrast

at the same time, on the contrary, in contrast, nevertheless, notwithstanding, nonetheless, conversely, like, unlike, even so, in the same way, as, unless, whether, though, even though, regardless, irrespective, otherwise, in comparison to, even when, to the contrary, but, or, nor, yet, inasmuch, contrary to, comparing, alternatively, rather, despite, ironically

Examples

for example, for instance, in the case of, in general, especially, if, specifically, in particular, generally, on this occasion, in this situation, to illustrate, to demonstrate, as an illustration, as a demonstration, unless, such as, provided that, once again, another example, a further example, a further complication, in such cases, in this way, in some of these cases, for these reasons, one way, another way, as discussed, using, particularly, that is, more specifically, except

Time

while, since, simultaneously, presently, meanwhile, thereafter, thereupon, afterward, at the same time, next, sometimes, in the meantime, eventually, following this, later, usually, occasionally, concurrently, preceding this, as, presently, at the time of this writing, often, rarely, throughout, by, at, during, continuing

Cause and effect

therefore, thus, consequently, as a consequence, for this reason, hence, accordingly, because, due to, in spite of, despite

Emphasis

surprisingly, indeed, interestingly, curiously, in fact, of course, naturally, evidently, certainly, clearly, obviously, apparently, fortunately, especially, significantly, perhaps, from my perspective, if possible, if so, basically, in reality, essentially

Concluding

finally, therefore, in summary, to conclude, in conclusion, to summarize, as I have shown, hence, thus, in other words, as said earlier, in any case, as a result, at least, as mentioned above, as said previously, thereby, in the present article, simply put

8.3 COHERENCE BETWEEN PARAGRAPHS

Coherence exists within a paragraph through the orderly succession of sentences. Yet, to create fluidity through the document and a lucid story for the reader, coherence must also exist through the orderly succession of paragraphs. Coherence between paragraphs is created through the same mechanisms discussed in Section 8.1, except on the paragraph scale using sentences, and occasionally words, as the transitioning elements. To demonstrate this coherence for a specific example, the first one or two sentences and the last sentence in the first six paragraphs of an article have been reprinted below, omitting the citations.

Introduction

[1] Single- and multiple-banded (hereafter, banded) clouds and precipitation are commonly observed in association with frontal zones in extratropical cyclones. . . . Indeed, some observational studies over extended periods of time show the presence of MSI [moist symmetric instability] in association with banded precipitating baroclinic systems to be rather common.

[2] Although we do not deny the likely existence of slantwise convection or the possible involvement of MSI in some precipitating systems in the atmosphere, it is our contention that CSI [conditional symmetric instability, a form of MSI] is frequently misused and overused as a diagnostic tool. We believe the following four reasons are responsible, in part, for the present situation. . . . Thus, for these four reasons, CSI is commonly observed yet often misinterpreted and misunderstood.

[3] The purpose of this article is twofold: to attempt to limit further misuse of the CSI paradigm by researchers and forecasters alike by highlighting common pitfalls, and to encourage future research explorations that are directed at the deficiencies in our understanding of MSI and slantwise convection. The remainder of this article is as follows. . . . Finally, Section 8 consists of a summary of main points, directions for future research, and a concluding discussion.

An ingredients-based methodology for slantwise convection

[4] Throughout this article, we wish to differentiate between *free convection* and *forced convection* as motions in the atmosphere that are associated with the presence and absence of instability, respectively. Unless otherwise specified, we use the generic term *convection* to imply free convection (gravitational or symmetric). [This paragraph is only two sentences long.]

[5] To clarify some of the confusion surrounding the concepts of CSI and slantwise convection, we find it useful to demonstrate parallels with the more familiar concepts of moist gravitational instability and convection. An exploration of these parallels begins with an ingredients-based methodology for forecasting deep, moist convection. . . . "Remove any one of these [ingredients] and there well may be some important weather phenomena, but the process is no longer deep, moist convection."

[6] For the purposes of this article, we adopt the same triad of ingredients from moist gravitational convection (instability, moisture, and lift) for the production of moist slantwise convection, where the requisite instability becomes MSI, rather than moist gravitational instability. . . . The ingredients-based methodology firmly labels CSI as the instability, clearly separate from the lifting mechanism.

Even with most of the central text within each paragraph omitted, the remaining text remains mostly readable. The reason is the effective coherence between the paragraphs. For example, enumeration was used within paragraphs 2 and 3. Repetition of "MSI" and "precipitating systems" was used between paragraphs 1 and 2, and repetition of "ingredients" was used between paragraphs 5 and 6.

Some improvement in coherence between paragraphs 2 and 3 could have been gained by repeating "misinterpreted and misunderstood" at the beginning of paragraph 3 with a slight revision: "The purpose of this article is twofold: to attempt to limit further *misinterpretation* of the CSI paradigm by researchers and forecasters alike by highlighting common pitfalls, and to encourage future research explorations that are directed at *correcting our misunderstandings* of MSI and slantwise convection." This revision shows more

SECTIONS AND SUBSECTIONS

Sections and subsections can be important to your paper for helping the reader see the organization of the paper more clearly. Sections and subsections allow readers to identify quickly the topics of interest to them and to skip the others. Section headings also provide some relief from whole pages of uninterrupted text, which can be imposing to a reader. However, creating subsections does not substitute for good transitional writing between the paragraphs (Section 8.3). Here are a few basic rules for creating sections and subsections:

- In general, at least two sectional headings are needed (e.g., Section 3.1 or 3a must be followed by a Section 3.2 or 3b). However, a minority of authors have argued that a single subsection within a section is legitimate. Creating a second subsection, they argue, would be forced, not natural.
- In general, some introductory text should exist between a major heading and a subheading (e.g., between the heading for Section 4 and the heading for Section 4.1). This material can be introductory material or a discussion of what will be covered within the section.

- Balance the number of headings, the number of topics to be discussed, and the length of the text under each heading. Too few headings and the corresponding text may be too long; too many headings and the corresponding text may be too short.
- Heading titles should have the same properties of a manuscript title, albeit much shorter: informative, accurate, clear, concise, and attention commanding (Table 3.1).
- Use descriptive titles, avoiding one-word titles (except for "introduction," "conclusions," etc.).
- Keep titles at each level parallel, if possible. If the titles are verb phrases (e.g., "Constructing the climatology," "Evaluating model performance"), do not intersperse noun phrases (e.g., "Comparison of control and no-flux simulations").
- Repeating the title in the body of the text shortly after starting the new section can give the readers comfort that you are going to address the topic that is described by the title.

Before submitting a manuscript, separate from the text and list all the section and subsection headings (e.g., table of contents, outline). Are the titles parallel (Section 9.4)? Does the organization of the paper as told through the outline make sense? See Section 4.15 for examples of effective paper organization.

clearly that the misinterpretations and misunderstandings of CSI will be addressed within the article.

8.4 LENGTH AND STRUCTURE OF PARAGRAPHS

In scientific writing, four to eight sentences per paragraph seems to be optimal in most cases. Although shorter paragraphs of two or three sentences can be used for emphasis from time to time, avoid single-sentence paragraphs as a general rule. Such paragraphs should be eliminated, merged in with another paragraph, or developed into a longer paragraph. On the other hand, coherent paragraphs much longer than eight sentences may be functional, but you may wish to break them up. Because the white space around paragraphs on the printed page serves partially as a visual break for the reader, long tracts of text can be imposing to the reader and are candidates for splitting into multiple paragraphs.

Paragraphing calls for a good eye as well as a logical mind. —William Strunk and E. B. White (2000, p. 17)

Within the paragraph, the sentences should vary in length and in rhythm, specifically in their construction or the location of the subject and verb within the sentence. Too many short sentences sound too sing-songy or elementary, whereas too many long sentences tire the reader. In the same way, the assemblage of paragraphs in the manuscript should also have variety in length and structure.

CONSTRUCTING EFFECTIVE SENTENCES

9

Well-written sentences convey information succinctly and precisely. Examples presented in this chapter guide authors toward improving their sentences. These improvements include such topics as subject–verb placement, overuse of passive voice, improper or inconsistent verb tense, and misplaced modifiers.

In the previous chapter, I said that paragraphs are the fundamental organizational unit of a paper. If this is the case, then sentences are the vehicle that delivers the message. Sentences composed of a series of disorganized words go nowhere. Whereas the construction of paragraphs focuses on coherence and unity of message, the construction of sentences focuses on concision and precision. In other words, sentences should say exactly what is meant in as few words as possible.

> A dog goes into a telegraph office, takes a blank form, and writes: "Woof woof woof. Woof, woof. Woof. Woof woof, woof."
>
> The clerk examines the paper and politely tells the dog: "There are only nine words here. You could send another 'Woof' for the same price."
>
> The dog looks confused and replies, "But that would make no sense at all."

Just like the dog's message, sometimes too many words can turn an otherwise clear sentence into nonsense. In this chapter, I present ways to improve the concision and precision of sentences. For some authors, applying the examples in this chapter will reduce the length of their drafts up to 20%, and, in the process, enhance clarity and precision. Although many of the examples to follow in this chapter will be grammatical, the present book is not intended to teach basic grammar skills. Nevertheless, some reminders about proper grammar usage will likely be useful.

9.1 ACTIVE VOICE VERSUS PASSIVE VOICE

One of the challenges facing scientific and technical writers is minimizing the use of passive voice and incorporating more active voice. Overuse of passive voice makes the manuscript dense to read and longer than necessary, so including more active voice generally strengthens manuscripts.

In active voice, the grammatical subject of the sentence acts upon the verb, whereas in passive voice, the subject is acted upon by the verb, which is a combination of a form of the verb "to be" (e.g., is, was, were) and the past participle (a verb with an "-ed" ending, commonly).

ACTIVE: I performed a simulation using a nonhydrostatic mesoscale model to understand the evolution of the squall line.

PASSIVE: A simulation was performed using a nonhydrostatic mesoscale model to understand the evolution of the squall line.

ARE FIRST-PERSON PRONOUNS ACCEPTABLE IN SCIENTIFIC WRITING?

Some teacher or professor in your past might have taught you to avoid the use of the first person (*I* or *we*), leading to a forced marriage with the passive voice. To appear disconnected from the research, common practice among authors of scientific and technical documents is to favor the passive voice, with the person who performed the simulation unstated and irrelevant. Such obtuse writing style has not always been the preferred style. Prior to the 1920s in the United States, active voice and first-person pronouns were quite common in scientific writing. Because science is done by individuals who make conscious decisions in designing, implementing, and communicating their research, such an air of impersonality, frankly, is disingenuous. We are intimately tied to our research and bias creeps in. The least we can do is acknowledge it.

Avoid first-person pronouns in the abstract—many journals do not allow it. However, most journals accept limited use of the first person in the body of the paper. I believe the first person can be quite effective when used sparingly and with purpose. Beware, however, that others may feel differently.

Avoid describing the rote methods of the research or the manuscript format almost entirely in first person or talking about yourself in the third person as "the author." Generally, you can use "this work" or "the present article" with the active voice to avoid first or third person.

DRAFT: I examined the events from Tables 1 and 2 for evidence of cloud-to-ground lightning. [sounds too conversational]
IMPROVED: The events from Tables 1 and 2 were examined for evidence of cloud-to-ground lightning.

DRAFT: We discuss the spatial distribution of the precipitation in northern Utah.
IMPROVED: The spatial distribution of the precipitation in northern Utah is discussed in the present article.

The subject of the sentence in active voice is "I," whereas the subject of the sentence in passive voice is "simulation." Because the first person "I" in this context is not generally used in a scientific document (see the below sidebar), passive voice dominates most scientific writing, even in situations where active voice would be preferred. Nevertheless, both active and passive voice are acceptable in scientific literature, although some authors would benefit from incorporating more active voice.

Here are three ways to change a passive sentence into an active one. First, put the object doing the action as the subject of the sentence (e.g., before the verb).

PASSIVE: Gamma or lognormal distributions commonly have been used to model *drop size distributions.*
ACTIVE: *Drop size distributions* commonly are modeled with gamma or lognormal distributions.

IMPROVED: The present article discusses the spatial distribution of the precipitation in northern Utah.

I use the first person very consciously to emphasize an action or decision that affects the outcome of the science being described.

DRAFT: Given option A and option B, the authors chose option B to more accurately depict the location of the front.
IMPROVED: Given option A and option B, we chose option B to more accurately depict the location of the front.

In the above example, because the results of the research may depend strongly on that choice, I want to make it clear to the audience that *we* made a decision to do something that impacts the outcome of the paper; two options were available, but *we* chose option B.

Similarly, the first person helps make sentences discussing speculation less awkward and more clear by indicating exactly who is speculating.

DRAFT: It is speculated that . . .
[Who is "it"? Who is speculating?]
DRAFT: The author speculates that . . .
[awkward]
IMPROVED: I speculate that . . .

If you feel that a sentence starting with "I" may sound too bold for many readers of a scientific paper, then move the first-person pronoun away from the start of the sentence with an introductory phrase: "Because the aerosol concentration increased dramatically, I speculate that. . . ."

Finally, I should comment about the use of "we" in a single-authored manuscript, or what is termed a *nosism.* Referred to by some authors derogatorily and incorrectly as *the royal we,* "we" in this context actually refers to "the author and the reader." Although some authors are comfortable with the nosism, others see "we" as condescending or patronizing. As with all language debates, exercise caution when employing contested language in your own writing.

Second, eliminate part of the verb.

> **PASSIVE:** Improved warnings *are perceived to be* an important safety benefit of weather radars.
> **ACTIVE:** Improved warnings *are* an important safety benefit of weather radars.

Third, pick a different verb.

> **PASSIVE:** A stationary snowband *was initiated* over southeastern Wyoming.
> **ACTIVE:** A stationary snowband *formed* over southeastern Wyoming.

Consider the following pair of sentences.

> **PASSIVE:** Light snow lasting four and a half hours was officially reported at Raleigh.
> **ACTIVE:** Raleigh officially reported light snow lasting four and a half hours.

Call for Papers from the Journal of the Passive Voice: A new publication has been started. It has been reported to be a sub-publication of Annals of Improbable Research *(AIR). The new journal has been named* Journal of the Passive Voice. *Articles written entirely in the passive will be seen to have been published in this new journal.* —Annals of Improbable Research, 4 (3), p. 15.

Both sentences are acceptable and would be welcome in a scientific document. How do you decide which to use? The answer depends on the desired emphasis, location within the document, and context within the paragraph (Table 9.1). Should you want to emphasize "snow," the sentence in the passive voice would be favored because its subject is "snow." On the other hand, if the sentence appears in a paragraph about the weather in Raleigh, the sentence written in active voice would probably be better.

Table 9.1   When to use active versus passive voice

Active voice is best used:
- ❯ to emphasize the subject of the sentence
- ❯ to emphasize the person or people doing the science ("I speculate that . . .")
- ❯ when describing figures or other work
- ❯ in declarative sentences, such as topic sentences
- ❯ to avoid sentences that begin with "there are" or "it has been shown that"

Passive voice is best used:
- ❯ when the subject of the sentence is unstated, unknown, or irrelevant
- ❯ to emphasize the object of the sentence
- ❯ within the data and methods section (to avoid first person)
- ❯ within abstracts (to avoid first person)
- ❯ for variety
- ❯ for coherence in the paragraph

To maintain the coherency of the paragraph through repetition (Section 8.2.1), you may need to choose one voice over the other to reverse the order of the sentence. In the first example below, choosing active voice in the first sentence means a similar structure to both sentences, which may sound elementary to some readers. In the second example, reversing the order of the first sentence by employing the passive voice results in coherency through repetition of "reduced dataset" in the stress position of the first sentence and in the topic position of the second sentence. Alternatively, both sentences could be combined, as in the third example, keeping active voice throughout.

FIRST SENTENCE ACTIVE: The reduced dataset consisted of stations that reported at least 80% of the possible surface observations. This reduced dataset consisted of 692,790 observations of nonfreezing drizzle from 584 stations.

FIRST SENTENCE PASSIVE: Stations that reported at least 80% of the possible surface observations were separated into a reduced dataset. This reduced dataset consisted of 692,790 observations of nonfreezing drizzle from 584 stations.

Articulate the action of every clause or sentence in its verb. —George Gopen and Judith Swan (1990)

COMBINED: The reduced dataset consisted of stations that reported at least 80% of the possible surface observations, resulting in 692,790 observations of nonfreezing drizzle from 584 stations.

In addition to writing in active voice, another way to make your sentences more potent is to choose verbs that emphasize action. Avoid weak verbs such as "occur," "see," "exist," and "observed"; favor stronger words that describe the relationship in the sentence rather than just saying the relationship exists. Too many sentences with "is," "are," "has," and "have" bore the reader (and the writer). As previously described, the reader looks to the verb in the sentence to see what the subject is doing, and passive sentences that lack action limit their ability to tell the story. Furthermore, selecting active verbs creates a more concise and precise sentence: "Brevity is a by-product of vigor" (Strunk and White 2000, p. 19).

DRAFT: An environment favorable for an airstream boundary *is* the result of the strong convergence and deformation associated with the surface cyclone.

IMPROVED: The strong convergence and deformation associated with the surface cyclone *creates* an environment favorable for an airstream boundary.

Do not be afraid to use the thesaurus. You do not have to write "Smith et al. (1995) ostended" when you mean "Smith et al. (1995) showed," but a little variety will improve your writing. Table 9.2 can help.

Choose active verbs rather than their noun forms. Avoid phrases such as perform a comparison, make a generalization, provide information, or reveal

Table 9.2 Some action verbs for scientific writing (augmented from Schall 2006, pp. 54 and 113)

acknowledge	compare	disagree	guide	list	recommend
admit	conclude	display	highlight	maintain	reiterate
analyze	consider	dispute	hypothesize	mean	report
argue	construct	distinguish	illuminate	measure	represent
articulate	construe	effect	illustrate	narrate	restrict
ascertain	contrast	elucidate	imply	neglect	reveal
assert	deduce	elude	improve	note	simplify
assert	define	employ	indicate	obtain	specify
assess	delineate	establish	infer	offer	speculate
attribute	demonstrate	estimate	inform	organize	state
believe	depict	evaluate	insist	postulate	suggest
calculate	derive	evince	interpret	predict	summarize
challenge	designate	exhibit	introduce	present	support
characterize	detail	explain	investigate	propose	surmise
clarify	determine	extrapolate	invoke	prove	synthesize
classify	devise	generalize	issue	provide	yield

a possible indication, when you can use more simple words such as compare, generalize, inform, or indicate. Similarly, often we add superfluous words when a more direct approach would suffice: acts to dry out → dries out; creates a moister environment → moistens; is used to denote → denotes; found to be → is; serves to introduce → introduces; and makes a measurement → measures.

9.2 SUBJECT-VERB DISTANCE

Consider the following sentence:

> **DRAFT:** Extratropical cyclones with two or more warm-front-like baroclinic zones over the central United States and southern Canada during 1982–1989 were examined.

Twenty words separate the subject "cyclones" from the verb "were examined." This distance keeps the readers in suspense, waiting to know what happens to the cyclone. Readers need understanding of what the subject is doing, and delays in receiving the second piece of information (the doing) inhibits comprehension. Words in between the subject and its verb are viewed as less important.

IMPROVED: Extratropical cyclones over the central United States and southern Canada during 1982–1989 were examined for the presence of two or more warm-front-like baroclinic zones.

9.3 VERB TENSE

Choosing verb tenses in scientific writing can also be confusing and not without controversy. The following guidelines appear to be generally held by most authors:

● Scientific fact is reported in the present tense: "The wavelength of maximum emission of solar radiation is 0.5 μm," "Ice pellets are frozen raindrops."
● Past events are described using the past tense: "On 12 December, 23 cm of snow fell," "An unusual climate shift occurred over the North Pacific Ocean around 1977."
● Present tense is used when referring to a figure, table, or calculation: "Table 3 shows," "the values are statistically significant."
● When the action started in the past and continues in the present, the present perfect tense (verb form of "have" and the past participle) is used: "the model has been developed."
● When the action started in the past continues in the present and will continue in the future, the present perfect progressive tense (verb form of "have" plus "been" and the present participle) is used: "the model has been developing."
● Future tense can be employed when referring to what will happen later in the paper, although concision argues for dropping the "will" and using the present tense: "Section 3 will discuss . . ." versus "Section 3 discusses. . . ."

Disagreements begin when considering the following situation. Should your own research (particularly the methods and results sections), as well as that of others, be reported in the past tense or in the present tense?

EXAMPLE 1: The simulation is/was run for 24 h, initialized from 1200 UTC 31 January.

EXAMPLE 2: Hansen (2005) derives/derived . . .

Most authors choose to write in the past tense because the work was done in the past. Furthermore, the use of past tense ensures that such a statement will remain true in the future, even if subsequent research comes to a different conclusion. These generalizations, however, are not supported by everyone.

Some authors argue that because published articles are in the past, their conclusions represent fact and should therefore be discussed in the present tense. (When past actions are discussed in the present tense, you can see why people get confused over verb tenses!) Nevertheless, others disagree, arguing reporting in the present tense "confers authority without substantiation." Ultimately, you must make up your own mind as to the verb tense you prefer in these situations. Whatever verb tense you choose, be consistent throughout the manuscript.

9.4 PARALLEL STRUCTURE

School teachers may have told you to mix up your writing by not repeating the same words and sentence structures. Although our teacher's advice may be appropriate for literary writing, repeating sentence structures, words, and phrases can be quite beneficial to readers of scientific documents (Section 8.2.1), especially in lists or when making comparisons. In performing experiments, scientists try to control as many variables as possible, changing only one variable at a time. Precise writing works the same way. Keeping structures parallel will help the reader follow your train of thought.

> **DRAFT:** The cyclonic path of the cold conveyor belt is represented by trajectories 21–23, whereas trajectories 24 and 25 resemble the anticyclonic path.
> **IMPROVED:** Trajectories 21–23 resemble the cyclonic path of the cold conveyor belt, whereas trajectories 24 and 25 resemble the anticyclonic path of the cold conveyor belt.

Similarly, words and expressions joined by a conjunction require the same form.

> **DRAFT:** Many of the standard statistical tests of differences assume independence of data points and that the underlying distribution of the sample is known.
> **IMPROVED:** Many of the standard statistical tests of differences assume that data points are independent of each other and that the underlying distribution of the sample is known.

> **DRAFT:** Given the gaps in our knowledge of the structure, evolution, and the dynamics of surface cold fronts . . .
> **IMPROVED:** Given the gaps in our knowledge of the structure, evolution, and dynamics of surface cold fronts . . .

9.5 COMPARISONS

Another form of nonparallel structure is incomplete comparisons.

> **DRAFT:** Surface confluence in west Texas with this surface pressure pattern was much smaller. [The reader is probably asking, "Smaller than what?"]
>
> **IMPROVED:** Surface confluence in west Texas with this surface pressure pattern was much smaller during the weak-dryline days than the strong-dryline days.

In the next example, the sentence suffers from both a partial comparison (what is being rigorously compared with theory?) and passive voice ("was incapable of being performed"). The revision solves both problems.

> **DRAFT:** A more rigorous comparison with theory was incapable of being performed because of the lack of theoretical studies on this complex situation with these three instabilities forced by frontogenesis.
>
> **IMPROVED:** Comparing these observational and numerical-modeling results with theory was not possible because theoretical and idealized-modeling studies are lacking for this complex situation with these three instabilities forced by frontogenesis.

Sometimes in our haste, we leave out words, shortening the sentence. Unfortunately, such omissions can convey sloppiness, ambiguity, or worse, inaccuracy. This is an example of how care should be taken with wording.

> **DRAFT:** Mammatus form in the four simulations initialized with soundings taken when mammatus were observed, whereas no mammatus form for the one no-mammatus sounding.

The second half of the sentence suggests that the mammatus form from soundings rather than in simulations. The revised sentence clarifies this inaccuracy.

> **IMPROVED:** Mammatus form in the four simulations initialized with soundings taken when mammatus were observed, whereas no mammatus form in the simulation initialized with the one no-mammatus sounding.

If the word "than" is present, check to see that the comparison is complete and that the structure is parallel. In the draft example below, the sentence reads as if the static energy of the subcloud air is being compared to the height of the cloudy air ("higher than the cloudy air"). The two proposed revisions make it clear that the *static energies* of the subcloud and cloudy layers are being compared. Thus, make sure that apples are being compared to apples and not to broccoli.

> **DRAFT:** The static energy of potentially warm, dry subcloud air is higher than the cloudy air above.
> **IMPROVED:** The static energy of potentially warm, dry subcloud air is higher than that of the cloudy air above.
> **IMPROVED:** The static energy of potentially warm, dry subcloud air is higher than the static energy of the cloudy air above.

The word "both" can be problematic, especially in comparisons. In the draft example below, whether the author meant the diabatic heating term was larger than the differential vorticity advection term and the Laplacian of the thermal advection term individually or the sum of the two terms is ambiguous, as shown by the two improved examples. Only the author knows which interpretation represents the correct meaning.

> **DRAFT:** The diabatic heating term dominated both the differential vorticity advection term and the Laplacian of the thermal advection term.
> **IMPROVED:** The diabatic heating term dominated *the two terms*, differential vorticity advection *and* the Laplacian of the thermal advection, *individually*.
> **IMPROVED:** The diabatic heating term dominated *the sum of* the differential vorticity advection term *and* the Laplacian of the thermal advection term.

Sometimes to describe a comparison, authors choose a sentence structure with parenthetical words or phrases. Such sentences, however, may be difficult to read and interpret. Often, such sentences are better written explicitly, even if they become longer. Revisions may also be possible by completely rewording the sentence.

> **DRAFT:** When temperature increases (decreases), relative humidity decreases (increases).
> **IMPROVED:** When temperature increases, relative humidity decreases, and when temperature decreases, relative humidity increases.
> **IMPROVED:** Temperature and relative humidity are inversely related.

A word pair that can slow down readers is "former/latter." Such words make sentences more concise, but often at the expense of requiring the reader to look backward in the text to remember the order.

DRAFT: . . . , where equation (1) is the continuity equation and equation (2) is the thermal wind equation. The implication of the former equation is . . .

IMPROVED: . . . , where equation (1) is the continuity equation and equation (2) is the thermal wind equation. The implication of the continuity equation is . . .

9.6 NEGATIVES

Negative information is more difficult for people to comprehend, often resulting in reduced understanding and reading rate because readers must first comprehend the statement, then negate it. Wording sentences positively improves their readability and tends to make them shorter. Eliminating the word "not" often makes a stronger sentence, as the examples below indicate.

DRAFT: This modeling study did not prove conclusive.

IMPROVED: This modeling study was inconclusive.

DRAFT: There did not appear to be any preferred geographical regions in which bow echoes developed from particular modes.

IMPROVED: Bow echoes showed no geographical preference to develop from particular modes.

Furthermore, increasing the number of negatives, especially words with a negative connotation (e.g., avoid, never, fail, unless, however), in a sentence further confounds comprehension.

DRAFT: At 1900 UTC, areas of drizzle across Pennsylvania were not associated with regions of higher visibility, unless fog was not present additionally.

IMPROVED: Areas of simultaneous fog and drizzle across Pennsylvania at 1900 UTC had lower visibility than areas of drizzle only.

9.7 MISPLACED MODIFIERS

Misplaced modifiers are also called dangling modifiers or dangling participles. As in the discussion of phrases starting with "using" on page 23, modifying words or phrases should be close to the words or phrases they modify. Not doing so often results in confusion or amusement for the reader. Phrases at the beginning of a sentence are especially problematic.

DRAFT: Inside the tornado, the model results show a rapid decline in wind speed. [The model is inside the tornado?]

IMPROVED: The model results show a rapid decline in wind speed inside the tornado.

DRAFT: Over the central United States, forecasters have found that castellanus clouds may mark the initial stages of elevated nocturnal thunderstorm development. [Forecasters are over the central United States? In hot-air balloons?]

IMPROVED: Forecasters have found that castellanus clouds over the central United States may mark the initial stages of elevated nocturnal thunderstorm development.

9.8 RHYTHM AND AESTHETICS

All the advice in this chapter means nothing if the sentence does not make any sense. After writing, read the sentences out loud. How do they sound? If you have to read the sentences twice to understand them, then your readers will have to read them three or more times. Look for a natural rhythm in your writing that helps the audience get comfortable when they read your work.

If something does not sound right, reword it. Reverse a word or two. Does that improve it? If not, try larger changes to the sentence. Perhaps, reverse the order of the sentence.

Where possible, avoid visually complex sentences, the visual equivalent of quicksand for readers. Too many of these will be tiresome for the reader. Things that add to visual clutter include equations, numbers, parenthethical phrases, too many phrases set off by commas, and symbols. Abbreviations with periods cause the reader to stop, as if at the end of the sentence, disrupting the flow in reading. Acronyms force readers to read all capital letters, which takes longer because they have less practice reading in all capitals. Text with many equations and not enough explanation in between is visually imposing. Follow the examples of authors who have written such articles well by interspersing text and equations to create a visual balance.

Sometimes you may be left with material that you feel compelled to include in the text, but you face difficulties in fitting the material into the structure of the paragraph while maintaining coherence. "A footnote!" you think. Footnotes serve a purpose, of course, but they should not take the place of an effectively written transition.[1] Avoid a large number of footnotes for ancillary, tangential, or unimportant material. If possible, either eliminate the footnoted material or include it in the text.

---

1. Readers expend time and effort searching out footnoted material. A large number of footnotes can be exhausting for the reader's eyes and can limit your ability to communicate your argument coherently.

USING EFFECTIVE WORDS AND PHRASES

10

How you choose your words can make the difference between text that is confused or clear, long-winded or lean, and ordinary or extraordinary. Eliminating redundant and complex words, trimming verbose and unnecessary phrases, and choosing precise and meaningful words engage the reader most effectively. Applying the lessons in this chapter to your writing will help convey your meaning to the reader concisely and precisely.

In the song "Open the Door, Richard," Louis Jordan calls out (followed by the crowd response in parentheses):

I met old Zeke standin' on the corner the other day.
That cat sure was booted with the liquor. (He was what?)
He was abnoxicated. (He was what?)
He was inebriated. (He was what?)
Well, he was just plain drunk. (Well, alright then!)

Jordan tries a colloquialism (booted with the liquor), a nonexistent word (abnoxicated), and a more tasteful word (inebriated), until he finally gives the crowd a word they understand.

As with sentences, words convey meaning, ideally with both concision and precision for audience understanding. If paragraphs are the fundamental organizational unit of a paper and sentences are the vehicle by which the message is delivered, then words are the sparkling new coat of paint on the vehicle that makes the sentences shine. Or, for poorly chosen words, the crud on the windshield that obscures a clear view down the road.

10.1 CONCISION

One of my teachers said, "do not use a $10 word in place of a 10-cent word." For example, authors often use the word "utilize" because it sounds more scientific, but "use" has the same meaning and is shorter. The same thing holds for the following pairs of complex–simple words: perform–do, initiate–start, facilitate–cause, and propagation–move. (The difficulties with "propagation" often go beyond it being a pretentious word. Page 361 has further discussion.) Authors might choose more complex words because they want to make their arguments more complex (and more impenetrable) or because they want to impress others with their large vocabularies. Whatever the reason, using complex words when simple ones would suffice generally makes the writing less clear.

Making writing concise is as much about reducing unnecessary words as it is reducing the complexity of the words. Minimize your use of phrases that have become intimately linked to one another so as to be cliché (e.g., meaningful dialog, time and time again, first and foremost).

Other phrases are simply redundant such as "smaller in size" ("small" already implies size) or "model simulation" ("model" and "simulate" are both similar terms). Use "smaller" or "simulation" (or "model results") instead. More examples are given in Table 10.1. Save some words, and be more creative.

Similarly, Table 10.2 lists words and expressions to avoid. One thing to notice about this table is the large number of phrases that begin with "it": "it has been noted that," "it is known that," and "it is clear that." These phrases are bad for two reasons. First, they add unnecessary length. Try removing these phrases from your sentences—the meaning of the sentence often will be unaffected. Second, the "it" is undetermined. "It has been hypothesized that enhanced deposition leads to more latent heat release." What does the "it" refer to? Who hypothesized this? If "it" is known, reword the sentence to incorporate the references or the first-person pronoun.

Like gratuitous variation, superfluous material can act as verbal camouflage. It can activate irrelevant connections in the reader's brain, and impede perceptual processing by making word patterns needlessly complicated. —Michael McIntyre (1997, p. 201)

Table 10.1   Redundant word combinations; words that could be eliminated are in parentheses

(absolutely) essential	(definitely) proved	(long) been forgotten	simply (speaking)
(already) existing	empty (void)	mix (together)	smaller (in size)
(alternative) choices	(end) result	(model) simulation	(solar) insolation
at (the) present (time)	(fellow) colleague	never (before)	(temporal) evolution
(basic) fundamentals	fewer (in number)	none (at all)	the (color) white
(completely) eliminate	first (began)	off (of)	the white(-colored) *noun*
(completely) false	(general) overview	(overall) summary	(time) evolution
(continue to) remain	(generally) tend to	past (experience)	variety of (different)
(currently) underway	introduced (a new)	period (of time)	(very) unique

---

On the other hand, the little word "about" deserves more respect than it gets (Table 10.2). Authors commonly step around this word with such verbosity as "approximately," "regarding," "with respect to," "more or less," or "in the vicinity of."

A different perspective on the phrases in Table 10.2 comes from Montgomery (2003, p. 9), who argues that such phrases may serve an important function such as pacing, flow, or transition. I agree that an occasional and purposeful use of such phrases can benefit the text, but I also caution that overuse of these phrases, which unfortunately occurs too often in scientific writing, runs the risk of wearing out the reader's patience.

Describe the science, not the figures. If the figure does not need an introduction, do not introduce it. This change not only reduces the number of

Table 10.2 Words and expressions to avoid and their shorter alternatives (partially adapted from Day and Gastel 2006, Appendix 2, and U.S. Air Force 2004, pp. 81–87)

Avoid	Alternative	Avoid	Alternative
a 15-min temporal basis	every 15 min	it should be noted that	(omit)
a greater number of	more	it was found that	(omit)
despite the fact that	although	it was/is noted that	(omit)
due to the fact that	because	more or less	about
for the purpose of	(reword)	note that	(omit)
in a number of cases	some	of particular interest	(reword)
in order to	to	on the order of	about
in reference to	about	over the Mongolia region	over Mongolia
in spite of the fact that	even though	summertime	summer
in terms of	by, in	temperature of −30°C	−30°C
in terms of stability	(omit or reword)	the period 1977–1999	1977–1999
in the context of	(omit)	the result indicates that	(omit)
in the event that	if	the results show	(omit)
in the matter of	about	the smallest values of lapse rate	the least stable
in the spring of 2008	in spring 2008	the southeastern part of Finland	southeastern Finland
in the vicinity of	near, about	the state of California	California
is equal to	is	through the use of	by, with
is shown to be	is	thunderstorm activity	thunderstorms
it appears that	(omit)	upward vertical velocity	ascent
it is apparent that	apparently	was acting to	was
it is contended that	(omit)	was found to be	was
it is important to note that	(omit)	was noted to	was
it may be expected that	(omit)	was observed to	was
it may be that	I think	with regard to	about
it must first be established that	(omit)	with respect to	about

words, but also shifts the focus from the figure to the science, which is where it should be. Further discussion of this point is found in Section 11.13.

> **DRAFT:** Figure 5 shows plots of surface temperature in the no-ice and control simulations, showing that the elimination of sea ice produced warmer arctic temperatures.
>
> **IMPROVED:** The elimination of sea ice in the no-ice simulation resulted in warmer arctic temperatures compared to that of the control simulation (Fig. 5).

10.2   PRECISION

The words we choose convey our thoughts. A carelessly chosen word can cause the reader to slow down, be confused, or even misinterpret the author's intended meaning. In addition, using excessive jargon and figurative language or anthropomorphizing inanimate objects fails to adequately describe the relevant science, and, to more careful readers, inadequately hides our lack of knowledge of the science. In this section, we look at how we can choose our words to achieve more precise meaning.

10.2.1   Denotation versus connotation

Words have two meanings—their *denotation*, the dictionary definition or literal meaning, and their *connotation*, the associated or implied meaning. Be aware of both meanings when writing. Use the dictionary to determine if the word you are considering has the exact meaning you intend. Sometimes similar words may have slightly different denotations. If a word is not precisely what you mean, use a thesaurus (along with a dictionary) to find a more precise word.

As an example of denotation versus connotation, authors commonly overuse "state" to mean "say," as in "Smith et al. (1996) stated the sky is blue." The primary definition of "state" in many dictionaries is "to declare definitively" as in legal proceedings (state your name) or in a scientific context (state a hypothesis or state the problem). This denotation is a much stronger and precise meaning than its connotation. Perhaps returning to this stronger meaning for "state" is something that we in science should strive for. Similarly, "claim" has the denotation of "say," but the connotation is that a person is not being truthful. Inappropriate use of "claim" can lead to implied bias against that person.

Sometimes words in common usage can be troublesome in scientific contexts. Consider "significance" (see also page 362), as in "a significant temperature anomaly." The scientific context implies that statistical tests have shown the results to be statistically significant, although the connotation is just "an impressive temperature anomaly." A selection of words that have scientific meanings different from their connotations are listed in Table 10.3.

Don't be afraid of elegant prose. Just as clothing can be utilitarian (keeping you warm and dry) and attractive at the same time, the best writing clearly communicates its message while providing a bit of aesthetic delight. Your prose doesn't have to be overly fancy to be pleasing. Like a classic tuxedo or black dress, a straightforward scientific paper can still sparkle with clarity and precision. —Bob Henson, writer/editor/ media relations associate, University Corporation for Atmospheric Research

Table 10.3 Words with troublesome connotations in a scientific context; see further discussion in Appendix B

accuracy/skill	correlate/correlation	severe storms
causing	observed/seen	severe weather
chaos/random	propagate	significance/significant
collaboration/coordination	resolution	theory

10.2.2 Jargon

Scientific writing cannot avoid *jargon*, the language that has been developed and has evolved to describe our science. Some jargon is specialized vocabulary, defined in scientific reference material, such as the *Glossary of Meteorology* (Glickman 2000), and essential for concisely conveying concepts between experts. Other jargon is scientifically incorrect, inappropriate, vague, or colloquial, as some examples in Appendix B demonstrate. In preparing your paper or presentation, jargon that is not likely to be understood by your audience should be defined or changed to a simpler language the audience will understand.

Sometimes multiple terms have arisen to describe the same thing. As an example, all the following terms refer to the same phenomenon: retrograde occlusion, back-bent, loop, broken-back, or bent-back occlusion, bent-back warm front, bent-back front, and secondary cold front. Part of good scholarship is not to create any more unnecessary terms, but to identify and clarify any discrepancies or confusion with existing terms. If multiple terms exist, consistency is key to communicating with your audience. For example, "gravity current" and "density current" describe the same phenomenon. Upon first mention of the phenomenon in your paper, introduce both terms, saying that both terms have been used interchangeably, but pick one term and stick with it throughout the manuscript. Even terms we think we may be familiar with, we may misuse. For example, "mammatus clouds" is incorrect, because mammatus are not clouds, but cloud forms.

Weather weenies, people who are passionate about the weather (see how I defined my jargon?), are a unique species. Online discussion groups about storm chasing have arisen, daily meetings in the weather-map room take place, and national forecasting contests challenge the best. Part of being a weather weenie is understanding the jargon, to be part of the in-crowd. Jargon can also intimidate others who are unfamiliar with that specialized jargon. But, more importantly, such jargon fosters sloppiness and a poor understanding of meteorological knowledge and atmospheric processes. In your writing, eliminate map-room jargon that is colloquial or obscures scientific meaning.

For example, do not refer to vorticity maxima in the jet stream as "energy." In fact, start in the map room, the area forecast discussions, the chat rooms, the weather blogs, and the mailing lists. Elevating the level of discourse will benefit your writing as well your scientific understanding.

> **DRAFT:** Upper-level support overran the surface low center resulting in bombogenesis.
> **IMPROVED:** Cyclonic vorticity advection increasing with height was associated with the rapid-deepening phase of the surface cyclone.

Accurately describe the process, making sure to not eliminate words or levels that may seem obvious to you but may make the wording unclear for the reader.

> **DRAFT:** Cold advection moved over eastern Texas.
> **IMPROVED:** A region of cold advection at 850 hPa moved over eastern Texas after 1200 UTC.

10.2.3 Unclear pronouns

The antecedent of the pronoun (the noun that the pronoun represents) should always be clear. Pronouns standing by themselves, not adjacent to a noun, are immediately suspect. To avoid problems, put a noun after each isolated example. "This" and "it" are especially abused. During revisions, search through the manuscript for "this" and "it," fixing instances in which the antecedent is vague.

> **DRAFT:** Frederick (1966) provided further support for *this* by showing the eastward progression of the warm spell across the United States, suggesting that *it* may be related to eastward-moving offshoots of the Aleutian low.

It is surprising how often repeating a noun works better than substituting a pronoun such as "it," "this," "them," "ones," etc., and it is surprising how seldom a repeated noun jars upon the reader. —Michael McIntyre (1997, p. 200)

Although the antecedent to "this" is in the previous sentence, which is not shown here, repetition (Section 8.2.1) would maintain the coherence from one sentence to the next and also define the unclear pronoun. Second, what does "it" refer to: "support," "evolution," "progression," "warm spell," or "United States"? The sentence can be reworded to define the antecedents for "this" (evolution) and "it" (progression). (The jargon "offshoots" is also dealt with as well.)

> **IMPROVED:** Frederick (1966) provided further support for this evolution by showing the eastward progression of the warm spell across the United States may be related to eastward-moving secondary low centers developing from the Aleutian low.

Here is another example where an unclear pronoun starts the sentence. Examples like this one happen when the author tries to refer to large parts of the previous sentence as the antecedent. Unfortunately, the audience may not recognize which parts of the previous sentence the author intended to be the antecedent.

> **DRAFT:** Such conditions lead to super-refraction of part of the radar beam, leading to the systematic underestimation being less than normal with increasing range. *This* means that the derived adjustment factors would be too large. [What is "this" referring to?]
>
> **IMPROVED:** Such conditions lead to super-refraction of part of the radar beam, leading to the systematic underestimation being less than normal with increasing range and derived adjustment factors being too large.
>
> **IMPROVED:** Such conditions lead to super-refraction of part of the radar beam, leading to the systematic underestimation being less than normal with increasing range. This range-dependent underestimation means that the derived adjustment factors would be too large.

An additional problem with pronouns is the *implicit "that,"* modifying phrases where the "that" may be omitted. Although omitting "that" may be common in writing, comprehension is sometimes limited by doing so. In addition, the implicit "that" is problematic for readers for whom English is a second language because the sentence structure is such that the words after the noun may not be recognized as modifiers. For clarity, include the "that," making it explicit.

> **DRAFT:** Cloud microphysical properties must be parameterized from the larger-scale fields the model can resolve.
>
> **IMPROVED:** Cloud microphysical properties must be parameterized from the larger-scale fields that the model can resolve.

This problem is not limited to "that"; read the sentence below with and without the "where."

> By knowing the ingredients needed to produce thundersnow, we can better explain the locations (where) thundersnow occurs.

For most readers, including the "where" is more explicit and clear.

10.2.4 Choosing the best words

Some words have been used many times or have definitions that are so vague that these words fit many circumstances. Unfortunately, such words are nearly worthless when precision is required. Consider the words in Table 10.4. These abstract words have lost their meaning in many scientific contexts. They are easy words to settle for when faced with a need to be explicit and precise. Rewrite sentences to eliminate these words when they are used ambiguously and choose more precise words. Consider the following examples:

DRAFT: Sounding analyses indicate the less stable nature of the lower troposphere over the surface occluded front.
IMPROVED: Soundings indicate the lower troposphere is less stable over the surface occluded front.

DRAFT: Situations that favor convective activity commonly occur in the spring and summer seasons.
IMPROVED: Convective precipitation commonly occurs in the spring and summer.

DRAFT: The coldest air is upstream of the trough axis, a favorable factor for further cyclogenesis.
IMPROVED: The coldest air is upstream of the trough axis, favoring further cyclogenesis.

DRAFT: A variety of factors appear to play a role in why the precipitation was so widespread in this storm.
IMPROVED: The precipitation in this storm was widespread for the following three reasons . . .

The vagaries of word choices sometimes lead to other interpretation problems as well. Consider the word "role" (P. A. Lawrence 2001). If a powerful tropical cyclone devastated Japan, how many different processes could be listed as playing a "role" in the cyclone's development? If something plays a role, what about other unmentioned items? How long would such a list be? Given all the ingredients required to produce a tropical cyclone, is it appropriate to say that any ingredient can play "the primary role"?

One way around the vagaries of these unfortunate words is to define the word precisely upon its first use. For example, "The high albedo of stratus strongly regulates the amount of incoming solar radiation. This role of stratus. . . ." Once defined this way, "role" can be used throughout the manuscript, referring specifically to the high reflectivity of stratus.

Nouns and pronouns are not the only parts of speech that can be less than meaningful. Adjectives and adverbs can also be empty and vague (Table 10.5). Obviously, what may be "obvious" to you might not be "obvious" to

Table 10.4 Potentially weak nouns when used in some circumstances. Not every use of these words is weak, but precision can be lost when using such words.

ability	degree	forcing	process
activity	development	influence	relationship
analysis	dynamics	interaction	role
approach	effect	issue	sense
case	element	level	situation
character	environment	manner	system
concept	event	nature	thing
context	factor	perspective	use

Table 10.5 More words to avoid

actually	feel	obvious(ly)	soon
basically	important	of course	still
certain(ly)	interesting	practically	type of
clear(ly)	kind of	quite	various
current(ly)	naturally	recent(ly)	very
extreme(ly)	now	regarding	wish

others. Different people may have different opinions about what constitutes "clear" evidence, so avoid irritating those people who have higher standards than you do. Rather than *stating* something is "interesting," explain *why* it is interesting.

Be careful about words that have meanings related to time. Words such as "recently" should be avoided. Is one year ago recently? Is ten years ago recently? "Now" can be also problematic as its casual use may confuse the literal reader. Is the author referring to "now" as in the time the paper was written? The time the case study occurred? The time when the paper is being read? Instead of "now" or "at this time," say "as of June 2006" or "at the time of this writing (March 2008)."

"Very" and "quite" are overused; eliminating most occurrences strengthens most sentences. Watch out for other verbal tics that add length not meaning, such as "basically," "practically," "various," "still," "really," and "kind of." Exclude any form of "feel" from your writing (e.g., "we feel the data show"); science is done with facts, not with feelings.

Quotes around colloquial or slang words can be distracting to readers and are usually unnecessary. Either find a more appropriate word, avoid colloquialisms or slang that ESL authors may not be able to interpret, or italicize a definition word.

10.2.5 Braggadocio and superlatives

Muhammad Ali said, "It's not bragging if you can back it up." Although boxers display braggadocio as part of their job, scientists who do so are generally viewed with skepticism, disdain, or worse. Be careful what you boast about in your papers, even if it is true. For example, some authors like to claim to be the first to do something. Before writing such flourish, think about whether it is really necessary and what others might think. Avoid phrasings such as "first," "novel," or "pioneering." Making such claims may be appropriate when

ASK THE EXPERTS

CREATING NEW SCIENTIFIC TERMINOLOGY

Mark Stoelinga, Senior Scientist, 3TIER, Inc.

From 1989 to 2003, I was part of a group at the University of Washington headed by Prof. Peter Hobbs that documented a characteristic set of frontal structures and associated precipitation systems in the central United States. Collectively, these structures became part of a new conceptual model for cyclones east of the Rocky Mountains, originally introduced as the Cold Front Aloft (CFA) conceptual model by Hobbs et al. (1990), and later expanded to the Structurally Transformed by Orography Model (STORM) by Hobbs et al. (1996).

Early in this research endeavor, it was clear to Hobbs's group that they were documenting frontal structures that did not conform to existing conceptual models and terminology for synoptic-scale structures, and so part of the research process involved the challenge of either conforming the new structures to existing terms, applying older terms that had fallen into disuse, or developing completely new terms for some of the features observed. As a member of this group, I share the following hindsight wisdom of what was done right and what was not, in the form of four

guidelines for developing terminology for new or modified concepts in science:

1. Use existing terminology whenever possible. Sometimes in science, we believe we have discovered a new process or described a new phenomenon that has not been documented previously. A careful search of the literature may indicate that essentially the same phenomenon has been described before, and existing terminology is sufficient. Even if the same phenomenon does not appear in the literature previously, many phenomena can be described using existing terminology. For example, in choosing the term "cold front aloft," Hobbs et al. (1990) properly acknowledged that the term was not new and that it was in fairly wide use among the U.S. operational forecasting community during the 1930s through 1950s to describe the same types of structures that our group was seeing fifty years later. Thus, Hobbs et al. (1990) were re-introducing the term, long after it had fallen out of favor.

2. Follow existing customs and conventions. Choose terms similar to existing terminology if the concept has some similarities to, or is a counterpart to, an existing concept. For example, to describe thunderstorm-induced straight-line winds, Hinrichs (1888) coined the term "derecho," a Spanish word meaning "direct" or "straight ahead." Hinrichs (1888) chose the

writing review articles about time-tested research or referring to people being honored at named symposia but are generally viewed negatively when describing your own articles.

In science, we expect evidence in support of claims. Superlative-laced writing demands similar supportive evidence. If you say that a particular cyclone was intense, provide evidence to indicate how intense it was. Do you have quantitative information that ranks this event relative to others?

Specific word choices may fuel trouble. Using "always," "never," "best," and other absolutes encourages readers to think of exceptions. Similarly, studies are rarely "comprehensive," and the level of detail in your "detailed" research is in the eye of the reader. (Shouldn't all our research be detailed?) You may say

word to be the straight-line counterpart to "tornado," derived from the Spanish word "*tornar*" meaning "to turn" (Hinrichs 1888; Johns and Hirt 1987).

3. Terms must be scientifically accurate, precise, and descriptive. If a new term is to be created, the principal task of the creator is to define a term that is scientifically accurate and precise, and sufficiently describes, albeit extremely briefly, the concept at hand. One example of a poorly defined term is "bent-back warm front," because such fronts rarely possess warm advection, the defining characteristic of a warm front. Also, part of the skill in creating a new term is to develop an appealing name. Creating a verbose name, even if accurate, can harm the chances of adoption. Many times a balance must be considered between conciseness, precision, and appeal.

4. Try to get terminology right the first time, and avoid subsequent changes. Perhaps the aspect of the evolution of the CFA conceptual model that caused the greatest consternation in the research community was the change, and subsequent repeal of that change, to the words that CFA stands for— a rather eggregious violation of this guideline. Our group received criticism for using a term implying a front can develop above the surface when classical frontogenesis theory dictates that fronts are strongest in the presence of a rigid or semi-rigid boundary such

as the ground or the tropopause (e.g., Hoskins and Bretherton 1972). In response to this criticism, our group changed the unabbreviated term to "cold frontogenesis aloft" in three papers published in 1995. However, reviewers and readers of these three papers were both confused by and critical of the new definition. In response, our group quickly reverted to "cold front aloft" again in 1996. In hindsight, the initial criticism could probably have been addressed without changing the term, particularly in light of subsequent research that identified the stable layer east of the lee trough as the missing lower boundary over which the front could advance aloft.

One of the principal challenges in creating new terminology is predicting how it will be received and used by the community, and how its definition might need to be adjusted in the future. Often while the initial study is underway, a research group may informally develop terms for convenience, to facilitate communication among the members of the group. Such terms can easily be revised and refined prior to their formal introduction via publication in the scientific literature. However, these terms must be carefully vetted (with consideration of the guidelines presented here) before submission. Once a term is introduced in the literature, modification or retraction may be impossible.

that "numerous" or "a limited number" of papers appear in the literature about a particular topic, but do you have a count? If so, provide a list of citations.

10.3 PROPER FORM

Avoid contractions, clichés, colloquialisms, and anthropomorphism. One reason is to assist readers who are not native English speakers. The other reason is that strong scientific writing generally does not contain these styles. This section addresses other types of proper form: abbreviations and acronyms, numbers and units, and adjective–noun agreement.

10.3.1 Abbreviations and acronyms

If a phrase is long and cumbersome to keep spelling out every time, then it is a candidate for a good acronym. Always define abbreviations and acronyms on first use. Spell out the word, then place the acronym in parentheses after.

> **CORRECT:** Model Output Statistics (MOS) surface temperatures from the Nested Grid Model (NGM) had a 1.4°C bias during near-surface temperature inversions.

Then, use the acronym throughout the rest of the manuscript—do not revert back. If you must introduce a lot of acronyms, consider a separate table defining all acronyms and variables. Often you do not need to introduce a relatively common acronym. Some abbreviations that are better known than their expanded forms (e.g., DNA, CAPE, NASA) should be defined upon first usage, but can be more commonly used in abstracts and titles without definition. Some journals may provide a list of acronyms and abbreviations that can be used without definition.

Often people will introduce acronyms as a shorthand for their own sake. "Jones and Stewart (2006)" becomes "JS06." Although convenient for you to avoid writing out "Jones and Stewart (2006)" all the time, readers may be frustrated, especially if you introduce more than a few acronyms. Does the acronym help the audience? Most are not necessary. If the acronym is only used a few times throughout the paper, consider whether it needs to be introduced at all.

You can also avoid creating new acronyms. If you introduce a long term or phrase that might require an acronym, you can minimize the number of times you need to use this term (and hence the acronym) in two ways. The first way is to structure a sentence (and surrounding sentences) so that you use a pronoun rather than the word or substitute a shorter more generic word in place of the longer word (e.g., "the cloud," "the dependent variable," "the model"). Alternatively, refer to the phrase by a shorter phrase. For example, after introducing a new type of two-moment cloud microphysical parameterization scheme, use the phrase "this new scheme" instead of an acronym.

10.3.2 Numbers and units

Although journals employ different styles, the following guidelines about numbers and units are consistent across many journals:

1. Numbers that are measurements, money, or decimals, should be written in numeric form.
2. Numbers less than or equal to ten should generally be spelled out, unless they are part of a list of numbers or quantities.
3. If a sentence begins with a number, spell out the number. Thus, avoid putting large numbers at the start of sentences.
4. Do not mix numbers in numeric and written form when used in a similar way, as in the example below.

> **CORRECT:** ... whether using a 2-layer, 10-layer, or 100-layer model ... [The "2" would normally be spelled out when describing a "two-layer model."]

Use the International System of Units (SI), wherever possible. Where measurements are taken in non-SI units and the measurement value is important, place the SI units in parentheses after the measurement.

> The 12-h accumulation of new snow at Albany, New York, measured 2 in. (51 mm) with a liquid equivalent of 0.11 in. (2.8 mm).

If you have used a statistical test to assess significance, include the following information: the name of the test, the statistic (e.g., t or F value), the degrees of freedom, and the probability of the statistic. If p is very small (e.g., 0.000056), writing $p < 0.001$ is sufficient. Furthermore, most common statistical tests assume the data are independent and identically distributed, and meteorological data are often neither, in time or space. Thus, has the time series been examined to ensure that samples have uncorrelated errors? If there is a correlation, has this been factored into the test (Wilks 2006, p. 144)?

10.3.3 Adjective-noun agreement

As Lipton (1998, p. 21) says, "Puppies are warm, not temperatures." Temperature is the quantitative measurement of the heat content of the air, whereas "hot," "warm," "cool," and "cold" refer to qualitative perceptions. Thus "warm temperature" is incorrect.

10.4 ELIMINATING BIAS

Sometimes you know who will read your articles; other times you do not. Knowing who your audience will be and catering to them can help the readability of your papers. For example, if you were working on South American cold surges, then many authors from South America are likely to read your paper. People who study cold surges in the Northern Hemisphere (North America, eastern Asia) may also read your paper. Therefore, avoid terms and phrases that are unlikely to be known by this audience. Choose simpler words rather than more complicated words that mean the same thing. An alternate view comes from some readers who say that the author's writing style should not be compromised to make it easier on those for whom English is a second language. Using the full range of English, they argue, helps such readers learn the styles and cultures of the authors. Despite that argument, I would prefer to reduce the burden on my audience by more carefully choosing my words.

10.4.1 Gender bias

The English language—like the Romance languages, which descended from Latin—has words to distinguish males from females: he versus she, him versus her. English does not have specific words for when the gender of a person is unknown or irrelevant, unlike Spanish and German that have a neuter gender. In contrast, the Finnish language has no gender at all—*hän* serves as "he" or "she." Unfortunately, in English, we have to be more clever to avoid language that favors one gender over another. So, how do we deal with something such as the following?

> **GENDER BIASED:** A master of the art of living draws no sharp distinction between work and play. His labor and leisure, his mind and body, his education and recreation, he hardly knows which is which. He simply pursues his vision of excellence through whatever he is doing and leaves others to determine whether he is working or playing. To him, he always seems to be doing both.
> —Wilfred Petersen

These are the best approaches to make your writing gender neutral:

- If possible, choose gender-neutral nouns: "chair" rather than "chairperson," "humanity" rather than "mankind/womankind," etc.
- Use both masculine and feminine: "he or she," "his and hers" (although this is awkward, especially when used many times in the text).
- Make the pronoun plural ("they"), and make other changes accordingly. Keep the sentence grammatically correct by ensuring subject–verb agreement.
- Rewrite the sentence to eliminate the pronoun.

Thus, the paragraph can be rewritten to incorporate these approaches.

GENDER NEUTRAL: Masters of the art of living draw no sharp distinction between work and play. Labor and leisure, mind and body, education and recreation—which is which is hardly known. These masters simply pursue their vision of excellence through whatever they are doing and leave others to determine whether they are working or playing. To themselves, they always seem to be doing both.

On the other hand, other approaches have been suggested that are less satisfactory:

1. The following constructions, albeit convenient, are awkward: "he/she," "s/he," or "(s)he."
2. You could use "one," but the result usually reads stilted, like when using "the author" to avoid the first person.
3. In some contexts, you could rewrite the sentence to say "you," but that would rarely work in a scientific context.
4. Some authors alternate "he" and "she" in examples, especially in longer texts like books. I argue for precision. If the gender is not known, do not force one upon the unsuspecting person.

Another mistake that can be made is to refer to a cited author as "he," when, in fact, the author may be a "she," or vice versa (e.g., Pat and Kelly could be either he or she; authors may only use their first initials; Kimberly Elmore is a he; the gender of authors from other countries may not be clear from their names). In such cases where you cannot be certain, it would be best to word the text in a gender-neutral manner.

10.4.2 Geographical bias

Despite globalization, the world is a big place. Although Americans may know where China is on a map, how many could name the rivers and mountain ranges there? What if the boot were on the other foot: Is it fair to expect a Chinese meteorologist to know the locations of the Ohio River Valley or Olympic Mountains? To aid in making manuscripts accessible to those not from your geographical area, define locations on a map. Some authors have a map of geographical place names as one of their first figures. Alternatively, annotating the figures or providing more description in the text of locations would assist others. Even if you are writing for a domestic audience, avoid general descriptions without meaning to outsiders (e.g., Golden Triangle, Capital District), unless you define them. Make your writing precise. Do not just say "the East Coast" without including "of the United States," at least for the first time.

Have you ever tried to read a paper about weather in the Southern Hemisphere? Reading "northerly flow" makes me think of cold air. In fact, northerly flow in the Southern Hemisphere is generally warm, having come from the equatorward, not poleward, direction. Thus, for our Southern Hemisphere colleagues, I recommend replacing "northward" and "southward" with their hemispheric-neutral words: "poleward" and "equatorward," respectively. Another example is to replace "positive vorticity" with "cyclonic vorticity." Such small efforts are worthwhile to those readers.

10.4.3 Cultural bias

Americans (of which I am one) tend to think the rest of the world understands us, or at least we like to think so. Sometimes our choice of words might be unfamiliar to people from other countries. Rewording those sentences or providing clarification can always help. I am not suggesting they be eliminated—I am suggesting that we choose words carefully and with purpose. Avoid metaphors and other colloquial expressions, especially those that may not translate well for the audience for whom English is a second language, such as "throwing out the baby with the bathwater."

10.5 MINIMIZING MISINTERPRETATIONS

The goal in scientific writing is to convey scientific content both accurately and precisely. Poorly structured text diverts readers' focus away from learning about your research toward trying to understand what the text means. By applying the guidance in this and the previous chapters, you can improve the readability of your manuscripts.

Do not write so that you can be understood, write so that you cannot be misunderstood. —Epictetus

Nevertheless, despite all your best efforts to be as precise and clear as possible, others may still misinterpret your writing. Once or twice a year I see my work miscited or misunderstood by others. Sometimes they have missed my point entirely, even contradicting sentences from my abstract. Although some of these instances may come from people not reading carefully and not citing my work properly, other situations may arise because of diversity in the scientific community. Specifically, the same piece of text can be interpreted a number of ways by different readers. In fact, Gopen and Swan (1990) argue that you can never completely eliminate alternative interpretations—the best you can hope to do is minimize interpretations other than your intended one by carefully structuring the text and considering all possible interpretations (and misinterpretations). Such approaches simply require practice, experience, and an open mind.

FIGURES, TABLES, AND EQUATIONS

11

The effective presentation of science goes beyond concise and precise writing. Figures, tables, and equations, if well constructed and well explained, can also convey science with both concision and precision. Creating effective figures, however, is more than just slapping the output from a commercial software package into the manuscript. Refinements and revisions are necessary to convert the working figures into their publication-quality descendents. Tables and equations can similarly be overused and underdeveloped, dumping grounds for inadequately explained data and science. Guidance provided in this chapter will help design, construct, and describe effective figures, useful tables, and helpful equations.

Although they may not hang in galleries or museums, beautiful and effective figures from scientific papers can be works of art, born from the artistic side that many of us scientists nurture. Instead of being in an exhibition next to a Renoir, O'Keeffe, or Picasso, such figures may appear in others' talks or be reprinted in textbooks. Moreover, the natural sciences are among the most figure-intensive sciences, with roughly a third of the article, on average, being graphs. Consider the importance of high-quality figures to effective communication through the following two examples.

EXAMPLE 1: In the early twentieth century, data from kite or balloon ascents were typically plotted on Cartesian graphs of temperature, relative humidity, wind speed and direction with height (Fig. 11.1). The Cartesian grid meant that even large differences in temperature or relative humidity between profiles did not appear to be very impressive. Even when Georg Stüve developed one of the first such thermodynamic diagrams (the Stüve diagram), the Cartesian grid remained. Adapting the pressure–volume diagram from classical thermodynamics to the atmosphere resulted in the innovation of the skew$T$–log$p$

Fig. 11.1 A typical plot of the vertical profiles of temperature, humidity, and wind in the time before widespread adoption of thermodynamic diagrams. (Fig. 1 in Clayton 1911.)

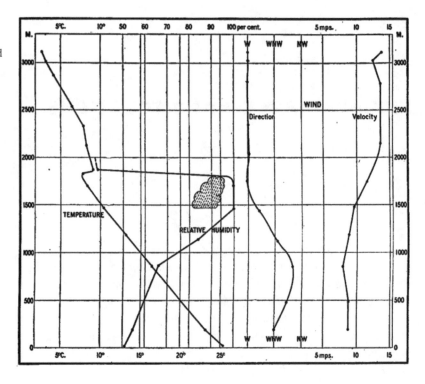

Fig. 11.2 A skew*T*–log*p* thermodynamic diagram with lines labeled. Sounding from 0000 UTC 20 July 2005 at Glasgow, MT. Temperature (solid lines), dewpoint (dashed lines), and horizontal wind (pennant, full barb, and half-barb denote 25, 5, and 2.5 m s⁻¹, respectively).

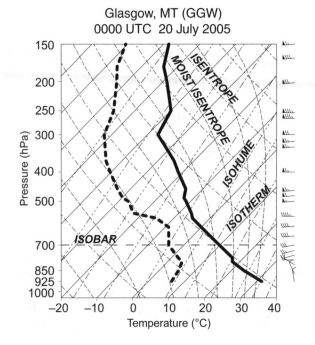

diagram. The clear advantage was that the oblique angle between temperature and the vertical coordinate, the logarithm of pressure, overlain with isolines of potential temperature, equivalent potential temperature, and mixing ratio, allowed one graph to show both the temperature and the moisture profiles and provided the necessary skewness to highlight small differences between profiles (Fig. 11.2). Furthermore, other quantities (e.g., wet-bulb potential temperature, lifting condensation level, level of free convection, convective available potential energy), otherwise difficult to obtain from Fig. 11.1, could be graphically calculated. Thus, the thermodynamic diagram allowed not only for a more concise presentation of the atmospheric profile, but advanced the science through the ability to calculate diagnostic quantities.

EXAMPLE 2: Before 1919, the structure of a typical extratropical cyclone was known only crudely (Fig. 11.3). Meteorologists in Bergen, Norway, eager to develop scientific forecasting methods based on observations, had constructed a dense observing network, revealing repeated patterns of temperature, wind, pressure, clouds, and precipitation associated with extratropical cyclones. This single figure (Fig. 11.4) encapsulated in schematic form the Bergen meteorologists' observations of numerous cyclones. The central image of Fig. 11.4 was the horizontal map of an extratropical cyclone showing the warm and cold air, the wind field, precipitation, and the direction of motion of the cyclone center. Below this map was a cross section through the cold front and warm front south of the cyclone center showing the thermal structure, airflow, cloud types, and precipitation. Above the map was a similar cross section through the elevated warm air north of the cyclone center. This three-paneled figure

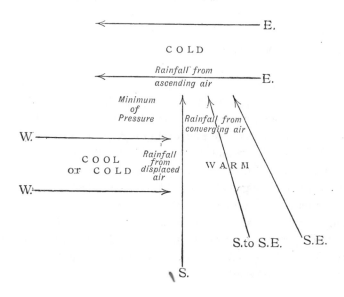

Fig. 11.3 Extratropical cyclone structure. (From Shaw 1911, his Fig. 96.)

Fig. 11.4 Extratropical cyclone structure. (From Bjerknes 1919, his Fig. 1.) Shaded areas represent precipitation.

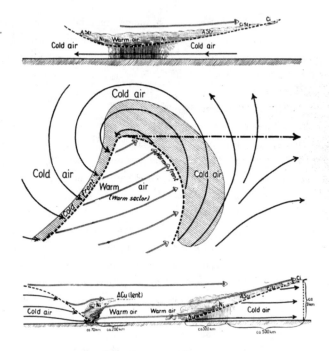

integrated the horizontal and vertical structure of the extratropical cyclone for the first time. Figure 11.4 was a vehicle for the Bergen meteorologists to educate forecasters about the typical structure of a cyclone, eventually contributing to the worldwide acceptance of the Norwegian cyclone model. When later results were being written up for publication to an international audience, many of the details and variety of cyclone structures were not presented in their conceptual model, in favor of simplicity. As Halvor Solberg, one of the Bergen meteorologists, wrote "the crystal clear drops [of water] seem more refreshing to a thirsty soul than a whole flood of muddy water."

Visual discourse adds variety for the eye and enhanced appeal for the mind. Does this seem trivial? It shouldn't: the psychology of reading is not a little complex. The living brain very much appreciates intelligence expressed in different forms. —Montgomery (2003, p. 114)

These two examples demonstrate how science benefits from high-quality figures, providing the backbone for effective scientific communication. Untitled graphics with poorly labeled axes and indeterminate lines (e.g., Fig. 11.5) may discourage readers from investing the time to read your manuscript, whereas graphics that are thoughtfully constructed and well labeled (Fig. 11.6) catch the reader's eye and encourage the reader to delve further into the paper.

Constructing eye-grabbing graphics, however, takes more time than you might think. Some figures may take half a day or longer to create, and several iterations may be necessary. Given that a picture is worth a thousand words, the effort can be quite worthwhile, though! On the other hand, in these days of cheap computer-generated figures, a paper with 50 figures does not imply an impact of 50,000 words.

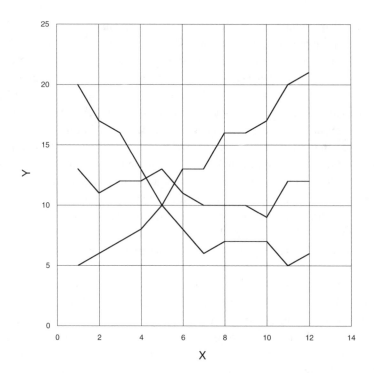

Fig. 11.5 A bare-bones line graph produced by a commercial software application using default settings. This figure is in desperate need of revision and annotation.

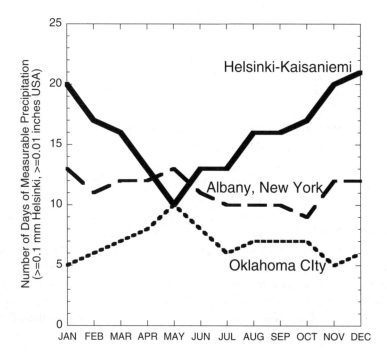

Fig. 11.6 Revised and annotated line graph from Fig 11.5. Average number of days of measurable precipitation per month at Helsinki, Finland; Albany, New York; and Oklahoma City, Oklahoma.

Such computer-generated figures have revolutionized scientific publications. Figure preparation is easy and changes can be implemented quickly. But, there is a big difference between the figures created during the research process (the raw computer-generated plots produced by a software application using default settings, the so-called *working figures*) and the figures submitted for publication (publication-quality figures, or simply *publication figures*).

The production of the working figure may be responsible for the "Eureka!" moment that makes science so enjoyable. Living beyond its creation, the working figure may appear in early drafts of the manuscript and serve as a placeholder for the eventual publication figure. The problem with the working figure is that it may not be the best way to present the data in a publication. Construction of publication figures often entails stripping everything off the working figures except the data, and rebuilding the graphic. The figure may be redrafted several times during the writing, revision, and peer-review process.

Figures 11.5 and 11.6 show the evolution of one such figure from working figure to publication figure. In the working figure (Fig. 11.5), the three data lines cannot be differentiated, the axes are labeled with variables instead of words, the tick labels are too small, the tick marks are too few, the borders of the graph are too thin, unnecessary grid lines compete with the data, and the data do not fill the domain of the graph. The revised version (Fig. 11.6) is easier to understand and more aesthetically pleasing.

The evolution from the working figure to the publication figure involves five checks, generally performed in this order:

1. **Design** of the type, shape, and layout of the figure
2. **Size** of the figure when published
3. **Aesthetics** to create a compact, self-contained, and visually pleasing figure
4. **Consistency** with text, other figures, and journal style
5. **Annotation** to enhance readability

The next five sections of this chapter discuss these five checks.

11.1 DESIGN

At the initial stage, decide on the type of graphic (e.g., scatterplot, line graph, horizontal map) to be produced. A discussion of some of the various options occurs later in this chapter. Decide on the style and layout of the figure. Will the figure involve color or half-toning (Section 11.6)?

Make the shape of the graphic appropriate for displaying the data. For example, square-shaped graphs should be used for graphics where both axes

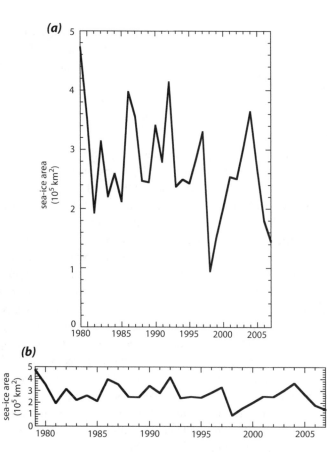

Fig. 11.7 The monthly averaged ice-covered area over the Canadian archipelago during the month of minimum Northern Hemisphere sea-ice area. Although (a) and (b) are the same graph, they are plotted with different aspect ratios.

are the same numerical values and the same units (e.g., 0% to 100% on each axis). Scale axes with the same units appropriately, if possible.

Small changes in the slopes of lines on a line graph are imperceptible if the slopes are near-vertical or near-horizontal. To maximize perception, the aspect ratio of the figure should exhibit the data in a way to approximate the slopes along a ±45° line whenever possible (called *banking to 45°* by Cleveland 1994, p. 70). For example, Fig. 11.7 shows the same data plotted on graphs with two different aspect ratios. Determining the period of most rapid decline in sea ice is nearly impossible from Fig. 11.7a, where the length of the *y* axis exaggerates increases and decreases. On the other hand, determining that 1997–1998 had the most rapid decline is straightforward from Fig. 11.7b, although reading the values of the individual peaks and troughs is more difficult on this short vertical axis.

If you have multiple figures of the same design, would combining them into a multipaneled figure be sensible to aid in comparing results? Such a combination also limits the total number of figures in the paper, which is easier on the audience, as well as the person doing layout. In a multipaneled figure, the panels [(a), (b), (c), . . .] should be ordered left to right, then down the page, as if reading a book. Label the panels prominently with larger font sizes on the upper left corner of each panel, in a title above the figure, or centered below the panel.

11.2 SIZE

Publication figures should be designed according to the size that they will likely appear in the journal. Whereas a working figure may occupy the whole sheet of paper (about 27 cm wide), the same publication figure may be as small as a few centimeters wide. Thus, knowledge of the intended journal is helpful. If the journal is published in two-column format (such as an AMS journal), determine whether the figure is likely to span one or both columns when published. The width of columns in the journal defines the maximum dimension of the figure: 7.9 cm (3.1 in.) wide for a single-column figure or 16.5 cm (6.5 in.) wide for a double-column figure in the AMS journals. When creating a multipanel figure, consider stacking two or three panels vertically to give the layout staff the opportunity to set the figure as a single-column figure.

Once the maximum dimension is determined, construct the figure so that all aspects (e.g., text, numbers, symbols) are legible. This task can be relatively straightforward given the ease of rescaling in most graphics programs. Most journals recommend font sizes of at least 8 point in figures after rescaling for publishing. Beware of dotted lines, which often do not survive reduction well. Print out the figure at the scale it will likely be published or experiment with various reductions using a photocopier as a final check.

Legibility during rescaling is enhanced by using sans serif fonts in figures. (*Sans* is French for "without," and serifs are the little flourishes on the ends of the strokes of the letters; e.g., the vertical lines on the ends of the "S.") Sans serif fonts (e.g., Arial, Geneva, Helvetica) are bolder and survive rescaling better than serif fonts (e.g., Palatino, Times), mostly because of the uniform width of the lines in the characters. Furthermore, sans serif fonts have the advantages of better surviving repeated photocopying and appearing more legible when viewed from afar (such as in a conference room).

11.3 AESTHETICS

This section should be subtitled "Take control of your figure preparation from automated software." I joke that GEMPAK (software for storing, calculating, and displaying meteorological data) has made my career possible. Without GEMPAK, I would have spent much more time performing calculations and writing computer code to make plots rather than doing science. GEMPAK and other software applications have given much power to us scientists. But, as

Peter Parker (Spiderman) is told by his Uncle Ben, "With great power comes great responsibility." We scientists, too, have a great responsibility not to abuse this power.

First, the power and ease of creating figure after figure from the computer must be tempered by the wisdom to include *only* those figures that contribute positively to the manuscript (Section 18.4.1). Second, this power also demands that the author be responsible for designing, creating, and revising the figure to maximize the audience's ability to understand it. Each element of the figure should exist for a purpose. A principal concern is to avoid what Tufte (2001, chap. 5) calls chartjunk—extraneous grid lines, annotations, moiré effects, and unnecessary graphical flourishes that detract from, compete with, or obscure the data, rather than supplement or enhance it.

In this section, we look to the design of the figure to enhance clarity, legibility, and aesthetic appeal. Here are some suggestions:

- ○ Make the data stand out from the rest of the figure. Data lines in line graphs should be thicker than borders, grid lines, etc., for clarity of the data. Make data lines black, make grid lines gray. Most default grid lines are unnecessary. The following ratios of widths is recommended: $1:\sqrt{2}:2$ or 1:2:4 for background grid lines:coordinate axes:data lines (e.g., Ebel et al. 2004, p. 432). For example, the data in Figs. 11.1 and 11.5 are barely distinguishable from the grid lines.
- ○ Minimize wasted space inside the figure. If the data have a maximum value of 34, making the maximum value of the axis 35 would minimize empty space in the graph. Starting the axis at zero is not necessary if the range of the data in a scatterplot is 31–34.
- ○ Including the figure-panel lettering [e.g., (a), (b)] on the inside of the figure panel will ensure that the figure is reproduced at its largest size and will not waste space around the outside of the figure.
- ○ Avoid unnecessary three-dimensional effects that many commercial software products can create for bar charts or pie charts.
- ○ Use font sizes proportional to the size of the figure.
- ○ Long blocks of text are easier to read if written in lowercase letters rather than uppercase letters. Varying the size of the text, and using italics, boldface, and color can go a long way to adding visual clarity to the figure. Too much variety, however, distracts the readers, so stick to just a few variations.
- ○ Axis titles should be accurate, concise, and include units. Avoid symbols, if possible. For example, "2-m air temperature (°C)" is better than "*T*." Axis titles can be in all capitals if short, in capitals only for the first word, or in headline style; that is, the first and other major words are capitalized. Whatever you pick, be consistent throughout all figures.

- Use a moderate number of tick marks: not too many, which would clutter the figure, and not too few, so that approximate quantitative information can be obtained from the graph. Tick marks can point outward from the graph so as not to clutter the interior of the figure or point inward to create a more compact figure.
- Try to put tick marks and labels on figures that represent something fundamental to the data (e.g., 3-hourly labels on hourly data for a 24-h period). Do not have tick marks for unphysical values (e.g., 5 minor tick marks in between the years 1987 and 1988 in a line graph that shows 20 years of annual temperature data). Avoid automated scales on axes that produce unusual intervals or maximum values (e.g., tick marks every 1.292°C instead of every 1°C).
- Make sure the tick-mark labels are positioned correctly relative to the tick marks. You may have to move the labels relative to the tick marks when the ticks represent categories rather than boundaries.
- Stick with standard units. Although many graphics programs may automatically determine the units and order of magnitude scalings, choosing values common in practice [e.g., geopotential height in decameters rather than hectometers, frontogenesis in K $(3 \text{ h})^{-1}$ $(100 \text{ km})^{-1}$ rather than 10^{-8} $\text{K s}^{-1} \text{ m}^{-1}$] is wise.
- Writing exponents in axis titles requires special care. The label "m × 10^{-3}" may confuse readers because the meaning of a value of "5" may not be clear. Does it represent 5×10^{-3} m or 5×10^{3} m? Omit the × and rewrite the label as "10^{-3} m." Such convention unambiguously yields "5×10^{-3} m." Additionally, say "contours are labeled every 10^{-3} m" in the caption.
- The scaling of axes can obscure or even misrepresent relationships between data. The use of nonlinear axes (e.g., logarithmic) also can help or hurt your goals. A logarithmic scale can be useful to avoid placing a break in the axis or when many large values skew the graph. When using logarithmic scales, label the axes as 10^4, 10^5, and 10^6 for clarity instead of 4, 5, and 6.
- Do not use ambiguous date labels such as 10/12/04 or 10.12.04: Does it represent 12 October 2004 or 10 December 2004?
- If plotting winds, make the units of the wind speeds and the plotting convention clear. For example, "pennant, full barb, and half-barb denote 25, 5, and 2.5 m s^{-1}, respectively."

11.4 CONSISTENCY

As work on your manuscript progresses, the working figures should be refined to be consistent with the text, other figures, and the journal style to ensure the whole package tells the same story. One of the most troubling errors reviewers can discover is inconsistency between the text and the data in the figures and tables. Such errors are unfortunately more common than they should be because different coauthors may be working on the manuscript, earlier versions of the figures may show slightly different results, authors may write from their memory rather than the figures, or authors may be careless.

Other less obvious, but still important, consistencies should also be double-checked. For instance, variable names, variable symbols, units, and terminology should be consistent between the text and the figures. Although convenient for computer-generated plots, an axis labeled in day of the year (not Julian date, see page 358) can be difficult for the audience to interpret. Convert day-of-the-year format to month–day format for consistency between the figure and the text, as well as ease for the reader.

Similar figures may occur several times within the manuscript. For such sets of figures, make each figure as consistent as possible with the others. Particular aspects to keep consistent include the size and shape of the figure, ranges on the axes, line type, line width, contour interval, and symbols.

Finally, be consistent with the style of the journal. Do not use casual or humorous fonts (e.g., **Comic Sans**), decorations, shadow effects, and bold annotations. The labels are one place to be wary of: many journals set scalar variables in italics and vectors in bold in the captions and text, so maintain this correspondence in your figures. Slashes as in "m/s" should be changed to exponents as in "m s^{-1}," if that syntax is consistent with journal style.

11.5 ANNOTATION

Even after applying the first four steps in effective figure design, most figures can still be improved by being annotated with useful information to enhance audience comprehension (Table 11.1). As we saw earlier (Fig. 11.5), the barebones line graph does not speak to the audience. The revised version (Fig. 11.6) is immediately understandable without reference to the figure caption, partly because of the annotation added to explain the graph to the audience.

Other examples of annotation include adding error bars, labeling geographically important features discussed in the text on a map, labeling the lines on a line graph (e.g., Fig. 11.6), or identifying the location of the 500-hPa cyclonic vorticity maximum responsible for the cyclogenesis with an "X." For example, a scatterplot between observed and modeled rainfall may benefit from a reference line with slope 45°, the so-called 1:1 line, to show the ideal relationship even if the data do not lie along this line (look ahead to Fig. 11.10). Describe all annotations in the figure caption.

Another reason for annotation is to make the figure self-contained so that the reader does not waste time searching through the figure caption for the relevant information. For example, if you can easily label the different lines on a line graph directly rather than creating a legend, you should do it. Another

Table 11.1 Annotations to enhance your figure

Horizontal length scale
Vector length scale
Reference lines (e.g., 0°C, time of frontal passage)
Geographical features cited in text
Locations of observing stations, radar sites, or other instrumentation
Insets to show detail to data, reference information, etc.
Grayscale or color legend
Title of the figure, if needed
Legends for data symbols or lines
Labels for each line or symbol, if possible
Figure-panel letters: (a), (b), (c), . . .
Error bars, standard-deviation bars, or confidence intervals
Circles, lines, arrows, or labels for important features described in the text

---

advantage of being self-contained is that the figure becomes easier to adapt into an electronic presentation (yours or someone else's), another manuscript, or even a textbook. Being self-contained will aid comprehension of your figure when it is separated from its context.

Computer-generated figures are often produced with a default legend, although this legend can often be moved to different locations on the plot. Some differences of opinion exist about where to place the legend. Some people prefer legends outside the plot box (so as not to clutter up the data field), whereas others prefer the legend inside empty space inside the plot box (because the legend may take up too much space outside).

Fig. 11.8 The inset shows details in the wind speed. In addition, the annotations 1 and 2 can draw attention to the features of interest (the two peaks in wind speed), which are described by these two numbers in the text of the parent article. (Fig. 11c in Schultz and Trapp 2003.)

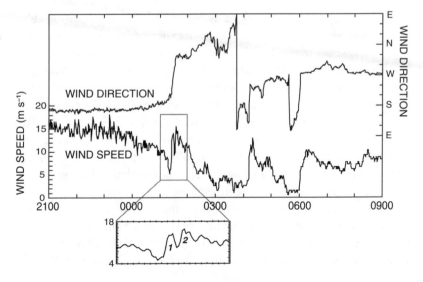

TIPS FOR PRODUCING SHARP DIGITAL GRAPHICS

In this era of electronic figure preparation, creating a figure from your software one way may produce a crystal clear graphic whereas other ways may produce pixelated or fuzzy graphics. Different software may produce different quality figures. Even software that converts between graphic formats may not yield the best quality. Try different approaches to obtain the best quality reproduction.

If you need to digitally scan photographic prints or other graphics for which no digital source exists, scan at a resolution between 300 and 600 dpi for color, grayscale, or continuous tone images, and at 1200 dpi for bitmap or line art images. The image should be scanned at the same size as you would expect to see in print, cropping white space around the edge of the image. Most computer monitors have screen resolutions of 72 dpi, so images that look sharp on a monitor may not be so when printed out or projected onto a screen.

When creating graphics, be aware of the graphic formats that the journal accepts. Some only want EPS (which can scale quite effectively) for line graphs and TIFF images for photos (highest resolution). Others are much more flexible. Journals may require figures in red, green, blue (RGB) format for the Web or cyan, magenta, yellow, black (CMYK) format for commercial printing. Stick to standard fonts (Times, Arial, Helvetica, and Symbol). Most publishers provide more detail on producing figures for their publications on their Web sites.

Figure insets are another form of annotation that can supplement the data in the figure or show detail that might otherwise be unseen. For example, Fig. 11.8 uses annotation to illustrate the detail in the time series of wind speed associated with a frontal passage. The time series of wind speed has two peaks (annotated 1 and 2 in the inset), the reasons for which are described more fully in the original article.

If more than one type of annotation is used (e.g., a map may be annotated with the names of cities and the names of observing stations), then distinguish them with different font types, sizes, or styles. The more aspects of the figure that can be annotated without cluttering up the figure, the less difficulty the reader will have interpreting the figure. Never allow annotation to overprint data—place such information in a legend, caption, or the text, if needed. Aim to place the annotation as close to its target as possible—long arrows extending away from the target may confuse the reader.

11.6 GRAYSCALING AND COLOR

Does your figure need grayscaling or color? Fields where the contours are quite detailed with lots of highs and lows (e.g., topography, cloud-top temperatures, radar imagery) benefit most from a color presentation. In contrast, grayscaling is much simpler to produce, more versatile for publishing, and less expensive to publish in some print journals. Although the cost of color figures may be prohibitively expensive for some, the costs can be minimized somewhat by creative grouping of different color images together, in some circumstances.

Although authors are making more use of color in scientific publishing, and indeed, many find it impossible to do without, there remains a strong incentive to stick to black and white, or at least design the figure with both considerations in mind. Color figures are ideal for many types of figures, but some readers will print your paper and its color figures on black and white printers or photocopiers. Features of remarkable clarity when presented in color may be uninterpretable when printed in black and white. Thus, just because you can employ color freely in an online journal, for example, does not mean that you should do so illogically and carelessly. Color should be used for clarity and emphasis, not decoration. Color emphasizes the data best when colored fields or lines are plotted against a light gray or muted background field. An example is color radar imagery plotted against a background of gray isobars.

The scales for grayscaling and colors can be continuous (hundreds or thousands of shades or colors that effectively appear as a smooth transition between gradations) or discrete (a few shades or colors with sharp distinctions between different gradations). Continuous color scales are effective for continuous fields such as satellite imagery, whereas discrete color bars are effective for fields that are usually contoured (e.g., 500-hPa geopotential height field). Avoid continuous color scales for fields where the audience might want to obtain quantitative information (e.g., precipitation). Specifically, grayscale resolution intervals should not be less than 20% because more than five grayscales limit the readers' ability to retrieve quantitative information (Fig. 11.9).

Use standard color schemes (such as with radar imagery), if possible. Avoid rainbow color schemes in most contexts. The colors may be too bold for the figure, and strong gradients in color distract the reader from a smooth distribution. Nevertheless, a number of ways exist to produce the color scales:

- Grayscaling and color scales can be presented as a uniformly increasing or decreasing intensity (e.g., visible satellite imagery from white for the highest albedo to dark for the lowest albedo). Such schemes do not bias the audience toward a particular color transition, as might a rainbow color scheme ranging from red to violet (e.g., the yellow to green transition may be particularly sharp).
- Abrupt transitions in scale take place at physically significant levels.
- A third approach, which is useful for plotting anomalies that have both positive and negative values, is a scale from light red to dark red for increasing positive values and from light blue to dark blue for decreasing negative values.

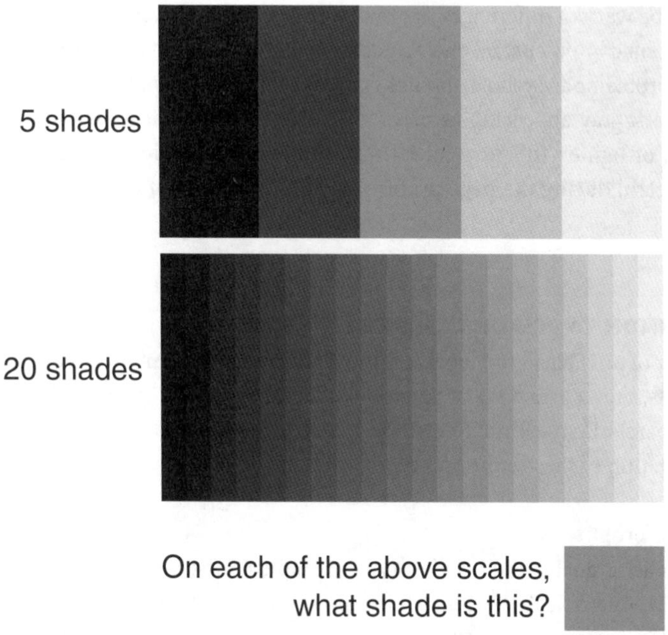

Fig. 11.9 Fewer shades are often better for determining quantitative information from your graphic.

5 shades

20 shades

On each of the above scales, what shade is this?

Whether you use grayscaling or color, figures generally benefit from a corresponding legend for the scale. As with other aspects of figure production, find what you like from what others have done, and emulate the best color schemes.

Color can also be used effectively for annotating the figure. By picking a distinct color for the annotation not present in the figure, a clear visual distinction between data and annotation can be created.

Warmer colors (red, yellow, and orange) will appear to jump out at the viewer relative to cooler colors (violet, blue, and green). Therefore, when constructing graphics, place warmer colors in the foreground and the cooler colors in the background. If two different colored lines cross, ensure the warmer color overprints the cooler color.

Another reason to be hesitant about automated color schemes is their ability to be interpreted by colorblind readers. Around 10% of males and 2% or less of females are colorblind. People with the most common form of colorblindness have difficulty distinguishing red and green (red–green colorblindness). Unfortunately, many common radar displays use red and green color schemes: reflectivity factor (red and green for high and low reflectivities) and Doppler winds (red and green for outbound and inbound velocities). Uploading your figures to www.vischeck.com will allow you to view them as a colorblind individual would. Such checks can be helpful before creating a color figure that may be uninterpretable by some in your audience.

Finally, be wary of differences in grayscaling or colors that may result between different computer screens (including projectors) and different printers. Particular problems are yellow and light greens fading in a background white field and dark gray appearing as black. Sometimes the publication process can darken or lighten the grayscales from what you originally intended. As such, be careful during the page proofs stage to inquire about the quality of the figures.

11.7   COMMON TYPES OF FIGURES

The number of possible figure types is limited only by the human imagination. Nevertheless, a few basic types of figures are commonly found in atmospheric science research. Below, I provide some advice to improve the construction and presentation of these common types of figures.

11.7.1   Line graphs

Line graphs are arguably the simplest type of graph, consisting of as little as a horizontal axis, a vertical axis, and a line representing the data (e.g., Figs. 11.6 and 11.8). Line graphs present the relationship between two quantities: the independent variable (what is selected to vary, the predictor or input variable, plotted on the horizontal axis, generally increasing rightward) and the dependent variable (what is measured, the predictand or output variable, plotted on the vertical axis, generally increasing upward).

If the line is created from connecting data points together, showing the location of the data points with a marker can be helpful to readers to see how much of the curve is interpolated or extrapolated from the available data. More than one line may appear on a line graph (e.g., Figs. 11.6 and 11.8). If so, the lines should be easily distinguished from each other by their symbols, line width, line color (even if grayscales), line type (solid, dashed, dotted, and dashed–dotted), or even their separation on the graph. One approach to illustrate the relative distinctiveness of multiple lines is to include the confidence intervals or error bars on the lines (defined, of course, in the caption so that the reader knows exactly what the bars mean). Furthermore, beware that the lines do not have sufficient overlap such that the individual lines cannot be distinguished from each other. One exception is a spaghetti plot or plume diagram from ensemble model output, where overlapping lines are common and are meant to show consistency in the individual forecast members.

When multiple lines exist that may represent different variables, different scales can be constructed for the left and right axes or the top and bottom axes. Such graphs are one way to compress the total number of figures and facilitate the comparison of two different graphs. For example, the double axes in Fig. 11.8 were constructed by plotting the wind speed from 0 to 40 m s^{-1}

(even though the largest value was less than 20 m s^{-1}) and plotting the wind direction from −90° to 450° (after adding 360° to all wind directions greater than or equal to 0° and less than 90°); both axes were appropriately relabeled. These rescalings allowed the wind direction and wind speed lines to not overlap and the wind direction line to appear as a single continuous line to avoid the otherwise nasty jumps between 330° and 30° that would happen with a graph ranging between 0° and 360°.

11.7.2 Scatterplots

As with line graphs, scatterplots can show the complex and nonlinear relationships between two quantities within a large dataset (Figs. 11.10 and 11.11). The difference between an effective scatterplot and a low-quality scatterplot is often related to the symbols representing the data. The choice of symbol, its size, and its color are options the author needs to control rather than accepting the default choices. Consider first a scatterplot with a single symbol, as in Fig. 11.10. In this figure, the size of the dots is large enough to be clearly visible, but small enough to minimize overlap, which would obscure the data. For a scatterplot with only a few widely spaced data points, the symbols can be larger than for a scatterplot with a larger number of overlapping data points. Data points in a scatterplot should never be too large and overwhelm the figure or too small and barely seen (or worse yet, eliminated upon reduction for publication).

Scatterplots with closed rounded symbols (e.g., ● ■ ▲) generally look better than scatterplots with open symbols (e.g., ○ □ △) or with edges (e.g., +, *, K). Open symbols often become closed upon reduction during printing, and small dots (such as periods) may not even be visible. When the data points overlap, closed symbols will partially hide the data. Thus, using open symbols where a lot of overlap in the data occurs better shows the variations in data density across the field. If open symbols cannot make your figure legible because of data overlap, then perhaps a density-contour plot of the field of points would be a more effective display.

For a scatterplot with multiple symbols (Fig. 11.11), choose symbols of roughly equal size and area to avoid introducing a visual bias in the scatter. For example, plots of plus (+) and minus (−) signs to represent cloud-to-ground lightning strokes of different signs on a scatterplot or map may be intuitive, but the larger area covered by plus signs dominates over that by the negative signs, potentially biasing the viewer to believe a larger number of positive lightning flashes or a larger areal coverage occurred than for negative lightning. (Might a vertical line be used instead of the plus sign?) Furthermore, multiple symbols must be distinct from each other (e.g., small bullets and squares may be nearly indistinguishable when reduced for publication). An intelligent use of different symbols is demonstrated by their large, but similar,

Fig. 11.10 Scatterplot comparing cloud-top temperatures determined from radiosonde data to cloud-top temperatures from satellite for a dataset of precipitating clouds. The thick solid line is the linear regression of the data, and the dashed line constitutes the perfect relationship. (Fig. 5 in Hanna et al. 2008.)

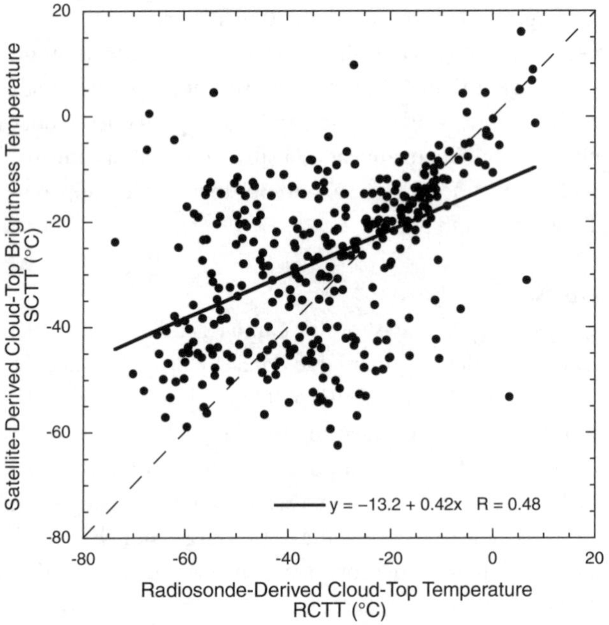

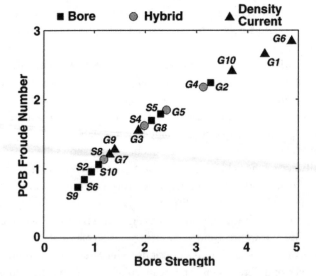

Fig. 11.11 Scatterplot of data from the formation of atmospheric bores from colliding density currents (bore is black square, density current is black triangle, and hybrid bore/density current is gray circle). The identifier next to each plot symbol indicates the type of post-collision boundary (G indicates gust front induced and S indicates sea-breeze front induced) and the case number (1–10). (Fig. 7 in Kingsmill and Crook 2007.)

size and grayscaling to distinguish the hybrid cases from the bores and density currents in Fig. 11.11.

Sometimes scatterplots are supplemented with line graphs showing the 1:1 line, theoretical relationships, correlation curves, or other derived quantities to help the audience better understand the relationships within the data in the scatterplot (e.g., Fig. 11.10). Such lines should be annotated either on the figure or in a legend, but definitely in the caption.

MISUSES OF LINEAR CORRELATION

A common diagnostic tool is to fit a least squares regression line to data in a scatterplot. A measure of how well the data fit that line is called the Pearson product-moment correlation coefficient (or simply, the correlation coefficient) r: 1 is a perfect correlation with a positive slope, −1 is a perfect correlation with a negative slope, and 0 is no correlation. The coefficient of determination r^2 represents the percentage of variance explained by the linear fit. Do not confuse r and r^2.

To demonstrate the potential misuses of linear regression, Anscombe (1973) presented four examples of statistical data, each with 11 points and each having the same regression line and correlation coefficient (Fig. 11.12). Only in the first example of Anscombe's quartet (Fig. 11.12a) is the resulting linear regression valid; the remaining three cases show common misuses of linear regression.

Before even considering calculating a linear fit, determine whether the data have an obvious relationship that may or may not be linear (Fig. 11.12b). More importantly, do you expect linear behavior from the data? For example, a graph showing a linear relationship between the mean raindrop size and radar reflectivity factor would be faulty because the equation relating the two is sixth order, not linear.

Beware of outliers, which may unduly influence the linear correlation. For example, an otherwise perfect correlation is ruined by one outlier (Fig. 11.12c), whereas a dataset with no correlation is afforded high correlation because of one outlier (Fig. 11.12d). If your dataset contains outliers, you can be honest with your readers by calculating the linear fit for the complete dataset and for the dataset with the outliers removed to show their influence.

Should you discover a large correlation in your dataset, avoid imposing physical links between data for which physical linkages are not apparent. As is often said, "statistical correlation does not imply causation" (discussed further on page 363).

Finally, a note on language. Limit use of the word "correlation" to situations with calculated correlation coefficients. Do not use "correlate/correlation" as a synonym for "relate/relation" or "correspond/correspondence" (page 353).

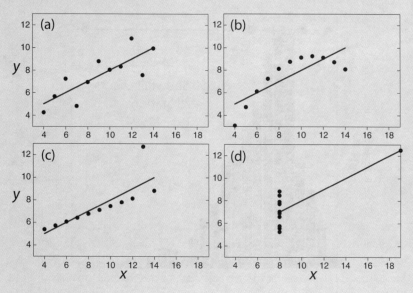

Fig. 11.12 Anscombe's (1973) quartet: four examples of datasets that have the same mean (7.5), same standard deviation (4.1), same linear regression line ($y = 3x + 0.5$), and same correlation coefficient (0.82). Only (a) is an appropriate use of linear regression.

11.7.3 Bar charts

Bar charts display the distribution of a dataset and are useful for comparing one category of data to another. A special type of bar chart, a *histogram*, shows the distribution of a dataset as a function of a quantity, plotted as number, frequency, or percentage (Fig. 11.13). Bar charts can be plotted with the bars oriented vertically (Fig. 11.13) or horizontally (Fig. 11.14). Choose a vertical orientation if the dependent variable is a quantity, number, or percentage; choose a horizontal orientation if the dependent variable is time, distance, or length—quantities that are more intuitively oriented along a horizontal axis. For Fig. 11.14, the horizontal orientation was chosen to make the names of the sources easier to read.

One of the secrets to creating an effective bar chart is to present the bars in the order that best illustrates the relationship you want shown. For most cases, this is fairly obvious (e.g., by chronologic order, by increasing numerical value). In other cases, the success of the bar chart depends upon this ordering. For example, consider the two bar charts in Fig. 11.14. The data are from a survey where Austin, Texas, residents were asked to identify all their sources of weather information and their most important source. The sources are listed in alphabetical order in Fig. 11.14a and in descending order from most cited source in Fig. 11.14b. The advantage of Fig. 11.14b is apparent. First, the order of the sources in Fig. 11.14b adds value to the figure, allowing the reader to immediately see the relative ranks of the sources. Second, with all sources ranked, a trend in the most-important-source category appears. Two of the most important sources do not follow the trend of decreasing percentage from

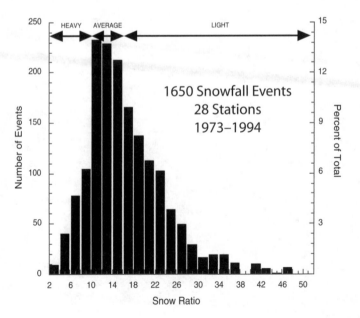

Fig. 11.13 Bar chart with vertical bars. (Fig. 2 in Roebber et al. 2003.)

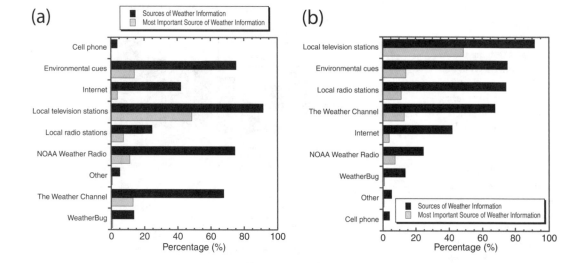

top to bottom (The Weather Channel and NOAA Weather Radio), suggesting that people prefer these two over other comparably ranked sources. Finally, the ranking of the sources in Fig. 11.14b provides enough space for the legend box to be moved to the inside of the figure, reducing the space in the journal required to publish the figure.

When more than one dataset is being plotted on the same graph, two approaches can be employed. The first approach is to place the bars side by side (e.g., Fig. 11.14b), although only a limited number of bars can be plotted effectively this way, depending on the shapes of the distributions and the number of categories, and whether colors are used effectively to distinguish the bars. If more categories need to be plotted, consider multiple panels of bar charts or use a line graph.

The second approach is to employ a stacked bar chart (Figs. 11.15a,b). Stacked bar charts are best used for categories that are subsets of a larger group (e.g., F0, F1, F2, . . . , F5 tornadoes as a subset of the total number of tornadoes). An effective use of this approach is to arrange the stacked categories in some logical order (e.g., by the size of the category, with the largest on bottom and smallest on top). Connecting lines between adjacent stacks can indicate relative proportions and help show trends better. Although stacked bar charts can be useful in some contexts, too many small categories in each stack may hinder easy interpretation of the results. In such cases, a line graph might be a better approach.

For an illustration of these points, Figure 11.15a is a stacked bar chart ordered from the largest category on the bottom, decreasing upward, whereas Fig. 11.15b is ordered from the smallest category on the bottom, increasing upward. With the smallest categories at the bottom (coincidently, also the

Fig. 11.14 The importance of thoughtful ordering of bars to produce an effective bar chart. (a) Alphabetical ordering of sources leads to an ineffective bar chart. (b) Ordering of sources by decreasing percentage produces a more effective bar chart. See text for further explanation.

Fig. 11.15 Stacked bar charts: (a) Ordering of the stacks from generally largest sector at the bottom (university) to the smallest sector at the top (student); and (b) ordering of the stacks from generally smallest sector at the bottom (student) to the largest sector at the top (university), which also happens to be the largest growing sector since 2001. Because of the difficulty in obtaining trends from (a), placing the largest or the fastest-growing bar at the top is more clear. (c) The same data as in (a) and (b) are presented in a line graph, showing the time trends of the five sectors much more clearly than either bar chart.

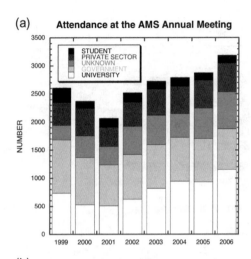

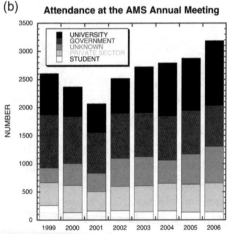

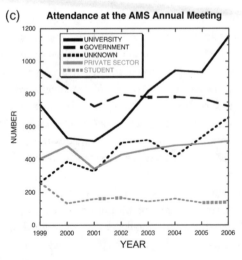

categories that exhibit few changes over time), the categories showing the largest changes are more apparent in Fig. 11.15b. Thus, we can see that the university sector at the top of Fig. 11.15b has the largest changes over time (a factor of 2) relative to the other categories. In contrast, with the largest categories on the bottom in Fig. 11.15a, all categories seem to be buoyed upward by the increase in largest categories after 2001, making determining trends in the smallest categories toward the top difficult.

When creating a bar chart, the bars should dominate the background. Plot the bars filled with black or grayscaling for maximum contrast with the background. For multiple bars, make sure the colors or shades of the bars are distinguishable. Shades in a stacked bar graph should go from lightest to darkest for maximum readability. Avoid cross hatching and other visually distracting filled patterns. Do not make the bars too thin or too thick—the bars should be wider than the space between the bars. If the bar chart displays a quantity that is continuous (e.g., temperature, wind speed, vorticity), the bars may be adjacent to each other with no intervening space. Otherwise, intervening space between the bars is appropriate for discrete quantities (e.g., number of occurrences of a phenomena).

The tops of the bars can be labeled with their values in two situations. First, extra precision may be required in some datasets. Second, one or more bars may be large and dominate the others. To show details in the lesser bars, rescale the axis to trim down the largest bars and label their values at the tops of the truncated bars.

For the independent variable (plotted along the x axis for a vertical bar chart), be aware of how the categories are labeled relative to the tick marks. In Figs. 11.15a,b, the years are labeled in between in the ticks in the bar charts under the bar, but in Fig. 11.15c, the years are labeled on the ticks in the line graph. For example, how should a bar chart with hail size labeled 2, 3, 4, . . . , 8 cm be interpreted? Does the category labeled "2 cm" imply hail sized only at precisely 2.0 cm or hail ranging from 2.0 to 2.9 cm? To avoid such ambiguity, either label the axes properly or include a statement in the figure caption "(the bar labeled '2' contains maximum hail sizes 2.0–2.9 cm)." Avoid overlapping categories (e.g., 0–5, 5–10, 10–15), either explicitly for convenience or implicitly by omission.

For the dependent variable (plotted along the y axis for a vertical bar chart), reserve the axis title "frequency" specifically for a number of events *per unit time*. In most cases, label the axis "number of events" instead.

Should you use a line graph or a bar chart? Line graphs and scatterplots function best with continuous data, whereas bar charts function best when used with discontinuous or countable data. Figures 11.15b,c show the same dataset in two different formats. At first glance, the data from the stacked bar chart (Fig. 11.15b) have several advantages: the overall shape of the data is more clear than in the line graph (Fig. 11.15c), the sum of the different sectors is easily determined, and the trend of the total attendance is more apparent. In contrast, the line graph (Fig. 11.15c) has the advantage of showing quantitative information about the different sectors more easily and trends are more visible within sectors (which can be seen in the stacked bar chart, but are more elusive). Thus, in Fig. 11.15c, we can readily see that attendance by the university and unknown sectors has increased since 2001, with university attendance increasing the most rapidly. Since 2001, the other three sectors (government, private sector, and student) have had comparatively less dramatic changes (Fig. 11.15c). Line graphs, therefore, are much more effective than bar charts for seeing trends, especially subtle ones.

11.7.4 Tukey box-and-whisker plots

Although the distribution of a single dataset is illustrated nicely by a histogram, histograms are less useful when comparing several datasets. Specifically, a common question asked is how similar two distributions are (i.e., could these two distributions have been sampled from the same population?). Box-and-whisker plots, also called box plots, are compact graphical representations of the histograms that provide information about the median, lower quartile, upper quartile, interquartile range (difference between upper and lower quartiles), and the outliers (Fig. 11.16). Whiskers typically represent the smallest and largest observations that are not outliers, those exceeding 1.5 times the interquartile range, although other conventions have been employed (e.g., minimum and maximum values of the whole dataset, 2nd and 98th per-

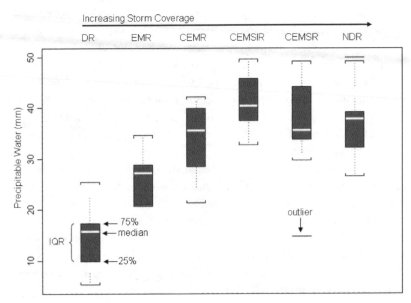

Fig. 11.16 Box-and-whisker plot of precipitable water for six synoptic regimes during the summer North American monsoon (DR, EMR, CEMR, CEMSIR, CEMSR, and NDR). IQR is the interquartile range, and the outermost brackets enclose values within 1.5 times the IQR. (Fig. 16 in Heinselman and Schultz 2006.)

centiles). Therefore, authors using box-and-whisker plots should define the characteristics of the box and whiskers in the caption or as an inset.

The benefit of box-and-whisker plots is that statistics of different distributions (e.g., quartiles, medians) can be more easily compared than the histograms. Where substantial overlap of boxes occurs, distributions cannot be distinguished statistically from each other. The larger the separation between medians and interquartile ranges, the more significant the difference between the distributions. For example, in Fig. 11.16, the large separation between medians and interquartile ranges of precipitable water for DR, EMR, and CEMR implies that these distributions are significantly different from each other, whereas the large overlap in the interquartile range of precipitable water for CEMSIR, CEMSR, and NDR implies these distributions are more difficult to distinguish statistically.

Some plotting packages display notches in the box near the median. These notches represent the variability of the median among samples derived from the populations. If the notches from two boxes do not overlap, then the two distributions have different medians at the 5% significance level.

11.7.5 Horizontal maps

Horizontal maps display the horizontal (or quasi-horizontal) distribution of a quantity, either from observations or models. Often the quantity is plotted on a constant-height surface, constant-pressure surface, constant-isentropic surface, or the dynamic tropopause. Examples include the surface map (Fig. 11.17), 500-hPa geopotential height map, precipitation anomalies over

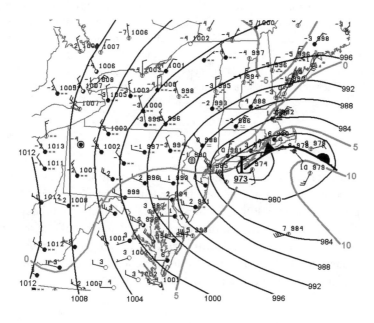

Fig. 11.17 An example of a horizontal map. (Fig. 3e in Novak et al. 2008.)

PHOTOGRAPHS IN SCIENTIFIC ARTICLES

Roger Wakimoto, Associate Director, Earth Observing Laboratory, National Center for Atmospheric Research

The atmospheric sciences are dominated by visual images. Not surprisingly, photographs strategically placed in an article can significantly enhance the presentation. Numerous styles of photographs appear in the literature, but I only choose a few that I have found particularly effective. Photographs presented in concert with an analysis that is based on remotely sensed data can add tremendous physical insight into a phenomenon (e.g., Fig. 11.18). An excellent example is combining satellite and Doppler radar analyses with photographs of thunderstorms, hurricanes, or tornadoes. Lidar analyses of pollution combined with visual images of the plume or layer can be striking.

Another effective approach is the before-and-after photographs. In climate studies, this approach might be an illustration of the dramatic retreat of a glacier or rising sea levels. In severe storm research, this approach is often two images of a building or structure before and after a devastating event. Record events are sometimes best shown in a photograph (e.g., largest hailstone, heaviest snowfall, record flooding, drought). A time series of photographs can effectively reveal the evolution of an event such as a building

being damaged in high winds, a hurricane landfall as viewed by a sequence of satellite images, and the explosive growth of a severe storm. Finally, photographs are often used in a complementary manner with design drawings to show an instrument or observational platform. There is a natural tendency to present the entire photograph in a publication; however, judicious cropping of the image should always be considered in order to minimize wasted space.

Most photographs are used in a qualitative manner; however, photogrammetric techniques can provide a wealth of quantitative information (e.g., Fig. 11.18). Photogrammetry permits labeling elevation and azimuth angles on top of the picture in addition to determining length scales. Superimposing an angular grid onto a photograph is not a new idea because it is analogous to placing a latitude and longitude grid onto a satellite image. Indeed, how a user could interpret a satellite image without including this photogrammetric information is difficult to imagine.

Placing a photogrammetric grid on top of a picture requires knowledge of quantitative information about the photograph: the focal length of the camera lens, the distance to the phenomenon of interest, the precise time the photograph was taken, and the exact location of the camera. Such information should be recorded at the time the photograph is taken. Finally, the horizon must be visible in the photograph for accurate placement of the grid. These additional requirements no doubt explain why such techniques have been underutilized in the literature.

the United States, a map of backward trajectories of air parcels, a horizontal map of isochrones of a front or convective system, and satellite imagery. Radar plan position indicator (PPI) plots are a form of horizontal map, although the quasi-horizontal surface is the surface of a slightly bulbous cone because of the radar-scanning geometry and Earth's curvature. By interpolating radar data onto a constant-height surface in a Cartesian coordinate system, radar data can be remapped onto a constant-altitude PPI (CAPPI).

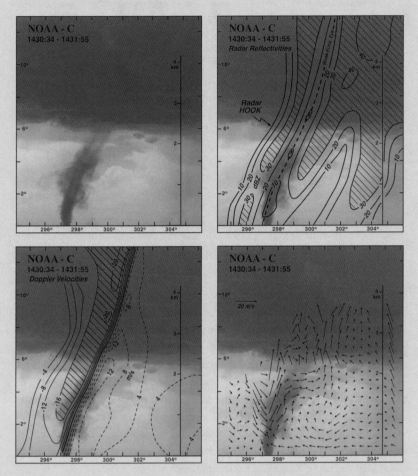

Fig. 11.18 Overlaying radar data on top of a tornado photo. (Fig. 9e in Wakimoto and Martner 1992.)

For traditional synoptic maps based on observational data (Fig. 11.17), two competing effects need to be considered: readability of the data and a large enough domain. For a large domain, such as the United States, not all surface data can be plotted and be readable in a journal-sized figure. One solution is to crop the figure so that only the relevant area of interest shows, although be careful of cropping the domain so much that relevant features that may move into the domain at a later time can be seen. Alternatively, the observational

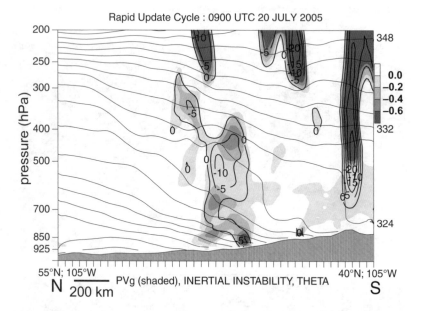

Fig. 11.19 Vertical cross section. (Adapted from Fig. 12 in Schultz and Knox 2007.)

data can be filtered so that only a fraction of the dataset is plotted. As the scale of the features being plotted decreases, more of the data should be plotted to show the mesoscale and microscale features. Whatever the solution, a proper balance needs to be struck between the plotted size of the station model, the density of the available data, and the size of the domain.

If geographical areas, observing stations, and instrument locations are described in the text, make sure that at least one of the horizontal maps is annotated with the names of these features. Include a horizontal length scale. Well-constructed titles with dates/times, quantities plotted, name of the model experiment, etc., help the reader understand the contents of the figure without resorting to reading a detailed figure caption.

11.7.6 Vertical cross sections

Vertical cross sections (Fig. 11.19) present data in a plane with the y axis representing the vertical dimension (e.g., height above the earth's surface, pressure). The x axis is typically a horizontal dimension, although time–height cross sections (Fig. 11.2) are also popular for showing the evolution of the vertical structure of the atmosphere (e.g., series of rawinsonde soundings, vertical wind profiles from Doppler radar). Radar range–height indicator (RHI) charts are also a form of vertical cross section.

Vertical cross sections often appear in conjunction with an accompanying horizontal map, showing the location of the cross section. Label the endpoints of the cross section clearly. Locations on the cross section indicating land–water boundaries, state or country boundaries, or mountains may be anno-

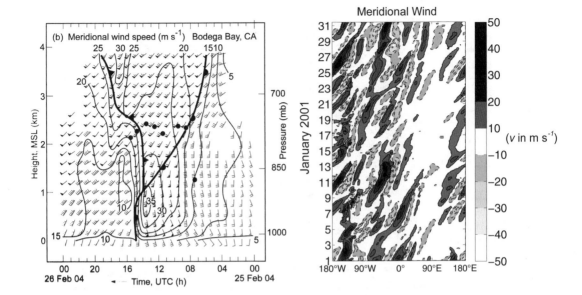

Fig. 11.20 Time–height cross section. (Fig. 2b in Martner et al. 2007.)

Fig. 11.21 The Hovmöller diagram shows the meridional wind velocity along the 2-PVU contour on the 320-K isentrope (*y* axis is the day in January 2001). (Figure courtesy of Olivia Martius.)

tated if it helps the reader to orient relative to the horizontal maps. Include a horizontal length scale (i.e., the 200-km long bar in Fig. 11.19) so that readers know the scales of the features in the cross section.

11.7.7 Thermodynamic diagrams

Thermodynamic diagrams are tools that graphically represent the relationships between pressure, temperature, mass, and water vapor as a function of height in the atmosphere (e.g., Fig. 11.2). There are several different types of thermodynamic diagrams, but the most useful are those that exhibit a proportionality between area on the diagram and energy (e.g., skew$T$–log$p$ diagram, tephigram, emagram). Ensure that the grid lines are thick enough should the figure be reduced for publication and that the data lines dominate these grid lines.

11.7.8 Hovmöller diagrams

A time–longitude plot to show the zonal movement of features averaged over a latitude band was originally developed by Hovmöller (1949) to track troughs and ridges in the jet stream. To accommodate tracking tropopause-based features along a meandering jet stream (where the features may not be easily tracked within a latitude band), Hovmöller diagrams have been refined by Martius et al. (2006; Fig. 11.21). In this figure, the positive (solid contours) and negative (dashed contours) meridional wind speeds show the mobile short-wave troughs along the 2-potential-vorticity-unit (PVU) contour on the 320-K isentrope during January 2001 for the Northern Hemisphere.

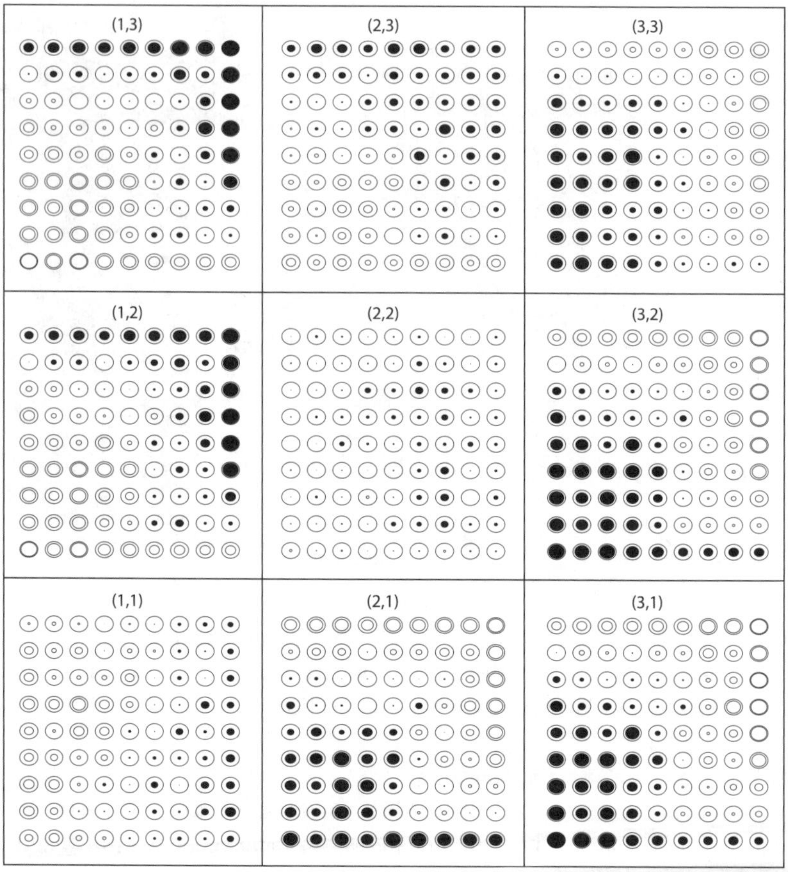

Fig. 11.22 An example of a small-multiples figure. In this figure, a 3 × 3 self-organizing map (Kohonen 1990) is constructed to show the different spatial and temporal evolutions of the intraseasonal oscillations of the Indian summer monsoon area-averaged rainfall anomaly over central India using large-scale circulation parameters. By using a higher-order 9 × 9 map, the event-to-event variability within each 3 × 3 map can be explored. (Fig. 8 in Chattopadhyay et al. 2008.)

11.7.9 Pie charts

Pie charts are circular graphs with the relative percentages of the whole broken up into slices. Generally, pie charts should be avoided in scientific publications as quantitative information is more easily obtained from bar charts or other figures. Furthermore, data from a pie chart can be presented more simply in a table or in the text. One situation where the simplicity of a pie chart can be quite effective, however, is in presenting a quantity (e.g., percentage of cloudy days) at many stations across a map (e.g., small multiples).

11.7.10 Small multiples

A figure employing the repetition of numerous small and nearly identical pictures is called a small-multiples figure (Fig. 11.22). Small multiples allow patterns to emerge from the dataset and the evolution of features in space or time to become apparent, as in this case illustrating the different spatial and temporal evolutions revealed from a 3 × 3 self-organizing map of the

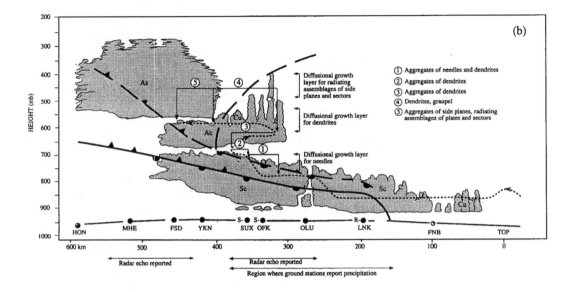

(b)

① Aggregates of needles and dendrites
② Aggregates of dendrites
③ Aggregates of dendrites
④ Dendrites, graupel
⑤ Aggregates of side planes, radiating assemblages of plates and sectors

Diffusional growth layer for radiating assemblages of side planes and sectors

Diffusional growth layer for dendrites

Diffusional growth layer for needles

Radar echo reported

Radar echo reported

Region where ground stations report precipitation

intraseasonal oscillations associated with the area-averaged rainfall anomaly over central India during the Indian summer monsoon (Fig. 11.22).

Fig. 11.23 A schematic figure combining synoptic, aircraft, microphysical, and visual cloud information through an arctic front. (Fig. 9a in Wang et al. 1995.)

11.7.11 Instrumentation figures

Figures showing a particular piece of instrumentation or nonstandard laboratory equipment can be either photos or line drawings. Photos have two weaknesses. First, photos may show the instrument sitting in a cluttered and dark laboratory space, in conjunction with other instrumentation as well. Such presentations and the inevitable shadows and lack of contrast do not allow the audience to focus on the important parts of the instrument. Second, most photos are unannotated, further confusing the reader. Instead of photos, many journals request that line drawings of the relevant instrumentation be presented. The line drawings enable the author to communicate more directly the functions of each component of the instrument, showing cross sections and exploded views of the interior, stripping away extraneous details that would be present in a photo. Paying a graphic artist to draw your instrumentation figure is generally worthwhile relative to the thousands of dollars you may pay in page charges. If not worthwhile, then consider whether the figure is even necessary to show.

11.7.12 Schematic figures and conceptual models

A conceptual model figure is an idealized figure synthesizing the results of the paper into a concise representation (Figs. 11.4 and 11.23). Papers that describe a complicated phenomenon, summarize a large number of cases, present a synoptic composite, or develop a conceptual model are amenable to such schematics. An effective schematic could be reproduced by others in review articles and textbooks, so ensure that the depiction is informative, accurate, clear, artistic, and aesthetically pleasing. Remember to make the schematic representative of the data in the paper. Do not overgeneralize or oversimplify. To ensure a high-quality graphic, enlist an experienced graphic artist.

11.8 FIGURES FROM OTHER SOURCES

Sometimes the best graphic is one that has been previously published by someone else. For example, the figures in a review article may come almost entirely from others. To achieve the best quality for the figure, try to obtain the digital file from the original author. If this is not possible, use a digital scanner to capture the image from the published journal. Experiment with different settings and scanners to obtain the highest-quality image. If the journal is online and you can screen-capture the image, expand the figure to the largest size possible without pixelation to obtain the best resolution during screen capture.

The quality of the scan or screen capture may be inferior, or the original figure may have extraneous material that may not be desired for your purposes. If so, the graphic may have to be redrawn. Such an approach can simplify or clarify graphics from older journals.

Before publishing others' figures, you may need to obtain permission of the copyright holder, who may not necessarily be the author (see Copyright on page 6). If you have received permission to reproduce the figure, identify the source of the figure and place a copyright notice in the figure caption "(Fig. 5 from Ackerman et al. 2009, © 2009 American Meteorological Society)."

Finally, if your figure or table comes from another source, properly cite the source in the caption: "From Jones (1995, her Fig. 5)." If the figure has been redrafted or otherwise altered, say so: "Adapted from Jones (1995, her Fig. 5)." If the changes to the figure are substantial or have affected the data, practice full disclosure: "Redrawn from Jones (1995, her Fig. 5) to emphasize cloud-top temperatures colder than −45°C." If the figure has not been previously published and is provided for use in your manuscript by someone not on the author list, list that person in the acknowledgments with "Figure 7 is courtesy of Sarah Jones" and place "Courtesy of Sarah Jones" in the figure caption.

11.9 TABLES

If care is taken in selecting which figures to include in a paper, even greater care should be taken when deciding which (if any) tables to include, an example of which is shown in Table 11.2. Presenting data in both figure and tabular forms wastes space. Pick one or the other, preferring figures. Many tables in published manuscripts could be better presented in a well-designed figure or eliminated altogether, although some tables, especially those with text, cannot be easily made into figures. Furthermore, tables take more time for the layout artist to construct, raising the costs of the article. Finally, tables can take more of the reader's time to interpret than a well-constructed figure or if the data were just presented as a list in the text.

Thus, the challenge of employing a table in your manuscript is to determine a) if the table is really necessary, b) whether the data in the table might better be presented as a figure or figures, and c) what is the most effective means to present the data in a table. These three topics are discussed in this section.

The first step is to decide whether to choose a table or a figure (or to include the material directly in the text). As discussed previously in this chapter, figures function best for communicating trends or spatial relationships between different quantities. In contrast, tables function best to:

- show precise quantitative information or a listing of numerical and alpha-numeric data;
- present data in a concise format that would otherwise be too repetitive in the text (e.g., timeline of events, lists);
- present a dataset that is too large to effectively communicate in a single fig-ure or even a group of figures (e.g., a list of dates and locations for weather events); and
- emphasize or organize important points that may not be readily apparent from the text. For example, many of the tables in this book summarize information spread out over several pages in the text. Collecting that infor-mation into the tables provides the reader with an easy-to-view list. Table 11.1 on page 114 is one such example.

You may not need a table in the following two situations. If the results are negative or not statistically significant, then consider whether the data even

Table 11.2 A sample table: examples of raindrop parameters at 700 mb (adapted from Table 1 in Shapiro 2005)

D (mm)	w_t (m s^{-1})	Re	C_D	λ (s^{-1})
1.0	−4.5	240	0.71	2.18
1.5	−6.3	510	0.55	1.58
2.0	−7.7	820	0.48	1.26
2.5	−8.7	1160	0.44	1.05
3.0	−9.3	1490	0.42	0.89

need to be presented in the manuscript. Could a sentence or two discuss the results instead? Second, if the data are not discussed in the text but are presented for completeness only, you may wish to consider whether they should be even be included in the manuscript at all.

If you have chosen to present your data in a table, then follow a few general rules about design to create an effective table:

1. Several smaller tables with more focused comparisons are preferable to larger tables. Alternatively, a multipaneled table with sections (a), (b), (c), etc. can be an effective means to present tables with similar structures.
2. Structure the table in the way that is helpful to how you think the audience will use the data. For example, most data, particularly numbers, look best when arranged in a column rather than in a row.
3. Arrange the rows and columns in some kind of context: alphabetical order, size, or order of importance. If the ordering scheme is unclear, describe it in the caption.
4. Design the table to avoid large white spaces between columns and rows.
5. Tables with too many columns to fit across the page may be printed in landscape orientation in the journal.

The final step in creating an effective table is to format the table elements (or data cells) to make a more readable table. More guidelines follow:

1. In general, I prefer the open layout that highlights the data, not restrains and obscures it with vertical or horizontal lines in tables. Many journals do not use such lines in tables, unless making distinctions between groups of columns or rows.
2. Each column must have a heading. Because column headings generally do not allow for much space, they must be concise, abbreviated, or broken up across multiple lines. Make all column headings have parallel structure (Section 9.4). Capitalize only the first word in the heading.
3. If a row or column repeats values, you may consider keeping those cells empty. Design your table to emphasize the entries that change, not the ones that stay the same.
4. As with the axes on a graph, specify units in the column heading.
5. Align columns of numbers by the decimal point. All decimals less than 1 must have a preceding zero (e.g., 0.23).
6. Indicate to the readers what data are most important, especially in large tables. For example, you might use boldface, italics, grayscaling, or other annotations (e.g., asterisks) to highlight statistically significant results. Or, you might put an asterisk next to data entries described in the text if they are not readily apparent to the readers.

7. Avoid creating an entirely new column for level of significance or other annotations that can be asterisked or included as footnotes. Use letters rather than numbers for footnotes to avoid footnotes appearing as exponents (e.g., 2^2 versus 2^a).

8. Abbreviations in the table must be defined in the caption, the body of the text, or a footnote.

11.10 CAPTIONS FOR FIGURES AND TABLES

In scientific writing, the text should discuss the results of the figure or table, and the caption should provide instructions for the reader on how to read and interpret the figure or table. Each caption should begin with a phrase that captures the essence of the figure or table. Make it unique from the other captions in the paper. The readers should understand the meaning of the figure from the caption alone. Many journals prefer discussion of the figure to occur in the body of the text rather than in the caption, so be clear on the particular format of the journal.

The most important characteristic of captions is that they must be complete. Never skimp on the caption. Every component of the figure and every panel in a multipaneled figure must be described in the caption, even if you think it is self-evident. Such completeness improves the ability of the reader to understand and interpret your figure. After writing a first draft that is complete, take care to write the captions clearly, then go through the captions again and try to make them more concise. Do not leave captions to when you are exhausted and nearing completion of the paper. Captions are too important to be neglected this way.

After a complete, clear, and concise caption is written, standardize formats for similar figures and tables within the manuscript. If you find a style that you like from an earlier manuscript, recycle the format for the present manuscript. As a final check, match the information in the caption (and figure) with that in the text.

11.11 DISCUSSING FIGURES AND TABLES IN THE TEXT

Once the figure and caption are completed, many authors think their work is done. A well-constructed figure should speak for itself. Wrong! The work is not done until the figure is adequately described in the text.

When discussing figures, discuss the most obvious aspect of the figure: the maximum in United States tornado frequency in the spring, the extensive stratocumulus west of California, and the positive correlation in the scatterplot. Thus, indulge the readers by noting the obvious, even if it is not the point you wish to make. Then, discuss the finer points or anomalies of the graph: the secondary maximum in tornado frequency in November, the ship tracks in the stratocumulus, and the outliers in the scatterplot in the lower-right corner of the plot. Not all figures need to be treated this way, but this approach is certainly sensible and satisfies the readers' needs.

The level of detail in the figure should be reasonably replicated in the accompanying text. If your figure is a complex flowchart with dozens of nodes, several paragraphs will probably be required to adequately discuss such a figure. In contrast, a simple line graph could be discussed in only one sentence. Do not waste space in the text explaining how to read the figure (e.g., "Figure 4 presents a scatterplot between the incoming solar radiation and the growth rate of new aerosol particles. Days with precipitation are represented by solid circles and days without precipitation are represented by open circles."). If your figure and caption are well constructed, such text is largely unnecessary. Tables also need to be discussed, but all the individual cells in the table do not need to be repeated in the text.

11.12 OVERSIMPLIFIED COMPARISONS

A common mistake that authors make when discussing figures in the text is to oversimplify the comparison. For example, papers that compare modeled to observed precipitation often broad-brush the comparison with a simplistic sentence such as, "a comparison between the observed and modeled precipitation fields shows remarkable similarity." In fact, the figures may not be similar, even on the most fundamental issue. Many reviewers will be rightly troubled by such text, which shows either that the author is being naïve by not critically discussing the results or that the author is being unscrupulous by trying to pass off a less-than-satisfactory comparison as satisfactory. Those who ignore or obfuscate obvious differences run the risk of being rightfully challenged.

For example, consider Fig. 11.24, a comparison between observed precipitation amounts and simulated precipitation amounts. At first glance, the model does remarkably well in capturing the essential feature of the observed precipitation, a maximum in eastern Missouri, albeit undersimulated (342 mm observed vs 300 mm simulated). But a research manuscript should not oversimplify such a comparison. Indeed, further inspection reveals potentially troublesome aspects of the simulation. Thus, a fair comparison should read something like this:

> Although the simulated precipitation amounts (Fig. 11.24b) are generally similar to the observed precipitation amounts (Fig. 11.24a), more careful inspection reveals some differences between these two fields. For example, the simulated maximum of precipitation is southwest of and less than the observed precipita-

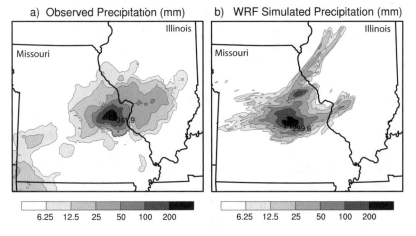

a) Observed Precipitation (mm)   b) WRF Simulated Precipitation (mm)

6.25 12.5 25 50 100 200   6.25 12.5 25 50 100 200

Fig. 11.24 Comparing observations and model—a fair comparison (see text). (a) Observed precipitation from the stage 4 product (1200 UTC 6 May to 1200 UTC 7 May 2000), and (b) modeled precipitation from a 3-km WRF run (0000–1200 UTC 7 May 2000). Precipitation in the southwestern corner of the domain in (a) occurs before the initialization time of the simulation in (b). (Figure courtesy of Russ Schumacher.)

tion maximum with more of the simulated precipitation extending farther west of the observations, the bulk of the observed precipitation in central Illinois is not simulated, and the tracks of simulated storms in northern Illinois are spurious. Despite these differences, we believe that the simulation is satisfactory for understanding the synoptic and mesoscale aspects of the maximum of heavy precipitation in eastern Missouri.

Although it takes more words to make the comparison this way, the description of the figures is more forthright. As discussed previously (Section 7.2), being honest about the quality of your research results is key to being respected as a scientist.

11.13 DIRECT VERSUS INDIRECT CITATION

Citing a figure or table in the text can be done in a direct manner or an indirect manner. The direct method is to describe the figure in the text: "Figure 2 shows the annual distribution of the number of tornadoes and tornado days per year in the United States based on data from 1973–2008." In contrast, the indirect method is to cite the figure parenthetically, at the end of material summarizing the results: "The numbers of tornadoes and tornado days across the United States are maximum in spring (Fig. 2)." The direct method should be used for complex diagrams that may take a paragraph to explain, whereas the indirect method is suitable for figures requiring minimal explanation. Many authors favor the indirect method, allowing the science to speak within the text rather than the figures. By placing the citation to the figure in parentheses, the indirect method also is a less wordy approach than other approaches that put the figure citation in the text (e.g., "As shown in Fig. 2, . . .").

"Table 3 is a list of," "Table 4 shows," or worse, "Table 6 demonstrates" . . . are unnecessary or incorrect (tables are inanimate and have never demonstrated anything). —Valiela (2001, p. 174)

When using the indirect method, place the citation at the end of the sentence where it is less intrusive. An exception to this guidance is when doing so makes what material is being referenced unclear. Consider the following two examples.

> **EXAMPLE 1:** While the surface cyclone explosively deepened to 971 mb, the convective line lengthened, ranging from northern Illinois to nearly the Gulf of Mexico in Texas, and was associated with a narrow axis of radar reflectivity factor exceeding 50 dBZ (Fig. 1d).
>
> **EXAMPLE 2:** While the surface cyclone explosively deepened to 971 mb (Fig. 1d), the convective line lengthened, ranging from northern Illinois to nearly the Gulf of Mexico in Texas, and was associated with a narrow axis of radar reflectivity factor exceeding 50 dBZ.

In Example 1, the sentence reads more smoothly with the citation to Fig. 1d at the end and implies that both the sea level pressure field and the radar imagery are included in that figure. Example 2 implies that only the sea level pressure field is cited and the radar imagery is likely not shown.

11.14 NUMBERING FIGURES AND TABLES

A few basic guidelines exist in describing figures and tables in the text. For simplicity, this discussion focuses on figures, although all of the guidelines below also pertain to tables as well:

- Number the figures in the order they appear in the text. Referring backward to figures already presented is certainly reasonable, and referring ahead to figures may be allowable in situations where grouping the figures together in certain situations makes more sense or for making a minor point on a figure to appear later. Regardless, organize the paper so as to avoid excessive jumping when discussing figures in the text. Try to present your argument in an organized linear manner. Too many references to previous figures may tire more careful readers.
- Every figure panel in the paper should be cited within the text. Otherwise, the figure panel is not necessary and should probably be deleted.
- Results from a figure referred to in the text should cite the figure. In some manuscripts, large blocks of text may describe results from figures, but never cite the specific figure numbers. Such a situation can be confusing to the reader looking to verify the text with the results in the figures. Authors can help by frequently citing relevant figures in the text. Such citations may seem unnecessary or excessive to the author, but readers rarely complain about too many citations compared to too few.

I select my figures as if I am a lawyer presenting a case in court. They are exhibits A, B, C, etc. My introduction is the opening statement telling the jury what I am going to prove. The figures are the evidence needed to make my case. My conclusions section is the final summation to the jury. They should have no choice but to believe me.
—*Robert Houze, University of Washington*

Another common mistake in submitted manuscripts is incorrect figure numbers. Whether these arise from carelessness during the editing process or the addition or removal of a figure during revisions, such errors during the review process can test even the most patient reviewer. Before submitting a manuscript, take just a few minutes to search out all figure references and make sure they correspond with the proper figure. At the same time you are checking for the correct figure and table numbers, make sure that "figure" and "table" are capitalized when referring to a specific figure, but in lowercase otherwise.

As a final admonition, despite the importance placed on figures on this chapter, I remind you that your manuscript should tell the story in the absence of figures. The text should be able to be read and understood without access to the figures. That does not mean that the figures are superfluous, as often the figures can be used to lay out the story of the paper (as discussed in Chapter 6 and by Robert Houze on page 168). Instead, both the text and the figures should be able to tell the story more or less independently of each other in a well-written article. Thus, the figure or table (along with its caption) should be self-contained.

11.15 PLACING FIGURES AND TABLES IN THE MANUSCRIPT

Refer to the style guide or Instructions to Authors about how to handle figures in the submitted version of the manuscript. Some journals want the figures placed within the text, whereas others want the figures accumulated at the back of the manuscript. When figures appear at the back of the manuscript, most journals expect each figure on a separate page.

The panels of a multipaneled figure must be joined together within a single file name. Some journals will not accept separate panels or may charge you for layout costs. One exception is if a figure or table is too big to be accommodated on a single page in the journal. Try to avoid splitting figures or tables across a page, but, if you must do so, split logically.

11.16 EQUATIONS AND CHEMICAL REACTIONS

Equations convey complex mathematical relationships. As important as equations (here to also include chemical reactions) can be to our common scientific language, a manuscript cannot stand on equations alone. Find the right balance between presenting every detailed step of a derivation and leaving out too many steps. Use the text to describe what was done to the equations, then provide the final equation. Avoid the overused phrase, "it can be shown that . . . ," which often is accompanied by too little detail about the steps involved.

The equations should be accompanied by text providing physical insight, a description of what the math means, perhaps even with some explanation under each term of the equation. After presenting an equation, sometimes a brief example can help the reader understand how the equation works by plugging in typical values or performing a scale analysis. Furthermore, many equations are derived under a set of assumptions. Those assumptions should be made explicit, especially if the relevant equation is just presented separate from its derivation.

The display of the equations can be as important as the actual equations themselves. Equations should be presented intermingled with the text, as if they were a seamless part of the text. Equations are generally treated as independent clauses, meaning that they could stand alone as complete sentences but never do. Most equations are listed on their own line, separate from the rest of the text. Brief equations may be included in the text. As with acronyms, do not start sentences with symbols or equations. Place punctuation after the equation, as if the equation were part of a sentence. Clear presentation and proper spacing of variables are essential. Nearly all scalars are italicized and vectors are boldfaced, even if included in the body of the text. Operators (e.g., sin, log, mod) are generally set roman. If you are wondering how to typeset your equations, find examples from your target journal or other high-quality scientific publications to understand press style.

As with abbreviations and acronyms, introduce all variables upon first usage. It is mathematically improper to use an equal sign to link a symbol to its text description (e.g., "c = phase speed" is incorrect; "c represents phase speed" is correct). Choose standard symbols wherever possible. A list of symbols in a document can be put into an appendix or table of symbols. Simple or common chemical symbols (e.g., CO_2, NH_3) generally do not need to be defined in the text. Other chemical compounds may go by their names or acronyms instead. For example, $(CH_3)_2S$ is dimethyl sulfide or DMS.

Equations and reactions are numbered sequentially, usually at the right margin. The numbers can be useful for referring to a certain equation in the text. For example, "Equation (6) presents the quasigeostrophic omega equation." Because different journals use different styles, follow the style of your target journal. Wherever possible, link the equation number to a specific name in the text: "Because of the continuity equation (3.1)" as opposed to "Because of equation (3.1)."

CITATIONS AND REFERENCES

Referencing previously published literature pays homage to those whose shoulders we have stood upon (or, for papers we may have a scientific disagreement with, those shoulders we wish to walk upon). This chapter discusses the mechanisms of citing and referencing, how to determine which literature to cite, how best to cite the literature, and tips for citing and referencing.

Given the importance of the previous literature to the content of an author's manuscript, authors should develop great skill in citing sources. To better understand citations and references in this chapter, let's ask the six questions of journalism: why, how, what, when, who, and where?

12.1 WHY CITE THE LITERATURE?

Among the many roles that we research scientists fulfill during our careers (e.g., teacher, collaborator, author), we are first and foremost scholars. One aspect of scholarship is the ability to read, evaluate, interpret, and critique previously published literature; and we demonstrate scholarship through the papers we write and the sources we cite. More specifically, we cite the published work of others to do the following:

- convince others we know our field, and we possess both breadth and depth;
- describe the history of the field;
- credit other authors for previously published ideas, research, hypotheses, and speculation;
- show the historical or intellectual development of our original ideas;
- distinguish our research from previously published work;

- critique previously published work; and
- cite direct quotations or figures from other sources.

A more practical reason for citing previously published work is to rebut reviewers and readers skeptical of your arguments. If you make a statement backed up by citing relevant, carefully chosen sources, then critical reviewers are compelled to show why each source is not relevant or is incorrect. Thus, a large number of well-chosen citations can be a shield to strengthen your arguments.

12.2 HOW TO CITE THE LITERATURE

There are two principal formats for citations and references in common use today. The first is the *author–date* system or Harvard reference system, which is the format used by the AMS and this book. Sources are cited in the text by author name and year, and are listed alphabetically in the reference list. The second is the *citation-order* system, or the Vancouver reference system, where sources are numbered based on their order of citation in the manuscript. A hybrid of these two systems is the *author–number* system, also called the *alphabet–number* system, which resembles the citation-order system in that sources are cited by number in the text, but resembles the author–date system in that sources are listed alphabetically in the reference list. The author–date and the citation-order systems are illustrated next.

The author–date system is as follows:

Blocking has also been shown to play a role in the modulation of the intensity of the Southern Hemisphere split jet (e.g., Trenberth and Mo 1985; Mo et al. 1987; Trenberth 1986, 1991). An early study by van Loon (1956) demonstrated that blocking in the Southern Hemisphere winter was favored in the southwest Pacific Ocean and to the southeast of Australia. More recent studies (e.g., Marques and Rao 1999; Renwick and Revell 1999) have confirmed the earlier findings and have established that the area near South America is an important secondary blocking region in winter and spring.

Marques, R. F., and V. B. Rao, 1999: A diagnosis of a long-lasting blocking event over the southeast Pacific Ocean. *Mon. Wea. Rev.*, **127**, 1761–1776.

Mo, K. C., J. Pfaendtner, and E. Kalnay, 1987: A GCM study on the maintenance of the June 1982 blocking in the Southern Hemisphere. *J. Atmos. Sci.*, **44**, 1123–1142.

Renwick, J. A., and M. J. Revell, 1999: Blocking over the South Pacific and Rossby wave propagation. *Mon. Wea. Rev.*, **127**, 2233–2247.

Trenberth, K. E., 1986: An assessment of the impact of transient eddies on the zonal flow during a blocking episode using Eliassen–Palm flux diagnostics. *J. Atmos. Sci.*, **43**, 2070–2087.

Trenberth, K. E., 1991: Storm tracks in the Southern Hemisphere. *J. Atmos. Sci.*, **48**, 2159–2178.

Trenberth, K. E., and K. C. Mo, 1985: Blocking in the Southern Hemisphere. *Mon. Wea. Rev.*, **113**, 3–21.

van Loon, H., 1956: Blocking action in the Southern Hemisphere. *Notos*, **5**, 171–177.

The citation-order system looks like this:

Blocking has also been shown to play a role in the modulation of the intensity of the Southern Hemisphere split jet.[1,2,3,4] An early study[5] demonstrated that blocking in the Southern Hemisphere winter was favored in the southwest Pacific Ocean and to the southeast of Australia. More recent studies[6,7] have confirmed the earlier findings and have established that the area near South America is an important secondary blocking region in winter and spring.

1. Trenberth, K. E., and K. C. Mo, 1985: Blocking in the Southern Hemisphere. *Mon. Wea. Rev.*, **113**, 3–21.

2. Trenberth, K. E., 1986: An assessment of the impact of transient eddies on the zonal flow during a blocking episode using Eliassen–Palm flux diagnostics. *J. Atmos. Sci.*, **43**, 2070–2087.

3. Mo, K. C., J. Pfaendtner, and E. Kalnay, 1987: A GCM study on the maintenance of the June 1982 blocking in the Southern Hemisphere. *J. Atmos. Sci.*, **44**, 1123–1142.

4. Trenberth, K. E., 1991: Storm tracks in the Southern Hemisphere. *J. Atmos. Sci.*, **48**, 2159–2178.

5. van Loon, H., 1956: Blocking action in the Southern Hemisphere. *Notos*, **5**, 171–177.

6. Marques, R. F., and V. B. Rao, 1999: A diagnosis of a long-lasting blocking event over the southeast Pacific Ocean. *Mon. Wea. Rev.*, **127**, 1761–1776.

7. Renwick, J. A., and M. J. Revell, 1999: Blocking over the South Pacific and Rossby wave propagation. *Mon. Wea. Rev.*, **127**, 2233–2247.

The author–date system is most advantageous to authors and editors, who do not need to renumber reference lists every time a change is made to the manuscript. Numbered systems are most advantageous to the environment by reducing article length by replacing names and years by numbers. Some readers say that articles using the numbered systems are easier to read, as they are

DEFINITIONS

Source. a document providing information, amplification, or context

Citation. the documentation, within the text, of an external source.

EXAMPLE: (Smith 1990, p. 303)

Reference. a source cited within the text and included in the reference list, with information on how to obtain the source.

EXAMPLE: Wernli, H., and H. C. Davies, 1997:

A Lagrangian-based analysis of extratropical cyclones. I: The method and some applications. *Quart. J. Roy. Meteor. Soc.*, 123, 467–489.

Reference list. a complete list of all references in the text

Bibliography. usually regarded as a complete list of all sources on a specific subject, some of which may not be cited in the text

Annotated bibliography. a bibliography with a written summary or abstract of each reference

not cluttered by citations, while other readers are annoyed by having to refer to the reference list frequently to know which number corresponds to which cited source. Because each journal adheres to its own style and expects authors to follow its style, identifying the target journal locks you into a referencing system for your manuscript.

12.3 WHAT LITERATURE TO CITE AND WHEN TO CITE IT

What constitutes a legitimate citation in the text of a scientific paper? Different well-meaning people may disagree on this point, and I hope to identify some of the subtleties within this section. Nevertheless, what information must be cited is very clear. Quoting Schall (2006):

- quotations, opinions, and predictions, whether directly quoted or paraphrased
- statistics derived by the original author
- visuals in the original
- another author's theories
- case studies
- another author's direct experimental methods or results
- another author's specialized research procedures or findings.

Be careful when employing long lists of citations parenthetically. Often, authors may want to demonstrate that a number of studies have been performed on a topic, so they may list five or more studies in a parenthetical note. When these studies are closely linked, this may be an acceptable practice. When the papers are quite disparate (e.g., a mixture of observational and modeling studies), breaking apart such lists into different groups may be preferable.

Although no formal rule appears to exist, standard practice is to list citations in chronological order. Not doing so will raise questions in readers' heads about why chronological order was not followed. If you want to highlight a certain paper from a list, add some text to indicate why this paper deserves special treatment. To put it simply, do not create exceptions that you do not explain to the reader.

Review articles can and should be cited in manuscripts, if relevant. If a well-written review article summarizes the main points the author wishes to make, the author could write, "A review by Keyser and Shapiro (1986) summarizes. . . ." If you use a review article for citing a general topic, write "(Keyser and Shapiro 1986, and references therein)," "(Keyser and Shapiro 1986, especially references within Section 4)," or something similar. However, such a citation should only be used when discussing the topic in general—for example, when so many relevant references exist that a complete list would be unnecessary or beyond the scope of the paper.

Although review articles or monographs may offer thorough literature syntheses, please cite the original sources. According to the Golden Rule or the ethic of reciprocity, treat others as you want to be treated. Had you performed some of the relevant research that is cited in a review article, you would feel slighted had another author not cited your original work in a later manuscript.

The cited sources in others' research articles will help you discover other research that may be relevant for you to cite. However, perform your own literature search to find relevant sources that have not been cited previously. No matter how thorough you are, you may inadvertently omit some references. This is natural. But, being unbiased, thorough, and accurate is the surest way to avoid potential omissions.

12.4 WHO TO CITE

Do not overreference yourself or your colleagues, particularly when other sources could and should best be cited. Unfortunately, this trap is easy to fall into as few people may know more about the topic at hand than you do, and your own work is what you know best. Not citing particular authors or research groups for personal reasons is also inappropriate. In situations where several groups have been working on a specific topic, providing at least one representative reference from each group is one way to not show bias. One-sided reviews of the literature that ignore alternative points of view, however, can be easily recognized by the audience, leading to a discrediting of your work as being biased and potentially offending the neglected authors (who might also be your reviewers!).

PRIMARY, SECONDARY, AND TERTIARY SOURCES VERSUS PEER-REVIEWED AND GRAY LITERATURE

Sources are classified according to their closeness to the origin of the information. Primary sources are the original source of the information. These include scientific journal articles, technical memoranda, dissertations, and datasets reporting original theoretical, experimental, or modeling results. Secondary sources derive from the original work and include review articles, biographies, bibliographies, monographs, and textbooks. Tertiary sources are those that are even further derived, taking their information mostly from secondary sources: encyclopedias, almanacs, and newspaper articles derived from a press release of a published article. The distinction between secondary and tertiary sources is sometimes vague, particularly for books. Fortunately, such distinctions are not often critical.

Cite primary sources to the greatest extent possible. If you are unable to obtain a copy of the original source yourself but you have seen it cited elsewhere, use the following convention: "(Sanders 1967, cited in Kessler 2008)." For opinions of others based on sources that you have no access to, use the following convention: "(Kessler 2008, discussing Sanders 1967)" or "Kessler (2008), in discussing Sanders (1967), said . . ."

Avoid citing secondary and tertiary sources when a primary source is available. Although such material may be useful for verifying facts, your audience will likely view such citations as being elementary. Such citations, however, can be quite effective to show the status quo or commonly accepted knowledge. For example, "Although Holton (1992, p. 208) said, 'The occurrence of inertial instability over a large area would be expected immediately to trigger inertially unstable motions,' new evidence indicates that this statement needs to be reexamined." Some textbooks, however, may be primary sources for some material or provide the most lucid explanation of the topic. If so, then these textbooks may be the most appropriate to cite.

Cite peer-reviewed sources wherever possible. Peer-reviewed sources are generally viewed with more authority. For example, if the same material appears in a conference extended abstract and a published article by the same author, cite the published article. Nonrefereed primary literature, such as dissertations, conference extended abstracts, and technical reports, is referred to as *gray literature*. Citing gray literature should be avoided and may even be prohibited in some journals. Some journals may expect gray literature to be footnoted or referenced parenthetically, as opposed to appearing in the reference list. Where citation of gray literature is appropriate, but prohibited by journal policy, citation to "(B. A. Colle 2006, personal communication)," "(Colle 2006, unpublished manuscript)," or "unpublished research results by Colle (2006) show . . ." may be acceptable. Manuscripts that are undergoing peer review but have not yet been accepted may also be similarly handled: "(Colle 2006, manuscript submitted to *Mon. Wea. Rev.*)." Citing secondary literature is more generally permitted where the secondary literature has made substantial or novel contribution, is widely recognized after years of "cult status," or contributes to the history of a particular discipline.

12.5 WHERE TO CITE THE LITERATURE

As with figure and table citations (Section 11.11), what is being cited and why it is being cited should be made very clear, both by the words surrounding the citation and the location of the citation within the sentence. Place the citation at the end of the sentence to avoid interruptions, unless doing so makes what material is being referenced unclear. Consider the following two examples.

EXAMPLE 1: Precipitation gauge undercatch, which can produce liquid equivalents that are 40%–70% less than snow collected and melted from snowboards, will introduce a bias toward larger snow-to-liquid-equivalent ratios (e.g., Peck 1972; Goodison 1978; Groisman and Legates 1994).

EXAMPLE 2: Precipitation gauge undercatch, which can produce liquid equivalents that are 40%–70% less than snow collected and melted from snowboards (e.g., Peck 1972; Goodison 1978; Groisman and Legates 1994), will introduce a bias toward larger snow-to-liquid-equivalent ratios.

Although Example 1 reads more smoothly with the citations at the end of the sentence, we interpret the location of the citations to indicate that the three references all report that undercatch leads to a bias in snow-to-liquid-equivalent ratios, which would be erroneous. Example 2 is more accurate in that the three references quantify only the precipitation undercatch, not snow-to-liquid-equivalent ratios.

The example below shows how specific attribution can be signified by avoiding a long list of parenthetical citations, a point also made in Section 12.3. In this example, three different types of studies (i.e., observational, numerical modeling, and idealized channel-model studies) show the prevalence of cold advection along what had been called bent-back warm fronts. To avoid confusion with warm fronts (which are associated with warm advection), the term *bent-back front* was used.

Because cold advection can occur in association with bent-back warm fronts [e.g., as noted in observational studies of oceanic cyclones (Shapiro and Keyser 1990; Neiman and Shapiro 1993; Blier and Wakimoto 1995), numerical modeling studies of oceanic cyclones (Kuo et al. 1991, 1992; Reed et al. 1994), and an idealized channel-model study of baroclinic development (Hoskins 1983, p. 18)], we refer to bent-back warm fronts as *bent-back fronts*.

What if your citation applies to an entire paragraph? If the paragraph opens with a well-written topic sentence (page 65), then a single citation following this topic sentence should indicate that the material that follows is related to the first citation. In cases where a more explicit statement is needed, the text could clearly say that the topic of the paragraph is discussed in more detail by the cited source.

12.6 QUOTATIONS

Use direct quotations in scientific literature sparingly, but effectively. Quotations without context or interpretation are unacceptable in scientific documents.

BLENDING SOURCE MATERIAL WITH YOUR WORK

Joe Schall, Health Communications Specialist,
National Institute for Occupational Safety and Health

Blending source material with your own work is a process of selecting the best material, extracting it from its original location without violating its intended context, and presenting it alongside your own work so that it supports your ideas rather than usurps them. You must labor to avoid the appearance—or the fact—of simply regurgitating ideas that others created. To become a skillful writer and researcher, it is important for you to develop your own assertions, organize your material so that your own ideas are the thrust of the document you create, and take care not to rely too much on any one source, or your content might be controlled too heavily by that source.

In practical terms, some unambiguous ways to develop your assertions and organize your material include:

- During the writing process, intentionally group your sources by some theme so that they blend, even within paragraphs. Your paper—both globally and at the paragraph level—should strive to reveal relationships among your sources and should also reveal the relationships between your own ideas and those of your sources.
- As much as is practical, make the paper's introduction and conclusion your own ideas or your own synthesis of the ideas revealed by your research. Use sources minimally in your introduction and conclusion, and choose from the most seminal sources for inclusion in these sections.

- In general, use the openings and closings of your paragraphs to reveal your work—that is, enclose your sources among your assertions, thinking of your own assertions as bookends for the sources. At a minimum, make it a regular practice to create your own topic sentences and wrap-up sentences for paragraphs, wording them so that readers intuit that they are yours in context.
- When appropriate, practice common rhetorical strategies such as analysis, synthesis, comparison, contrast, summary, description, definition, hierarchical structure, evaluation, hypothesis, generalization, classification, and even narration. Even when we read a literature review, we should have a sense that the author is managing the material rather than vice versa, and a well-placed transition, a simple enumeration of points, or a brief definition composed by the author will reinforce that necessary authorial control. In short, prove to your reader that you are thinking as you write.

To effectively blend source material with your own work, you must also clarify where your own ideas end and the cited information begins, and your very wording can, in effect, neatly fill this gap and create context for the cited information. A phrase such as "A 2002 study revealed that" is an obvious announcement of a citation to come. Another common technique is the insertion of the author's name directly into the text to announce the beginning of your cited information, in particular if that author is prominent enough to warrant repeat citation. Finally, when you compare the work of one author to another, you can create context for your narrative through a simple phrasing devoted to advancing the theme you are discussing, such as "A follow-up paper by Watkins et al. (2002) expanded on the radiative effects of clouds on climate, by investigating. . . ."

Avoid using quotations for general points that could have been said by just about anyone. Remember that some quotations may require you to obtain permission from the copyright owner.

Direct quotations must always be enclosed in quotes and cited. The quotation must be written identically to the original, except where italics are added for emphasis (say "[emphasis added]"), material is deleted (use ellipses, "..."), or any words are added (as for clarifying pronouns, "it [the mesocyclone] had a remarkable signature in the radial velocity field"). Page numbers should generally be included in the citation to aid readers who may wish to read the quote within its original context.

Placing "[*sic*]" (Latin, "thus,") in a direct quotation is a signal to the reader of a misspelling, variant in spelling or phrasing from current usage, or an error in the source. Use "[*sic*]" at the location in the quotation where the error is made to avoid giving the appearance of having made a mistake yourself.

12.7 CITATION SYNTAX

Syntax is the set of rules by which we construct our language. American humorist Will Rogers once said that syntax "must be bad, havin' both sin and tax in it." Here is a collection of advice related to the syntax of proper citations.

Abbreviating articles. Some authors choose to abbreviate an article by the initials of the authors and the year if the article gets cited multiple times in the manuscript (e.g., McKay and LaTour 2007 becomes ML07). As discussed on page 98, avoid such abbreviations, unless absolutely essential and used many times throughout the manuscript.

Article/paper/study. When citing literature, make your writing more concise by eliminating the often unnecessary *article*, *paper*, or *study*, as in the following: "The Johnson (2001) article demonstrated...." or "The study by Johnson (2001) demonstrated...." These can be said more simply as "Johnson (2001) demonstrated...."

e.g. This is an abbreviation for the Latin *exempli gratia*, which means "for example." Many style guides recommend using this expression inside parentheses only, preceding an incomplete list. For example, "Stratus clouds are an important control on the radiation balance of the atmosphere (e.g., Harrison et al. 1990; Stephens and Greenwald 1991; Hartmann et al. 1992; Klein and Hartmann 1993)." Clearly, listing every single paper that made that claim would not be feasible for such a simple sentence, but some references are needed, perhaps the most important ones, a relevant review article, or textbook. If only one reference is needed, then "e.g." is not needed. Dr. Richard Tyson of Newcastle University says the following about using "e.g." before lists of references: "Unless used thoughtfully, deliberately, and appropriately, it certainly does not help." Always put a comma after "e.g." in American English. In British English, the trailing comma may be omitted.

et al. From the Latin *et alia* ("and others"), "et al." indicates that the source has more than one author (e.g., Garrett et al. 2005). The period after the "al." is essential, unless you mean to imply the coauthor is named Al. Depending on the style of the journal, a comma may or may not follow "et al.," and "et al." may be italicized.

Figures and tables. If you include figures or tables from other sources in your manuscript, include a citation to the specific figure number or page in parentheses. If you wish to reproduce the figure caption, place it within quotes and cite that, too: "(Figure and caption from Hakim et al. 2002.)" If you have altered the figure in any way, add the phrase *adapted from* preceding the citation: "(Adapted from Hakim et al. 2002)." Make sure to distinguish between figures from other sources and figures in your current paper [e.g., "(Fig. 10 in Parker 2000)" or "(Parker 2000, his Fig. 10)" is less ambiguous than "(Parker 2000, Fig. 10)"].

Footnotes. Although preferred in the arts, humanities, and some social sciences, do not use a footnote to list a citation in scientific writing.

i.e. This is an abbreviation for the Latin *id est*, which means "that is." Many style guides recommend using this expression inside parentheses only to mean "in other words," to expand upon words and phrases. Do not use "i.e." for citations when you mean "e.g." Always put a comma after "i.e." in American English. In British English, the trailing comma may be omitted.

Initials, when needed. Occasionally you may wish to cite two papers published in the same year by two different people with the same surname. Use initials of their first and middle names to distinguish them in citations: C. Schumacher et al. (2008) and P. N. Schumacher et al. (2008).

Page numbers. Page numbers, section numbers, or chapters should be provided for books and other such citations, unless making a general statement that refers to the whole book. You may also wish to include page numbers with citations where it may not be obvious where the citation originates from (e.g., for quotes). Page numbers should be added to citations after the year "(Martin 2006, p. 123)." A single page number is indicated "p. 34," and a range of pages is indicated "pp. 1–45." Always use an en dash (page 348) between numbers when indicating a page range (entered in LaTeX as two hyphens; in Microsoft Word for Mac as Option-hyphen; or in Microsoft Word for Windows, by going to Insert, Symbol, and choosing the en dash from the Symbols tab, though it is recommended to create a keyboard shortcut for yourself).

Personal communication. The AMS (2008) defines a personal communication as "a completed manuscript that was never published, or an informal discussion, or written communication with researchers," and is cited "(L. Wicker 2006, personal communication)." Avoid citing opinions or common knowledge in this way. If the name of the person is not obvious from the context of the paper by using just the initials, include the full name and affiliation in the acknowledgments section. If possible, get the information in written form, not verbal communications, to protect you and the person being cited. Always ask sources for permission to publish by showing them the precise text as it will appear with their names.

Repeating the year. One of the common questions is how often to continue listing the year after the author's name when discussing the paper multiple times within a paragraph, a section, or the entire manuscript. I err on the side of continuing to include the year, even if it means being slightly repetitive. That way, there is never any confusion that I am referring to an article or book, rather than the name of a particular person. Including the year helps avoid confusion should there be more than one paper by a particular author in the reference list.

See. Citations do not need to be prefixed with "see," as in "(see Mudrick 1974)."

12.8 REFERENCE LISTS

Before the advent of the personal computer, many authors documented, annotated, and stored their references on index cards. In the computer era, many authors use an electronic database for references. Such databases allow easy creation of reference lists from entries already prepared by the author. Many Web sites for journals will export references for their articles in various formats, thereby facilitating the preparation of your personal database.

Regardless of how the reference list is created, follow the format of the citations and reference list given by the style guide for your target journal. For sources in which the style guide does not provide a format, check reference style guides such as the most recent version of *The Chicago Manual of Style*. For journals, class projects, or other writing assignments where the referencing style is unstated, select one of the standard referencing styles (e.g., author–date system, citation-order system) and maintain consistency with that style throughout your manuscript.

Incomplete and inaccurate references can be frustrating to your readers. Accurate citations are also required for proper attribution in citation services and cross-linking in online databases. Before submission, authors should perform two checks to ensure the completeness and accuracy of the citations and reference list.

The first check is that all citations within the text have references and all references are cited within the text. I do this by printing out the reference list and then electronically searching for all occurrences of "19" or "20" in the manuscript file (these numbers being the prefix to nearly all years that might be cited in my manuscript). When found, I mark off the reference from the list. Going through the whole file this way, I can ensure the completeness of the reference list and identify any inconsistencies between the citations and the reference list (e.g., wrong years, misspellings of author names).

Second, all material in the reference list is verified for accuracy with a paper or electronic copy of each source (e.g., page numbers, volume numbers, correct spelling of authors' names). Maintaining a file with verified reference lists can help speed this process. Much of the time spent copy editing journals is in correctly formatting and verifying the reference list, so authors can help keep the cost of page charges down through providing accurate references. Although the journal will often ask for clarification on incomplete or inaccurate references, the author should not rely on the editing staff to do this.

12.9 CITING DIGITAL MATERIALS

Digital materials are being increasingly cited in scientific work. Published materials, such as CD-ROMs, peer-reviewed electronic journals, and online government documents, are primary literature and should be cited where appropriate. However, care should be taken with numerous other online documents, such as Web sites and electronic online presentations, which are gray literature and generally should not be cited. The particular format will depend on the referencing style adopted by the journal, as presented in the style guide or Instructions to Authors, but a list of the type of information to include is found in Table 12.1. Indicate when any information is unavailable or unknown: "publisher unknown."

Table 12.1   Information to include when referencing digital or online sources (adapted from the Monash University Language and Learning Online Web site)

Author or editor
Title of the Web site (if one exists)
Title of the host Web site
Date that the page was last updated or copyrighted (if known)
Name of database or type of medium (e.g., CD-ROM)
Date the information was accessed
URL of the page or the distributor
Identifying number:
    DOI, ISBN, citation number, document identification number, or access number from an online archive

---

12.10 A FINAL ADMONITION

Authors should always read the articles in their reference list. The truth is that most authors have not read the papers they cite or cite them for the wrong reason—Simkin and Roychowdhury (2003) estimate that only about 20% of all sources are read by the citing author. There are very practical reasons for completely reading the sources you cite rather than relying on other published research. Would you rely on what Smith et al. said about Sanders (1955), or would you rather interpret Sanders (1955) in your own way? Furthermore, how many times have we failed to remember important passages of papers, or have our ideas changed about the papers since we last read them two, ten, or twenty years earlier? Revisiting previously read literature is valuable to our professional development and can help ensure the accuracy of our citations. In addition, obtaining a copy of each cited paper and corroborating with the citation ensures that you have the year and page numbers in the reference correct.

EDITING AND FINISHING UP

13

Once the first draft of the manuscript is completed, emphasis shifts from primarily writing to primarily editing. Editing existing text is often easier than writing it. Eventually, multiple drafts of the manuscript are written and rewritten. Days of focusing on the manuscript have led to exhaustion, but also the satisfaction that the manuscript is ready to submit. This chapter discusses the editing process and how to bring the manuscript to a close. These steps include approaches to making revisions, getting feedback from others, dealing with minor formatting and syntactical issues, and making the final edits.

Short of seeing your article in print, producing the first complete draft of your manuscript is probably one of the most satisfying experiences in writing. I call this point the *hit-by-the-bus moment*. If I were hit by a bus on my way home with that manuscript saved on my computer at work, my coauthors would be able to retrieve that manuscript, and, with a reasonable amount of revision, be able to submit it in posthumous tribute to me. Reaching the hit-by-the-bus moment is an important milestone in publishing your manuscript. Celebrate a little. You deserve it.

Unfortunately, your work is not done yet. So look both ways before crossing the street (especially in England!) to avoid saddling your coauthors with the burden of finishing the paper. The next section describes a process to make these revisions go more smoothly.

13.1 THE PROCESS OF REVISION

In movies, writers are often depicted at a manual typewriter, not a word processor. They finish each page, progressing until their book is completed. I have wondered if this were Hollywood's unrealistic portrayal of authors or whether writers exist that are talented enough to crank out perfect prose without any revisions. My question was answered, in at least one case, when I heard that Jack Kerouac's *On the Road* was typed on a scroll of paper over 36 m long so that he could maintain his train of thought without changing sheets of paper. For nearly all of the rest of us, we need to perform often laborious editing to produce near-perfect text.

Indeed, editing is time consuming. Editing involves going through the text with strict attention to detail at a much slower pace than writing. Many passes through the manuscript are often required for the author to recognize all the revisions that need to be made. In fact, most authors I know and admire create dozens of drafts (50 drafts would not be unreasonable) before their manuscripts are ready for submission.

Unfortunately, most manuscripts received at journals would benefit from a more rigorous revision process. Rather than laziness on the part of the authors, I believe that most manuscripts simply do not receive the detailed editing they need because the authors focus on writing text, rather than editing it. Upon reaching the hit-by-the-bus moment, many authors may immediately submit their manuscript, thinking they are done. To the contrary, this transition between the first complete draft and the submitted version is critical to delivering a high-quality manuscript. At the hit-by-the-bus moment, the role of the author must evolve from one dominated by writing to one dominated by editing. When writing, the author worries about creating valid scientific arguments, ensures the sections of the manuscript are properly organized, and creates all the necessary figures at least in draft form. When editing, emphasis shifts toward getting the most impact from the writing through effective sentence structure, clear and precise word choice, and concision.

To help during this transition, I offer the following organized approach to editing. Although some may find this approach overly prescriptive, others may welcome its formalism. As discussed previously during the writing process, as the paper reaches its first draft, the writing shifts from large-scale issues on the writing/editing funnel (Fig. 7.4), such as the flow between paragraphs and between sentences, to smaller-scale issues, such as word choice, misspellings, and typos. Editing proceeds in the same manner. By starting with the largest-scale issues first before worrying about sentence- and word-level problems, you can save yourself considerable effort.

Once the organization of the text at a certain scale is determined (such as at the sentence level), Schall (2006, pp. 42–44) recommends making revisions in three stages that he calls CPR: concision, precision, and revision. In the concision stage, trim unnecessary text. In the precision stage, make the writing more clear. Finally, revisions sharpen the transitions, as the coherency techniques discussed in Section 8.2 are designed to do. By first applying the writing/editing funnel to the organization of the text, then following the CPR technique, the manuscript goes through a rigorous top-down edit. Before illustrating this editing approach with an example, two more techniques need to be introduced, one to help organize jumbled text, the other to help trim it down.

13.2 LOSING YOUR WAY

As writing and editing proceeds on your manuscript, you may lose your path. You may know what you want to say, but the writing just somehow cannot represent it. Perhaps your writing looks like what Strunk and White (2000, p. 25) call a "succession of loose sentences," sections of text devoid of organization. Such a situation typically arises when authors sit down and type directly into the computer without a well-structured outline. Furthermore, because manuscripts take more than one day to write, the internal coherence that might develop if it were written in a single day is lost. Or, seeing the forest through the trees may be difficult as the words become too familiar. Alternatively, some authors have difficulty producing a manuscript with a logical progression, and they need help in determining when their train of thought derailed. Whatever the reason, your text may contain great ideas and will contribute to a strong document eventually, but, in its present form, lacks structure and organization.

If you feel this way, let the work sit overnight or for several days. If the respite does not help, look at the big picture you are writing about. What is the logical progression of ideas? Are you first presenting model output, then the supporting observations? Would it help to present the observations first to motivate the modeling simulations? Do you jump around between scales of motion, presenting mesoscale and microscale observations interspersed among the synoptic-scale discussion? Has the structure of logical arguments (Section 7.3) been violated, leading to confusion?

One way to visualize how the text is organized is to go through the confusing section and label the topics of the text in the margins of the paper, grouping the topics into similar themes. Classify statements, then look for common themes to group together. The weaknesses in the organization of the text will likely become apparent, and you will see your way toward improving the text. If you need to, print out the text, cut out the different sections from the manuscript with scissors, and try possible arrangements of the strips of paper on a table.

The paragraph-level organization of the writing/editing funnel can be one of the most difficult stages of writing a scientific document. Once the paragraphs (clearly defined by their topic sentences) are in place, sentences and words follow much more easily.

13.3 CONDENSING TEXT THROUGH PRÉCIS

The next step is to make the text more concise. When we write, especially the first draft, we often write like we think or talk, throwing in unnecessary verbal baggage, duplicate phrasings, colloquialisms, and other phrases we hear commonly in conversations. Or we may express tangential thoughts arising from our nonlinear thinking. Unfortunately, we may be so close to our own writing that we cannot see its faults.

One exercise that helps to identify baggage and tangents is précis (pronounced *pray-see*), the process of shortening existing text to its bare essence. In other words, what is the minimum number of words that can convey the same meaning? To précis a section of text, rewrite the text by eliminating unnecessary words, but do not omit any principal points nor alter its meaning. Précis differs from paraphrasing because many of the words in the original text are still used in précis. In contrast, paraphrasing uses different words to convey the essence of the original text.

Greetings, my friend. We are all interested in the future, for that is where you and I are going to spend the rest of our lives. And remember my friend, future events such as these will affect you in the future. —Criswell, the psychic in the movie Plan 9 from Outer Space

Even well-written text can be a target for précis. For example, the trimming of words required in précis can be an effective tool for meeting stringent length requirements for abstracts. Below we take an example piece of published text and reduce it through précis.

ORIGINAL TEXT: The National Weather Service (NWS) is now in the midst of a major paradigm shift regarding the creation and distribution of its forecasts. Instead of writing a wide array of text products, forecasters will make use of an interactive forecast preparation system (IFPS) to construct a 7-day graphical representation of the weather that will be distributed on grids of 5-km grid spacing or better (Ruth 2002). To create these fields, a forecaster starts with model grids at coarser resolution, uses "model interpretation" and "smart" tools to combine and downscale model output to a high-resolution IFPS grid, and then makes subjective alterations using a graphical forecast editor. Such gridded fields are then collected into a national digital forecast database that is available for distribution and use. The gridded forecasts are finally converted to a variety of text products using automatic text formatters.

There is little question that the NWS must trend toward graphical forecast products if it is to remain effective and relevant. First, only graphical/gridded distribution can effectively communicate the detailed spatial/temporal information that is becoming available as model resolution increases, knowledge of local weather features advances, and observing systems improve. Second, gridded forecasts are required for effective distribution over the Web and through the media. Third, many new forecast applications (such as transportation applications and automated warning systems) require a digital/gridded forecast feed.

Although graphical tools clearly have a major place in the forecast office of the future, the current implementation of IFPS by the NWS has major conceptual and technical deficiencies that threaten to undermine the institution's ability to provide skillful forecasts to the public and to other users. This paper will examine some of these problems and will provide some suggestions regarding the forecast preparation system of the future.

PRÉCIS: National Weather Service (NWS) forecasters will use an interactive forecast preparation system (IFPS) to construct a graphical representation of the weather on high-resolution grids. A forecaster downscales model output and makes subjective alterations to a high-resolution IFPS grid. Such gridded fields are collected into a national digital forecast database and are converted to text products automatically. The NWS must trend toward graphical forecast products to remain effective and relevant. Graphical/gridded distribution can communicate detailed information, be delivered over the Web and through the media, and serve many new forecast applications. Although graphical tools have a place in the future forecast office, IFPS currently has deficiencies preventing skillful forecasts. This paper examines these problems and provides suggestions for the future.

Although the original 292-word text is already reasonably compact, notice how the 119-word précis contains only essential content. You might try your own précis of the original text and see how many words you can eliminate.

13.4   AN EXAMPLE OF THE EDITING PROCESS

To illustrate how to employ the writing/editing funnel and CPR approaches to revise text, consider the following draft abstract sent to me by first-time author Jari Tuovinen. He had worked on it as much as he could and needed some guidance to make further revisions.

ORIGINAL DRAFT: The spatial and temporal occurrence of large (at least 2 cm in diameter) hail in Finland was studied using many different methods to collect observations. The study period covered summers from 1930 to 2006 containing months from May to early September (first half of a month) each year. The maximum hail size in a single hail fall was mainly less than 4 cm in diameter (65% of cases). The number of observed cases decreases as hail size increases, yet number of nonsevere, under 2 cm hail cases seems to be the most common hail size. In extreme cases, even 7–8 cm (baseball size) hailstones have been observed and photographed.

Altogether, 240 severe hail cases were found in this study all over the country, the northernmost being located near latitude 68.5°N. So far, this case might be the northernmost large hail observation in the northern hemisphere. The under-reporting of hail, large or small, is great in Finland due to low population density, vast forest or lake areas and the nature of mesoscale event itself. The era of advanced technology and more widespread interest in severe weather events among the general public and media since 1990's is seen in the dataset of large hail observations as an increasing trend of observed cases. According to seasons' 1997–2006 data, a yearly average of 8–12 cases is expected during four to six severe hail days. Most of the observed large-hail cases (84%) occurred from late June through early August. July was the peak hail month with almost 66% of cases.

The peak of diurnal distribution was observed mainly during afternoon and early evening hours. For larger hailstones (4 cm or above), the peak time of occurrence was a little later (1600–2000 LT) compared to smaller, 2–4 cm, sized hailstones (1400–1800 LT). The largest density of cases was observed in an agriculture-intensive area of western Finland whereas the proportion of over 4 cm hail cases was bigger in the eastern part of the country. The number of observed cases in northern Finland is the smallest. The average synoptic pattern associated with 16 large hail cases included a low pressure centre or a through of low over western Scandinavia which enabled the southerly or southeasterly rush of warm air mass to Finland.

Let's begin with the largest-scale issues. Does this text flow smoothly from one theme to another? Although individual sentences may read well, the abstract as a whole does not read clearly. The text jumps from one theme to another. To show how the organization of the abstract was affecting its clarity, I classified the sentences into different themes in the margin, resulting in the following.

Purpose

DRAFT WITH MARGINAL NOTES: The spatial and temporal occurrence of large (at least 2 cm in diameter) hail in Finland was studied using many different

Dataset

methods to collect observations. The study period covered summers from 1930 to 2006 containing months from May to early September (first half of a month)

Size

each year. The maximum hail size in a single hail fall was mainly less than 4 cm in diameter (65% of cases). The number of observed cases decreases as hail size increases, yet number of nonsevere, under 2 cm hail cases seems to be the most common hail size. In extreme cases, even 7–8 cm (baseball size) hailstones have been observed and photographed.

Dataset

Altogether, 240 severe hail cases were found in this study all over the coun-

Spatial distribution

try, the northernmost being located near latitude 68.5°N. So far, this case might be the northernmost large hail observation in the northern hemisphere. The under-reporting of hail, large or small, is great in Finland due to low population

Changes in dataset over time

density, vast forest or lake areas and the nature of mesoscale event itself. The era of advanced technology and more widespread interest in severe weather events among the general public and media since 1990's is seen in the dataset of

Annual cycle

large hail observations as an increasing trend of observed cases. According to seasons' 1997–2006 data, a yearly average of 8–12 cases is expected during four to six severe hail days. Most of the observed large-hail cases (84%) occurred from late June through early August. July was the peak hail month with almost 66% of cases.

The peak of diurnal distribution was observed mainly during afternoon and early evening hours. For larger hailstones (4 cm or above), the peak time of occurrence was a little later (1600–2000 LT) compared to smaller, 2–4 cm, sized hailstones (1400–1800 LT). The largest density of cases was observed in an agriculture-intensive area of western Finland whereas the proportion of over 4 cm hail cases was bigger in the eastern part of the country. The number of observed cases in northern Finland is the smallest. The average synoptic pattern associated with 16 large hail cases included a low pressure centre or a through of low over western Scandinavia which enabled the southerly or southeasterly rush of warm air mass to Finland.

Diurnal cycle
Size/diurnal cycle

Reporting issues/spatial distribution

Synoptic patterns

Notice how text on the dataset, hail size, and its spatial distribution each appear in two separate locations within the abstract. Having this material closer together would make more sense. After this annotation step, the way to reorganize the abstract became more clear:

1. purpose of the paper
2. dataset
3. size of the hail
4. annual cycle
5. diurnal cycle
6. spatial distribution
7. reporting issues, which follows from the spatial distribution
8. changes in the dataset over time
9. synoptic patterns

Putting like material next to like material produces shorter, smoother-flowing text, as demonstrated in the revised abstract below.

DRAFT WITH SENTENCES REARRANGED: The spatial and temporal occurrence of large (at least 2 cm in diameter) hail in Finland was studied using many different methods to collect observations. The study period covered summers from 1930 to 2006 containing months from May to early September (first half of a month) each year. Altogether, 240 severe hail cases were found in this study. The maximum hail size in a single hail fall was mainly less than 4 cm in diameter (65% of cases). The number of observed cases decreases as hail size increases, yet number of nonsevere, under 2 cm hail cases seems to be the most common hail size. In extreme cases, even 7–8 cm (baseball size) hailstones have been observed and photographed. Most of the observed large-hail cases (84%) occurred from late June through early August. July was the peak hail month with almost 66% of cases. The peak of diurnal distribution was observed mainly during afternoon and early evening hours. For larger hailstones (4 cm or above), the peak time of occurrence was a little later (1600–2000 LT) compared to smaller, 2–4 cm, sized hailstones (1400–1800 LT). The northernmost hail case was located near latitude 68.5° N. So far, this case might be the northernmost large hail observation in the northern hemisphere. The largest density of cases was observed in an agriculture-intensive area of western Finland whereas the proportion of over 4 cm hail cases was bigger in the eastern part of the country. The number of observed cases in northern Finland is the smallest. The under-reporting of hail, large or small, is great in Finland due to low population density, vast forest or lake areas and the nature of mesoscale event itself. The era of advanced technology and more widespread interest in severe weather events among the general public and media since 1990's is seen in the dataset of large hail observations as an increasing trend of observed cases. According to seasons' 1997–2006 data, a yearly average of 8–12 cases is expected during four to six severe hail days. The average synoptic pattern associated with 16 large hail cases included a low pressure centre or a through of low over western Scandinavia which enabled the southerly or southeasterly rush of warm air mass to Finland.

This reorganized text is structurally more sound. The next step is to make the text more concise. Two big changes to the abstract included (i) deleting the material on synoptic patterns because that material was later deleted from the manuscript and (ii) deleting the material on the northernmost hail report because we did not consider it important enough to include in the abstract. Further concisions can be seen from just the first two sentences: "spatial and temporal occurrence" was changed to the simpler "climatology" and "summers . . . May to early September (first half of a month)" was changed to "the warm seasons (1 May to 14 September)." Later in the abstract, we deleted the phrase "number of nonsevere, under 2 cm hail cases seems to be the most common hail size," which seemed unnecessary. Also, the last sentence was reworded from passive to active voice. These, and other revisions (shown in italics), were applied to make the text more concise.

DRAFT AFTER CONCISION: *A climatology* of large (at least 2 cm in diameter) hail in Finland was studied using many different methods to collect observations. The climatology covered the *warm seasons (1 May to 14 September)* during 1930–2006. Altogether, 240 severe hail cases were found [*deleted text*]. The maximum hail size [*deleted text*] was mainly less than 4 cm in diameter (65% of cases). The number of observed cases decreases as hail size increases [*deleted text*]. In extreme cases, even 7–8 cm (baseball size) hailstones have been *reported*. Most of the [*deleted text*] large-hail cases (84%) occurred from late June through early August. July was the peak hail month with almost 66% of cases. The peak of diurnal distribution was observed mainly during afternoon and early evening hours. For larger hailstones (4 cm or above), the peak time of occurrence was a little later (1600–2000 LT) compared to smaller, 2–4 cm, sized hailstones (1400–1800 LT). [*deleted text*] The largest density of cases was observed in an agriculture-intensive area of western Finland whereas the proportion of over 4 cm hail cases was bigger in the eastern part of the country. The number of observed cases in northern Finland is the smallest. The under-reporting of hail [*deleted text*] is great in Finland due to low population density, vast forest or lake areas and the nature of mesoscale event itself. The era of advanced technology and more widespread interest in severe weather events among the general public and media since 1990's is seen in the dataset of large hail observations as an increasing trend of observed cases. According to seasons' 1997–2006 data, *Finland experiences a yearly average* of 8–12 severe-hail cases during four to six severe-hail days. [*deleted text*]

Having shortened the text, next we attempted to make the text more precise. The following were some of the issues that were addressed at this stage:

- In the drafts so far, the terms "severe hail" and "large hail" were used interchangeably. We standardized all usage of the term to "severe hail," which is consistent with the definition as applied in the United States (page 362).
- The expression "at least 2 cm in diameter" was changed to "2 cm in diameter or larger" to be inclusive of hail exactly 2 cm in diameter.
- The "many different methods to collect observations" was made more specific: "newspaper, storm-spotter, and eyewitness reports."
- The period 1930–2006 is now preceded by "77 years" so that readers do not have to do mental subtraction.
- "The number of observed cases in northern Finland is the smallest." Smallest compared to what? Reworded to "Most severe-hail cases occurred in southern and western Finland, generally decreasing to the north, with the majority of the cases near population centers."
- "Under-reporting of hail is great," which is ambiguous, was changed to "underreporting of hail is a particular problem . . . due to. . . ." Also, the journal's format is to not hyphenate "underreporting."
- "Nature of mesoscale event itself" was changed to "relatively small hail swaths" to be more precise about what made the hail cases mesoscale events.
- "The era of advanced technology" was replaced by an exact listing of the specific technologies that have led to better reporting.

- The mean number of hail cases and hail days is given as a single number rather than a range.
- Why was the 1997–2006 data important for determining the averages? The answer was simply because it was the last ten years, a nice round number that captured a period of relative homogeneity in the dataset. The reason is made more explicit in the revised text.

These and other changes in italics make the abstract more precise.

DRAFT AFTER PRECISION: A climatology of *severe* hail *(2 cm in diameter or larger)* in Finland was *constructed by collecting newspaper, storm-spotter, and eyewitness reports.* The climatology covered the warm season (1 May to 14 September) during the *77 years* 1930–2006. Altogether, 240 severe hail cases were found. The maximum hail size was mainly *4 cm in diameter or less* (65% of cases). The number of observed cases decreases as hail size increases. In *a few* extreme cases, even 7–8 cm (baseball size) hailstones have been reported. Most of the *severe*-hail cases (84%) occurred from late June through early August. July was the peak hail month with almost 66% of cases. The peak of diurnal distribution was observed mainly during afternoon and early evening hours. For larger hailstones (4 cm or above), the peak time of occurrence was a little later (1600–2000 LT) compared to smaller, 2–4 cm, sized hailstones (1400–1800 LT). *Most severe-hail cases occurred in southern and western Finland, generally decreasing to the north, with the majority of the cases near population centers.* The largest density of cases was observed in an agriculture-intensive area of western Finland whereas the proportion of over 4 cm hail cases was bigger in the eastern part of the country. The *underreporting of hail is a particular problem across much of Finland* due to low population density, vast forest or lake areas, and the *relatively small hail swaths. Since the 1990s, a greater interest in severe weather among the general public and media, a storm-spotter network, improved communications technology, and an official Web site for reporting hail* have increased the number of reported hail cases. *During the most recent ten years (1997–2006),* Finland experiences an annual average of *ten* severe-hail cases during *five* severe-hail days.

Finally, in the revision stage, we critiqued the text even further, enhancing the transitions, checking for grammatical errors (e.g., missing "the"s and hyphens, changing "due to" to "because of"), and generally cleaning up the text. The version of the abstract that was submitted to the journal is below.

IMPROVED: A climatology of severe hail (2 cm in diameter or larger) in Finland was constructed by collecting newspaper, storm-spotter, and eyewitness reports. The climatology covered the warm season (1 May to 14 September) during the 77 years 1930–2006. Altogether, 240 *severe-hail* cases were found. The maximum *reported severe-hail* size was mainly 4 cm in diameter or less (65% of *the* cases), *with the number of cases decreasing as hail size increased.* In a few extreme cases, *7–8-cm (baseball-sized)* hailstones have been reported *in Finland.* Most of the severe-hail cases (84%) occurred from late June through early August, *with July being the peak month (almost 66% of the cases). Most severe hail fell during the afternoon and early evening hours 1400–2000 local time (LT). Larger hailstones (4 cm or larger) tended to occur a little later (1600–2000 LT) than smaller (2–3.9 cm) hailstones (1400–1800 LT).* Most severe-hail cases occurred in southern and western Finland, generally decreasing to the north, with the majority of the cases near population centers. *The proportion of severe hail less than 4 cm in diameter is largest over the agricultural area in southwestern Finland where crop damage caused by severe hail is more likely to be reported.* The underreporting of hail is a particular problem across much of Finland because of *the vast forest and lake areas, low population density, and relatively small hail swaths.* Since the 1990s, a greater interest in severe weather among the general public and media, a storm-spotter network, improved communications technology, and an official Web site for reporting hail have increased the number of reported hail cases. During the most recent ten years (1997–2006), Finland *experienced* an annual average of ten severe-hail cases during five severe-hail days.

13.5 NEARING A FINAL VERSION OF THE MANUSCRIPT

As revisions continue and the manuscript reaches a stage where it could be submitted, the following steps can be implemented in this final push toward completing the manuscript:

As a reviewer, I see a lot of papers that are sent in with the idea that they will do the final editing after the reviews (or perhaps that the reviewers will provide what they need to edit to final form). My personal view is that when you submit a paper it should be in final form and that you should be comfortable with the paper going directly to press as is. It is a waste of time for all of us to review anything less. —Jim Steenburgh, University of Utah

- Set the manuscript down for a while. Clear your brain. Do you ever work unsuccessfully all afternoon to debug a computer program only to immediately see the error in the code first thing the next morning? If so, then you know that being too focused on the manuscript can blind you to otherwise obvious typos and inaccuracies.
- Are you in the mood to edit? If not, do not force yourself. Sometimes words seem to pour out, and you do not want to stifle that creativity by editing minutae. Other times, you can have much more focus and clarity and be a much better editor. Some days you just want to go kayaking.
- Print the document and crank up the intensity of editing a notch. If any part of the text is unclear, is inconsistent with other parts, lacks justification, or needs better transition, do not hesitate to revise it. If you have to

read a sentence twice to understand it or you feel uncomfortable with a figure, readers likely will, too. I have noticed that when I submit a manuscript despite having some minor sneaking suspicions about a piece of text, reviewers almost always pick up on my concerns. So, I should have fixed those revisions myself before submission, and saved the reviewers the trouble.

- Plan multiple passes through the manuscript, focusing on a different goal each time. For example, one pass might stress the evidence for the argument, another pass might stress transition, and yet another pass might stress grammar.

- Start editing the manuscript from the last page and work to page 1. Looking at your writing out of order will provide a different perspective. Plus, doing so will provide the often-overlooked figures, tables, captions, and references some deserved attention.

- Read the manuscript backward sentence by sentence. Although this, too, may sound extreme, doing so will enable you to focus on the sentence structures and words. Reading a paper consistently in the forward direction may make you overly familiar and comfortable with the text, and unable to see the problems with it.

- Read the manuscript out loud to yourself or others. Does it sound like it makes sense? Do you stumble over certain sentences? Are words and punctuation omitted?

- Evaluate your working title and abstract to make sure that they still represent the manuscript accurately. If not, revise them.

- Look for consistency between the abstract, introduction, body of the text, and conclusion. Do all the main results appear in each? Prof. Robert Houze of the University of Washington recommends going through the manuscript with a highlighter and marking the main points of the manuscript, confirming their consistency throughout the manuscript.

- Recognize your weaknesses and work to improve them. Do you commonly misuse certain word pairs (e.g., that/which, whereas/while, because/since)? Do you tend to write with phrases starting with "it"? Maintain a list of your foibles, either on a piece of paper near where you write or in a file on your computer. Refer to your list and search for those weaknesses throughout the manuscript. Recognizing, listing, and fixing weaknesses in your manuscripts will improve your writing over time.

How is one to seriously review a paper that the authors don't seem to have read carefully even once? If this were to get published as is, the authors would make a joke of themselves. —Peter Houtekamer, Environment Canada

13.6 MINDING THE LITTLE THINGS

Take care to mind the little things in the manuscript, such as correct spelling, proper use of abbreviations, appropriate use of commas, the difference between hyphens and dashes (both en and em dashes, discussed in Appendix A),

FINAL CHECKS OF YOUR MANUSCRIPT

Final checks of your manuscript (partially adapted from an AMS document "Final questions to ask yourself about your completed manuscript").

☐ Title page is complete and includes date and corresponding author address.
☐ Abstract and conclusions cite the most important results and are consistent.
☐ Consistent terminology is used throughout.
☐ All acronyms are defined upon first usage.
☐ All citations appear in the reference list, and all references are cited.
☐ References have been double-checked for accuracy and correct format.
☐ All section numbers, figures, and tables are numbered in sequential order.
☐ All citations to figures and tables refer to the correct figure and table numbers.
☐ Spell check and grammar check have been performed.
☐ Pages are numbered.
☐ Lines are numbered in margin (if required by the journal).
☐ Your personal list of common weaknesses has been checked.

grammar, format of references, and so forth (see the sidebar "Final Checks of Your Manuscript"). Unfortunately, some authors treat these aspects as unnecessary and inconvenient.

When I have reviewed others' work and have commented on the little things, I have often heard, "That is the technical editor's job." Wrong! You are the author. Do you trust someone else with your manuscript? Although technical editors are competent people, they can make mistakes. The best way to avoid mistakes in your manuscript is for you not to make them in the first place. Also, because editors are busy preparing your manuscript for publication, having them fix mistakes you could have easily fixed wastes their time. Being inconsiderate slows down the publication process and increases publishing costs for everyone. Specifically, Ken Heideman, director of publications for the AMS, says that a manuscript where the author has taken care to follow the *Authors' Guide* (American Meteorological Society 2008) only takes three hours to edit, whereas one in poor shape can take more than seven hours.

The author has the responsibility for submitting a proper manuscript. Most journals have Instructions to Authors and a style guide. Follow the directions! Some journals supply a template for authors to follow. Failing to use their template may result in the unreviewed manuscript being returned to you.

Minding the little things can be worthwhile for other reasons, too. These little things mean a lot to some people. After submission, your manuscript will go to an editor and several reviewers. Do you want them to know that you are sloppy? Sure, many people may not comment on your omitted commas and misspellings, but many will notice. Reviewers who have to wade through such messiness lose patience more easily and are more likely to recommend rejection.

Most importantly, not minding the little things often means not minding the big things. Manuscripts with lots of little errors often contain big errors in the science as well. Carelessness often has no bounds. For all of these reasons, taking care of the little things instead of leaving it to the technical editor will help ensure the eventual publication of your manuscript.

13.7 RECEIVING FEEDBACK

Days, weeks, or months of working on a manuscript blinds you to your own writing. Many times I have heard frustrated authors say, "I cannot do anything more with this paper, submit it and let the reviewers have it." If you find yourself in this position, let the work sit for several days to give you a fresh perspective on the manuscript. Being in a state of panic is no way to deliver a good product.

Ask a trusted colleague to look at the relevant sections *before* submission. No matter how carefully you revise your manuscript, others will make different suggestions than you will. In fact, some laboratories and organizations require a formal internal review process. Colleagues who have a penchant for being tough reviewers are also valuable to review your manuscript. Least helpful are syncophants who are "yes men" and "yes women," people who return a manuscript with but a few red marks. Strive to find people who intellectually challenge you and hold you to high standards. Furthermore, you may also invite nonexperts to read your manuscript. They may pick up on terms that should be defined for a more general audience. Other good reviewers are those who may not agree so readily with your conclusions. No one is more apt to find the flaws in your manuscript than someone who disagrees with you scientifically. You may disagree with their concerns, but at least you will be aware of what some issues might be and can revise your manuscript by taking their criticisms into account.

If possible, say what kind of feedback you are expecting from your informal reviewers. Are you worried about your interpretations being correct? Are you concerned that the paper does not flow well, or that it is not targeting the right audience? Do not expect every reviewer to fix all your grammatical mistakes, but if that is what level of detail you need, be sure to request that kind of feedback.

Resist the temptation to send a manuscript to others without doing your best to clean it up. First, you want to offer the best product to others. Second, you do not want others to spend time wading through your poorly edited manuscript. Third, you need to develop these editing skills for yourself. An exception is if you give specific instructions: "Don't worry about the details of the introduction, it needs a lot of small-scale editing. What I really need from you is advice about whether the material makes sense to you." Or, "Don't bother with the grammar and spelling—I need the most help with the organization of the paper at this time."

Besides colleagues, there are other sources where you can get help. Your university likely has a writing center where you can get free advice on your manuscript from a qualified individual. There are also professional manuscript editing services that can help improve your manuscript for a fee, although the quality of these services may vary (see the sidebar "Professional Manuscript Editing Services" on page 200).

Other approaches to get feedback involve posting your article on the Web, on your own home page, or on an Internet archive (e.g., arXiv), if appropriate. You might also give a seminar at your home institution. Go on tour and present your work at other institutions, especially those where you think you would get good feedback from the audience. Anything you can do to get feedback before submission allows you to make revisions that improve the manuscript.

13.8 THE NEED FOR CONCISION

As a final plea for editing to produce a shorter manuscript, I include this section. An ever-increasing amount of literature is being published in an increasing number of journals by an increasing number of scientists, yet the time scientists can devote to reading the literature is finite. Keeping track of the current literature, let alone the previous literature, is a difficult task. All authors need to do their part by writing shorter manuscripts.

There is a saying that a paper should be as long as it needs to be, but no longer. The AMS, which until 1991 did not have a limit on the length of papers, has twice dropped the maximum length of papers not requiring editor approval to the present 7500 words (about 26 double-spaced pages). This is not to say that longer articles will not be put into the review process, but they may face additional scrutiny for excessive length by editors. So, do your best to submit the most concise manuscript you can. Here are some reasons why:

- The audience is more likely to read a shorter paper.
- Shorter papers are generally quicker and easier to write.
- Getting small bits of research published is easier than publishing one all-encompassing piece of work.
- Shorter papers usually garner more favorable reviews.
- A few shorter papers over several years will keep your name in the spotlight more than one long paper will.
- Shorter papers usually have fewer coauthors, and hence assigning credit to the appropriate authors is easier.

- Shorter papers are less likely to overgeneralize the research results.
- A long manuscript may be excluded from many journals that have strict length requirements.

Authors looking for inspiration to make their papers shorter can find it in *The Elements of Style*, by Strunk and White (2000, p. 23), whose admonition to "omit needless words" has been the editor's rallying cry:

> Vigorous writing is concise. A sentence should contain no unnecessary words, a paragraph no unnecessary sentences, for the same reason that a drawing should contain no unnecessary lines and a machine no unnecessary parts. This requires not that writer make all his sentences short, or that he avoid all detail and treat his subject only in outline, but that every word tell.

On the largest scale (chapters of dissertations and sections of journal articles), the best way to maintain focus is by having a well-defined purpose to the document. Anything that deviates from the focus should be a strong candidate for removal. In an upcoming section on writing case studies of weather events (Section 18.4.1), I discuss how writers typically think that every detail of the case study is important to describe to the reader. This is wrong. Only include what is needed to tell the story. One or two tangents makes for entertaining reading, but repeated insertions of tangential and extraneous material tests the patience of the reader.

As a final exhortation in this long-winded section on being concise, Daniel Oppenheimer of Princeton University won the 2006 Ig Nobel Prize in Literature for his article "Consequences of erudite vernacular utilized irrespective of necessity: Problems with using long words needlessly" (Oppenheimer 2006). His acceptance speech was the following: "My research shows that conciseness is interpreted as intelligence. So, thank you."

13.9   THE RIGHT LENGTH

Although more concise papers are generally favored, some would argue that shorter papers may allow an author to get credit for multiple publications containing relatively little new knowledge. How do you know when your manuscript is the right length?

Inescapably, authors, reviewers, editors, and publishers will never stop arguing about the appropriate length of scientific manuscripts. The least publishable unit (LPU) or the quantum value of publishable material (publon) will vary among scientists. Reviewers may want you to do more analysis, but shorten the paper, something that seems contradictory. Arbitrary word limits imposed by journals may unnecessarily constrain lengthy, but otherwise novel, manuscripts, although I believe that journals have the right to determine their own requirements. Finally, some 10-page manuscripts are too long, and some 35-page manuscripts are not long enough.

For yourself, you will know when your manuscript is the right length when you have:

- made solid arguments in support of your evidence,
- avoided tangential arguments and figures,
- made sentences concise and precise, and
- eliminated redundant and verbose words and phrases.

When these conditions are met, you have just the final edits to perform before the manuscript is done.

13.10 THE FINAL EDITS

Once you can see the end is near, you may feel like a runaway train, eager to reach the destination. Take this feeling, and go with it. But, do not rush the manuscript out the door.

The rush to complete the manuscript may cause you to start skimping on the last steps. Maintain your cool, and work to complete the final revisions. Always perform near-final edits on single-sided paper, which among other things allows you to easily compare text and figures appearing on adjacent pages when checking for internal consistency. With all the writing and revising that was done on the computer, the printed words will look different, allowing you to spot errors more easily. In fact, several versions of the manuscript should be revised on paper.

Perform near-final edits when you are fresh and undistracted. For me, the best time to revise is first thing in the morning, before I eat breakfast, before I read my e-mail, and before my mind starts preparing for the day. Others find that evenings in the library are when they are least distracted or after the children and spouse are asleep. Do whatever works for you.

Refer to your list of common writing weaknesses, searching the manuscript for possible examples that need to be fixed. Use a spell checker or grammar checker to catch obvious errors, but do not expect perfection from this software. One way to maximize the utility of these checkers is to customize your own settings (e.g., those settings in Microsoft Word are under "Preferences"). For example, my version of Word allows a check for subject–verb agreement, which is turned on because it functions reliably and is useful for catching my mistakes. On the other hand, the passive-sentence check is turned off because I find the constant reminders of sentences I have chosen to write in the passive annoying.

How do you know you are done editing? If you are still making substantive changes to the sentences, you still need at least one more round of edits. If you average one minor revision per page or less, then the manuscript is probably ready to submit. If you find yourself making a change during one round of revisions and undoing that change during the next round, submit the damn thing, will ya?

AUTHORSHIP AND ITS RESPONSIBILITIES

Nearly everyone who publishes will collaborate with others on their research. Working with others can be a satisfying or a frustrating experience. One potential difficulty can be determining who will be listed on the paper as coauthors. This chapter describes guidelines for determining authorship and authorship order, the responsibilities of the corresponding author, and the responsibilities of all coauthors.

One active scientist can typically write one or two papers a year. In contrast, a group of people can increase this output tremendously. Therefore, the opportunity to collaborate with people on research can be good for your career and productivity.

Science is becoming increasingly interdisciplinary, perhaps because of the increasing complexity of the problems needing to be solved. One measure of this rise in interdisciplinarity is the increasing number of authors per article over time. For 19 atmospheric science journals, Geerts (1999) found that the average number of authors per article increased from 1.2 in 1950, to 1.5 in 1965, to 2.0 in 1980, to 2.9 in 1995. Articles with tens or even hundreds of authors are common in some disciplines such as biology, medicine, and high-energy physics. For example, the first papers published by the members of the Human Genome Project announcing that they had sequenced the human genome had over 200 coauthors. One can imagine the headaches of coordinating 200 different authors for such an article—sometimes coordinating with just one coauthor is problematic enough!

Authorship is one of the most significant decisions that may be made about a manuscript. The author list is the first item in the citation and the reference, and people who contributed the most to the research should receive the most credit. For example, at some journals (e.g., *Proceedings of the National*

Does this liquified colleague have a right to be listed as a coauthor? Cartoon by Nick D. Kim.

Academy of Sciences), the role of each author to the creation of the manuscript is published on the front page of each article. Unfortunately, authorship is one of those things that is rarely openly discussed among the contributors. What are the rules for determining the author list and its order?

14.1 DETERMINING AUTHORSHIP

In principle, determining authorship should be quite simple, yet no formal rules exist across all scientific disciplines. One codification of these rules was provided by the International Committee of Medical Journal Editors (ICMJE) in 2003, who stated that all authors of a manuscript must satisfy all three of the following criteria:

1. Substantial contributions to conception and design, or acquisition of data, or analysis and interpretation of data;
2. drafting the article or revising it critically for important intellectual content; and
3. final approval of the version to be published.

ICMJE (2003) continue, "Acquisition of funding, the collection of data, or general supervision of the research group, by themselves, do not justify authorship." This statement by ICMJE (2003) is arguably the most concise and clear definition of authorship. These criteria can also be used in reverse, too. If a person aims to be a coauthor on a scientific paper, he or she must contribute to the scientific content of the manuscript, help draft or revise the manuscript, *and* approve the final version. Someone failing to be involved at all three levels should be removed from the author list.

Unquestionably, two people trading favors by adding their names to each others' manuscripts to increase their publication statistics is not acceptable. Equally inappropriate is adding a prominent name to an author list to elicit greater attention to the manuscript. Furthermore, scientists who think that their names have been added to manuscripts for which they did not contribute work at the level discussed above should demand their names be removed from the manuscript. Unfortunately, such an action may have to occur after the manuscript has already been submitted. Scientists should use such situations to educate others about the rules of authorship, hoping to avoid similar future occurrences.

The morality is clear. *If a coauthor is willing to take credit for the article, that coauthor should be prepared to accept responsibility for it as well.* In fact, all authors listed on an article should be prepared to accept responsibility for everything within the article, not just their own contributions. If there are parts of the paper in which you have not directly participated, it is incumbent upon you to learn more about them and the techniques and methods involved, even if you never rise to the level of expert on par with your coauthors. (After all, sharing expertise is one of the joys of collaborating with individuals with different skills than you have.) You may even ask a trusted colleague, who is not a coauthor, for comments on the paper if you lack confidence in the material. Regardless, such informal peer review can only strengthen the paper.

To understand better why these issues of authorship should be taken so seriously, consider the following situation. Suppose you are fifth author out of six on an article published two years ago. Allegations surface that the lead author had manipulated data to arrive at a better linear correlation in the principal figure in the article. Although the figure looked bizarre to you when you read a draft version of the manuscript before submission, you were too busy to raise the issue with the lead author who was eager to submit. Although the lead author was wholly responsible for the unethical behavior, *all authors* suffer under the same cloud of discredit. Consequently, the legitimacy of all your articles may be questioned. To avoid such scenarios, all coauthors must take their role seriously and only commit to manuscripts that they can express total confidence in.

14.2 DETERMINING AUTHORSHIP ORDER

Determining authorship order can be almost as contentious as who is on the author list. Imagine if the issue was whether you would be first author or second author in a three-author paper. Would you rather see for perpetuity the paper listed as You et al. or Someone-Else et al.? Even two-authored papers can be challenges. In one article published by the AMS, a footnote on the first page of the article read, "the authors contributed equally to this study."

How to deal with the order of the authors on multiple-authored papers can be difficult. Let's begin with the lead author. Lead authorship could result if an author meets one or more of the following criteria:

- Outstanding contributions—the lead author has demonstrated leadership during the study to make the manuscript come to fruition.
- Major intellectual input—the lead author had the scientific insights to make the manuscript possible.
- Active participation in work—the lead author did the most work throughout the course of the study.
- Most contribution to writing—the lead author did most of the writing.
- Major feature of the manuscript—the lead author developed the principal feature of the research.

Given that more than one author may have contributed to the paper on these levels, several schools of thought exist in determining author order. The most common interpretation is that the first author is the one that did the most work, the one that wrote the majority of the paper, or the one that oversaw the group developing and writing the manuscript. Subsequent authors are those that did progressively less work.

A second approach occurs in some laboratories where multiple-authored papers are commonplace. The last name on the author list, rather than being the person who did the least amount of work, is reserved for the leader of the laboratory (assuming, of course, that the laboratory head also satisfies ICMJE's three criteria for authorship). After the first two positions on the author list for such papers (usually a student and the direct supervisor), the last position is actually regarded as one of the most prestigious.

A third approach occurs in some papers where the first few authors are the ones that did all the work, then at some point, the author list proceeds alphabetically to indicate that the effort of the remaining authors is comparable. For example, such an alphabetical list may appear in some field program reports to indicate the people involved in the planning and execution of the field program, but played a relatively small role in the manuscript.

A fourth approach is to perform a quantitative assessment of each person's contributions in several different categories such as project design, implementation, writing (e.g., Schmidt 1987; Ahmed et al. 1997; Devine et al. 2005; Tscharntke et al. 2007). Numerical ranking of the scores can then indicate the author order.

Clearly, many different models for authorship exist, and each research group must decide on their own approach. Sometimes the author list or author order may change as work proceeds on the paper, responsibilities evolve, people leave or join the research group, or substantive comments from colleagues affect the research or writing. Trying to implement a uniform standard for authorship order across science, let alone just atmospheric science, is simply not feasible.

One issue that frequently arises is how to deal with coauthorship on articles resulting from a student thesis written up for formal publication by the advisor. This scenario is common for students who wrote their thesis and graduated, but did not continue in science, yet the advisor wants the research published. In such scenarios, some advisors will assume lead authorship because they performed the bulk of the effort required to produce the manuscript, which otherwise would not have been published. Although a reasonable supposition, others may interpret this scenario as the advisor stealing the students' work. This perception is avoided by advisors who are adamant that, because the research was done by the student, the student should be the lead author, even if the advisor was responsible for the production of the manuscript. In all situations, students and advisors should openly discuss publication issues early during the collaboration. Students are often understandably uncomfortable discussing this issue. The advisor therefore needs to initiate the discussion.

Because of the different scenarios for authorship and the intensely personal feelings that may arise from these issues, I suggest the following rules about authorship be involved in each multiple-authored paper:

1. Authorship should be discussed among all those involved. The lead author, corresponding author, most senior person on the author list, or head of the research group should explain why all authors are listed on the paper in the proposed order, being open to concerns from all authors.
2. Whatever rules of authorship are employed should be consistent throughout the research activities of the group or the series of papers on the particular topic.

As your career evolves and you consider a new job opportunity, ask the supervisor about their group's authorship standards, inquire from the other employees about their experiences, and seek out the group's publications to see that appropriate credit is given. If the standards of this group do not meet yours, consider a different position. Your ability to have the career you want depends on you receiving the credit you deserve for the work you did.

14.3 OBLIGATIONS OF AUTHORS

With the list and order of authors determined, each author has responsibilities to the manuscript. The American Geophysical Union (2006), emulating a similar document by the American Chemical Society, developed the following list of such obligations for authors:

1. An author's central obligation is to present a concise, accurate account of the research performed as well as an objective discussion of its significance.

2. A paper should contain sufficient detail and references to public sources of information to permit the author's peers to repeat the work.

3. An author should cite those publications that have been influential in determining the nature of the reported work and that will guide the reader quickly to the earlier work that is essential for understanding the present investigation. Information obtained privately, as in conversation, correspondence, or discussion with third parties, should not be used or reported in the author's work without explicit permission from the investigator with whom the information originated. Information obtained in the course of confidential services, such as refereeing manuscripts or grant applications, cannot be used without permission of the author of the work being used.

4. Fragmentation of research papers should be avoided. A scientist who has done extensive work on a system or group of related systems should organize publication so that each paper gives a complete account of a particular aspect of the general study.

5. It is unethical for an author to publish manuscripts describing essentially the same research in more than one journal of primary publication. Submitting the same manuscript to more than one journal concurrently is unethical and unacceptable.

6. An author should make no changes to a paper after it has been accepted. If there is a compelling reason to make changes, the author is obligated to inform the editor directly of the nature of the desired change. Only the editor has the final authority to approve any such requested changes.

7. A criticism of a published paper may be justified; however, personal criticism is never considered acceptable.

8. Only individuals who have significantly contributed to the research and preparation of the article should be listed as authors. All of these coauthors share responsibility for submitted articles. Although not all coauthors may be familiar with all aspects of the research presented in their article, each should have in place an appropriate process for reviewing the accuracy of the reported results. A deceased person who met the criteria described here may be designated as an author. The corresponding author accepts the responsibility of having included as authors all persons who meet these criteria for authorship and none who do not. Other contributors who do not meet the authorship

criteria should be appropriately acknowledged in the article. The corresponding author also attests that all living coauthors have seen the final version of the article, agree with the major conclusions, and have agreed to its submission for publication.

14.4 OBLIGATIONS OF CORRESPONDING AUTHORS

Because each coauthor has responsibility to read and approve the manuscript at each step in the submission process, the corresponding author has the following additional obligations to *all* coauthors:

- Tell all coauthors that they are being considered as an author as early in the research process as possible. (Do not laugh—manuscripts have been submitted and authors did not know they were listed as such.)
- Provide all coauthors a reasonable amount of time to comment on the manuscript *before* it is submitted.
- Get the signatures of all coauthors on the copyright forms, if required.
- Send reviews to all coauthors.
- Inform all coauthors of the manuscript status throughout the process.
- Involve all coauthors in responding to the reviewers.
- Manage all comments by coauthors and resolve differences among them, if needed.
- Tell all coauthors the final deadline for comments on the final version of the manuscript before submission.
- Offer all coauthors an opportunity to comment on the page proofs.
- Send reprints (either digital or paper copy) to all coauthors when the article is published.

Being corresponding author may mean balancing differing viewpoints or different levels of attention to detail. Simply put, the corresponding author should ensure that papers are held to the highest standard among all the coauthors.

After publication, the corresponding author is often the person contacted from readers interested about the manuscript, wanting data, asking questions, etc. Thus, the corresponding author should be prepared to accept these requests as well.

15

SCIENTIFIC ETHICS AND MISCONDUCT

As scientists, we pay homage to the truth. Unfortunately, this search for the truth can be tainted by some unethical individuals who steal research and text from others, publish nearly identical papers in multiple journals, fabricate data, or manipulate images. Some are caught—many others are probably not. This chapter discusses ethical issues for researchers.

Consider the following cases:

Case 1. In 2006, the *International Journal of Remote Sensing* retracted three articles on satellite detection of biomass burning written by a network of four authors. These three articles "substantially reproduced the content" of five articles written by different authors and published in other journals.

Case 2. Editor A received an e-mail from Author 1. Author 1 alleged Author 2's published derivation in Editor B's journal "used liberally and verbatim material" without citing Author 1's earlier article in Editor A's journal. Although Author 2 agreed to publish a correction, Editor B said that the journal did not print corrections to published articles.

Case 3. *Advances in Atmospheric Science* published an article by Author 3 in 2004. Author 4 contacted the journal because the model used by Author 3 in the article was developed by Author 4's group, but was not cited as such in the article. Previously, Author 3 had been a visiting scholar to Author 4's group. *Advances in Atmospheric Science* retracted Author 3's article.

Although the news occasionally reports on misconduct in hot fields such as genetic engineering, medicine, or nanotechnology, these three examples from atmospheric science illustrate that, unfortunately, our discipline is not immune from scientific misconduct either. Such ethics violations may be more common than reported, but the most common problem is authors submitting work that has been published previously. Perhaps "publish or perish" has caused some scientists to be careless, greedy, or unlawful. Regardless of the reason, such misconduct wastes the time of authors and editors, hurts careers, and ruins the credibility of scientists in the eyes of the public.

In a manner similar to other scientific organizations, the National Science Foundation (NSF) has defined misconduct.

> Research misconduct means fabrication, falsification, or plagiarism in proposing or performing research funded by NSF, reviewing research proposals submitted to NSF, or in reporting research results funded by NSF.
>
> (1) Fabrication means making up data or results and recording or reporting them.
>
> (2) Falsification means manipulating research materials, equipment, or processes, or changing or omitting data or results such that the research is not accurately represented in the research record.
>
> (3) Plagiarism means the appropriation of another person's ideas, processes, results or words without giving appropriate credit.
>
> . . . Research misconduct does not include honest error or differences of opinion.

This chapter addresses these three types of misconduct.

15.1 FABRICATION AND FALSIFICATION

I would hope that this book does not have to address fabrication of data. Simply put, making up data or results is unacceptable, and such cases are generally straightforward to prove. Falsification, however, can be more subtle. For example, consider observational data from an innovative, but noisy, instrument. By overly filtering the data to eliminate the noise and to allow the weak signal to appear, could you be accused of publishing falsified data if you fail to describe your filter?

Ideally, all data are good, although, unfortunately, bad data exist. Responsible scientists know how to address bad data. Quality control measures to address bad data are commonly employed in most studies. If the quality control measures to eliminate potentially bad data are properly and thoroughly described, others may argue with your choices, but you cannot be accused of fabricating or manipulating data.

A growing form of misconduct is alteration of digital images. For example, in January 2006, *Science* retracted two papers by South Korean scientist Hwang

Woo Suk because of evidence that human stem cells were cloned, not in the laboratory, but using computer software such as Photoshop. Unfortunately, occurrences of image manipulation are frighteningly common. The *New York Times* reports that the U.S. Office of Research Integrity found that 44.1% of their allegations of fraud involved image manipulation.

To avoid such problems at the *Journal of Cell Biology*, managing editor Mike Rossner formulated a list of checks for image manipulation for all submitted manuscripts. Of 1300 manuscripts, 14 (1%) were rejected because the images were manipulated deliberately to mislead the reader. More significantly, 20% of those 1300 manuscripts had one or more images manipulated so fundamentally that editors asked authors to resubmit the images. The offense could have been as simple as altering the contrast to eliminate fainter features that were not the focus of the image.

Should such touch-ups of imagery be allowed? As Rossner and Yamada (2004) argue, misrepresenting your data deceives your colleagues who trust that you are accurately presenting your results. Moreover, images may contain information that may be noise to the author, but signal to someone else. Thus, leaving the images in their original state as much as possible is preferred. The following list provides some general guidelines for handling images:

- Creating or eliminating data within an image is scientific misconduct.
- Increasing the resolution of the original image is unacceptable because, in essence, new pixels (i.e., data) are created. Decreasing the resolution of an image is acceptable because the resulting pixels are merely averages of the originals.
- Small adjustments to gamma (brightness) generally do not qualify as misconduct if these adjustments do not obscure (or white out) data.
- Composite or inset images are acceptable, if properly denoted by sharp boundaries around the individual components.
- Any manipulations performed (e.g., false color imagery, filtering) should be reported in the figure caption.
- Annotating figures with text to aid the readers' understanding is generally not considered misconduct.

Rossner and Yamada (2004) conclude:

Data must be reported directly, not through a filter based on what you think they "should" illustrate to your audience. For every adjustment that you make to a digital image, it is important to ask yourself, "Is the image that results from this adjustment still an accurate representation of the original data?" If the answer to this question is "no," your actions may be construed as misconduct.

15.2 PLAGIARISM

Because the largest number of alleged misconduct cases involve plagiarism, the majority of this chapter focuses on it. Plagiarism can be intentional or inadvertent. Plagiarism examples are numerous, likely because of the ease of plagiarism, the sometimes subtle distinctions between proper citation and plagiarism, and different cultural norms.

Plagiarism is using another person's intellectual property as if it were your own. Intellectual property is not only the text and figures in a scientific paper (Case 1 on page 183), but may also be equations (Case 2), computer code (Case 3), ideas, hypotheses, speculations, or calculations. Someone else's scholarship can also be plagiarized, a problem that review articles are particularly susceptible to. For example, a published article may track the discrepancy between two differing viewpoints over time. Later authors who use this article as a pathway to their own exploration of the previous literature should cite the original article as the source of their scholarship.

Using data or figures from the Web and not providing proper credit is plagiarism. The advent of many Web-based applications for plotting meteorological data (e.g., soundings from the University of Wyoming, the National Severe Storms Laboratory Historical Weather Data Archive, reanalysis data from the Earth System Research Laboratory Climate Analysis Branch) has made it

DISCLOSING CONFLICTS OF INTEREST

David P. Jorgensen, Research Scientist, National Oceanic and Atmospheric Administration (NOAA)/ National Severe Storms Laboratory, and Publications Commissioner, American Meteorological Society

Several embarrassing revelations in the mass media in recent years illustrate how potentially biased science can result by undisclosed financial conflicts of interest. Although *conflict of interest* most often refers to financial relationships, it can also involve personal, professional, ideological, political, or religious views. Here, I focus on financial conflicts of interest.

Why is it necessary to disclose author financial arrangements with sponsors? Is peer review not enough to ensure that biased results are weeded out of the publication process? The answer, unfortunately, is no.

Studies document the positive association between sponsors' interests and the outcomes of research—in other words, sponsors get what they pay for (i.e., favorable research results). Might good science still be done even if a sponsor's interests are served? Certainly. However, undisclosed financial arrangements, when uncovered following publication, could cloud the reader's judgment about the results and embarrass the journal, even if authors believe that their conduct has been above reproach.

The simplest way to inoculate yourself against charges of bias is to require disclosure of financial arrangements within the body of the work. Yet, academic and government investigators operate under varying institutional rules, and journals have not as yet operated with a uniform policy for disclosure.

Although the AMS does not, as yet, have a policy for author disclosure of relevant financial conflicts, authors should be voluntarily open and honest about

easier to access such data and create graphics for research articles, but proper credit must be given in publications and presentations. Often these Web sites will provide a recommended statement for acknowledging their site.

Given the often thin line between citation and plagiarism, how does one cite accurately without plagiarizing?

- ◗ Because plagiarism is stealing someone else's ideas, the best way to avoid plagiarism is to start with your own ideas. If you develop your own thoughts, then use others' work as supporting or refuting evidence. It is more difficult to prove plagiarism of ideas.
- ◗ If your ideas originate with others, build upon others' work, or are suggested by their work, then provide a citation to demonstrate the intellectual route through which you developed your own thoughts.
- ◗ In your handwritten notes and in text documents, make a clear distinction among material derived from sources (either direct quotations or paraphrasings), your own interpretations of the source, and your own thoughts. Never copy text verbatim from the source document into your own document without placing quotes around the material and citing the reference. Omitting this step may lead to problems at a later time when distinguishing your own ideas from borrowed material is impossible.

such potential conflicts. The safest course would be to err on the side of greatest disclosure, although still recognizing that some relationships are clearly irrelevant. Best practices dictate that any relationship with a sponsor that has a direct stake in the contents (or results) of a submitted paper, whether or not that relationship relates to that paper, should be disclosed. In other words, common sense should help guide authors' disclosures. A rule of thumb: if disclosure of an apparent conflict would cause the author, or publisher, embarrassment if disclosed following publication, then the conflict should be disclosed prior to publication.

Such statements of disclosure usually appear in the acknowledgements. For example:

This research was supported by the following grants: National Science Foundation Grant ATM-1234567, Carnegie-Mellon Cooperative Agreements XX12-456LM and XX12-987ZZ, TRMM NASA Grants NAG5-9876 and NAG5-1234, and EOS NASA Grants NAG5-9999 and NAGW-8888. The lead author has been a paid consultant during 2006–2009 with SpyCrafters Aerospace, Inc., which manufactures the satellite microwave sounder instrument used in this work.

How far in the past should one go in determining relevant conflicts? A few journals require a conflict be current, yet most say one to five years. Given the long lead time between research and final publication, best practices say a three-year statue of limitations is a reasonable minimum standard. Finally, is there a minimum dollar amount below which need not be disclosed? Most journals say no. Any relevant conflict that could appear to influence a researcher's objectivity should be declared, regardless of how small.

- Instead of direct quotation, read the source and summarize in your own words. Do not start with the source and rewrite it. Avoid borrowing too much of the sentence structure and language from the original source.
- Phrases invented by others (even if only one or two words) should be cited with direct quotations.
- Larger pieces of relevant text can be directly quoted and cited. Direct quotations should be used sparingly and effectively (Section 12.6).
- Cite all sources faithfully, whether directly quoted, paraphrased, summarized, or interpreted. Clearly distinguish your thoughts from those of the source.

Avoiding plagiarism can be even more challenging for those for whom English is not their native language. In one highly publicized case, the online archive arXiv withdrew 65 articles by a group of 14 Turkish authors "due to excessive reuse of text from articles by other authors." One of the authors, a professor and dean, defended his plagiarism saying "using beautiful sentences from other studies on the same subject in our introductions is not unusual. . . . I aimed to cite all the references from which I had sourced information, although I may have missed some of them" (Yilmaz 2007). Unfortunately, such defenses are simply not acceptable in science.

Be inspired by specific words, phrasings, and sentence structures that you like from other sources on different topics. Emulate, but do not copy, these sources. Combine the emulated sentence structure with your own ideas to make the writing your own. More suggestions for authors for whom English is a second language may be found in Chapter 16.

Even more difficult to avoid is cryptomnesia, or when someone believes they are creating original ideas, words, or phrasings only to find that the inspiration comes from the past (e.g., something previously read or heard). Perhaps the most famous case of cryptomnesia is the lawsuit alleging George Harrison's 1970 song "My Sweet Lord" plagiarized the Chiffon's 1963 song "He's So Fine." The judge ruled that the melodies of the songs were nearly identical, but that Harrison's plagiarism was subconscious and unintentional. Nevertheless, Harrison had to surrender hundreds of thousands of dollars in royalties. I am unaware of prosecuted cases of cryptomnesia in science, although we are all influenced directly or indirectly by what we have previously heard and read. Perhaps the best defense against cryptomnesia is being aware of the literature and maintaining thorough notes.

15.3 SELF-PLAGIARISM

Can you plagiarize yourself? Indeed you can. Given the ambiguity, apparent paradox, and misunderstanding of the term self-plagiarism (also called autoplagiarism) and the different techniques of self-plagiarism, I prefer to be more specific by using the following four terms from Roig (2006): duplicate publication, salami-slicing, text recycling, and copyright infringement.

Duplicate publication can occur in one of two ways: submitting the same manuscript to two different journals at the same time or submitting largely similar manuscripts with slightly different interpretations or only minor differences. Nearly all journals reject this practice. Nevertheless, as many as 8% of biomedical articles are believed to be duplicate publications. Submitting the same abstract to different conferences, however, does not necessarily constitute duplicate publication, nor does submitting a non-peer-reviewed conference abstract or article to a peer-reviewed journal. Some conferences in other fields may require that presentations must be original or not presented elsewhere, especially if copyrighted proceedings are produced, but that does not seem to be common in atmospheric science.

Salami-slicing is breaking apart a larger study into two or more smaller publications for the sole purpose of obtaining more publications. To avoid the perception that studies are being salami-sliced, authors should list all other relevant publications in the manuscript and in the cover letter to the editor.

Text recycling is when similar or identical text appears in different manuscripts. Text recycling usually occurs with the data and methods section of the manuscript. For example, authors who have written a compact description of their modeling system may reuse this text from paper to paper. In general, most scientists would not find a problem with such text recycling, as long as new results are presented in each separate manuscript. Nevertheless, such duplication of text is a form of self-plagiarism.

Finally, authors who self-plagiarize by any of the three methods discussed previously may also be guilty of copyright infringement. *Copyright infringement* is reproducing material (even if properly cited) from a published article that is copyrighted. Thus, not all situations of copyright infringement are plagiarism. Authors may self-plagiarize by taking sections of a paper published by one journal and republishing the same material in a different journal with a different publisher without seeking permission, as for text recycled for the modeling system description. Remember that they may be your words, but the publisher may own the copyright to your articles. (For more information on copyright, read the Ask the Experts column by Ken Heideman on page 6.) Reusing material between articles, especially with different publishers, could be copyright infringement.

15.4 CONSEQUENCES OF MISCONDUCT

Although honest mistakes do occur in the sciences, these mistakes can usually be corrected with published corrections or corrigenda. With more serious errors, the manuscript may be voluntarily withdrawn or retracted by the author. If misconduct is determined to have occurred, the problematic paper may be forcibly retracted. The author, the author's affiliated institution, or both the author and the institution may be disciplined by the journal.

If proven, plagiarism can destroy your career. In Case 3 at the beginning of this chapter, *Advances in Atmospheric Science* barred any submissions by the author for at least three years. Other instances in other disciplines have resulted in jobs being terminated, awards being stripped, and grants being cancelled. Even fighting accusations of plagiarism can cost money, time, and your reputation.

Although not flawless, commercial tools exist for school teachers and professors to test for possible plagiarized class assignments. Similarly, journals have partnered with such companies to develop tools for detecting plagiarism of submitted scientific articles. Such tools scour the Internet and search online databases of journal articles to detect duplicate text that may already have been published. By identifying plagiarized manuscripts during the submission process, journals hope to avoid the negative publicity from having to retract already published, but fraudulent, articles. Despite the obvious advantages of such tools, however, the ultimate responsibility for submitting manuscripts free of plagiarism lies with the author.

GUIDANCE FOR ENGLISH AS A SECOND LANGUAGE AUTHORS AND THEIR COAUTHORS

Given the challenges that face scientists in communicating their work effectively in their native language, authors who make the attempt in a second (or third or fourth) language must surmount even greater obstacles. This chapter addresses the concerns of nonnative English speakers: common pitfalls to avoid, steps toward improving language skills, and advice from nonnative English speakers. Furthermore, this chapter offers guidance for native English speakers when working with nonnative English–speaking colleagues.

Until thirty years ago or so, most scientists receiving Ph.D.s in the United States had to demonstrate proficiency in a foreign language before graduating. To keep abreast of important advances being published in other countries and in other languages, scientists needed to read somewhat proficiently in one or more foreign languages. Some well-known atmospheric scientists were fluent in many of the classical languages of science, even if not their native tongue. Tor Bergeron spoke seven languages fluently and knew some of three others. Over time, however, English has become the dominant international language of science. And when scientists adopt English, they also adopt a way of looking at the world.

Inherent in every language is a way of thinking, but all authors—whether or not native English speakers—must demonstrate their skill in perceiving and performing science in a manner that is respected within English-speaking cultures. Although a journal may reject a manuscript for significant weaknesses in writing alone, such flaws do not solely explain the lower acceptance rate for manuscripts authored by English as a second language (ESL) scientists. My experience as an editor suggests that reviewers who recommend rejection place greater emphasis on scientific errors than the mechanics of grammar.

Thus, the difficulty for ESL authors in gaining acceptance for their work goes beyond their use of the language. By being aware of these cultural differences, ESL authors can ensure greater success as a scientist.

16.1 CULTURAL DIFFERENCES REQUIRE DIRECT COMMUNICATION

English-speaking countries (e.g., Australia, Canada, England, United States) tend to be composed of people from diverse cultural and ethnic backgrounds, more so than other countries that have a more homogenous culture (e.g., China, France, Japan, Russia). This diversity means that value and belief systems vary more among members of English-speaking societies than among other countries. As a consequence, Americans, for example, have to be more explicit and direct with each other to communicate more effectively. Such a culture is referred to as a *low-context culture*.

In contrast, other countries that are more homogenous ethnically and culturally share strong common bonds among their members (a *high-context culture*). Hence, a greater part of communication between individuals in this type of society can be implicit because a common set of values and beliefs more likely underscores communication. For example, relative to English standards, French writing may appear sophisticated, Italian writing may appear flowery, Mexican writing may appear emotional, Hindi writing may digress, and Japanese writing may be imaginative and beautiful. In the context of scientific writing in English, such styles may not seem appropriate to other readers, particularly native English-speaking authors whose culture requires a more explicit approach.

Communication in science has evolved to be more explicit, partly because of the dominance of English. But because science is done by research groups all over the world, each with a different culture (or cultures), communication must be explicit and direct. As I have emphasized throughout this book, a scientist's job is to communicate so that the reader can understand, and an explicit approach most likely ensures being understood. Knowing that effective communication in science goes beyond vocabulary and grammar and incorporates this cultural context is an important step to becoming a better communicator, regardless of whether you are a native English speaker or not.

Other differences in culture may also be relevant to the last chapter's discussion of proper scientific conduct. In some cultures, memorization and imitation are acceptable ways of communicating others' results, but, in science, the preferred way is through paraphrasing and direct quotations. Duplicating others' science without attribution, either their work or their words, is unethical (Section 15.2).

16.2 COMMON WEAKNESSES IN MANUSCRIPTS WRITTEN BY ESL AUTHORS

I assembled a list of common weaknesses often found in manuscripts by ESL authors (Table 16.1). Although these problems are certainly not unique to ESL authors, the frequency with which they appear in manuscripts authored by ESL scientists suggests some validity to this list. You may recognize some of the items from this list in comments on your manuscripts from reviewers and coauthors.

The first group of weaknesses in Table 16.1 relates to the formulation of the research question. Manuscripts consisting merely of descriptions of case studies or model simulations are especially prone to this problem. New scientific results may be limited, and any unique aspects of the research are omitted or barely mentioned. Many times these manuscripts apply previously published methods to new areas, which may be a worthy contribution, but the author fails to describe the uniqueness and importance of the study. Whether warranted or not, manuscripts with these weaknesses often get rejected for publication.

The second group in Table 16.1 is closely related to the first group because authors may not be fully aware of the previous literature. Understandably, authors who have not written much in English are less likely to have read much in English. Nevertheless, as argued in Section 4.7, knowing the literature is an essential part of being a scientist. When the author does not display knowledge of the literature, reviewers question whether the research is sufficiently innovative to be published. If the author has a good grasp of the depth and breadth of the previous literature on a topic and what the current challenges are, and focuses research toward unresolved topics, the author can make a contribution to the science and receive a greater recognition by others, even if the mechanics of the language are not perfect.

A further manifestation of failing to adequately explain the purpose of the research is seen in the third group of Table 16.1. ESL authors often do not describe their data, methods, results, and reasoning in sufficient detail for readers to understand the study. These concerns are often independent of the quality of the science, but readers may interpret such omissions as the authors not understanding what they were doing or why they were doing it. Remember that to communicate in a low-context culture, you must be more explicit than you may think is necessary.

Finally, the last two groups in Table 16.1 relate to the ability to communicate effectively. Transition between thoughts helps the reader maintain focus while reading, especially within paragraphs. Help on topics more related to grammar can be found online as well as in many textbooks for the English language. Web searches often can be useful to identify correct sentence or

Table 16.1 Common weaknesses often found in manuscripts by ESL (and native English–speaking) authors, and chapters in this book that address these issues (in parentheses)

1. Formulation of research (Chapter 2)
   - ▶ Purpose of research is not clearly stated.
   - ▶ Uniqueness or utility of research is not clearly stated.
   - ▶ Research contains little, if any, new scientific results.

2. Relevance to previous work (Chapters 2, 4, and 12)
   - ▶ Author fails to understand the relevant scientific issues of the present day.
   - ▶ Previous literature section is cursory, incomplete, outdated, or lacks synthesis.
   - ▶ Author fails to compare new results with previous research results.

3. Writing and presenting the science (Chapters 4–7, 11, and 18)
   - ▶ Data and methods are not described in enough detail or with sufficient clarity.
   - ▶ Reasons for doing experiments are not explained.
   - ▶ Descriptions of thought processes are not transparent enough.
   - ▶ Equations alone do not convey the underlying theory and physical interpretation.
   - ▶ Results are presented with too much confidence (e.g., "models and data agree perfectly") or with too little explanation.
   - ▶ Results are overinterpreted to a level not warranted by the data.
   - ▶ Results are not critically considered.
   - ▶ Limitations of the data are not discussed.
   - ▶ The text contains inconsistencies, often involving the figures.

4. Organization and flow (Chapters 8–9 and 13)
   - ▶ Transition is lacking between paragraphs and sentences.
   - ▶ Sentences lack a good rhythm, and are choppy, too short, or too long.

5. Language and style (Chapter 10 and Appendices A and B)
   - ▶ The style of the journal is not followed.
   - ▶ Phrases and words are used incorrectly.
   - ▶ Articles are improperly used or missing (*a*, *an*, and *the*).
   - ▶ Verb tense is incorrect or inconsistent (Section 9.3).

Rewrite frequently. Ask someone to check your writing. Do not be afraid of iterating on the same writing over and over again. I always find I can improve both grammatically and scientifically after each iteration. —Fuqing Zhang, Pennsylvania State University

word structures. Commonly used phrases will have more hits, and the context of the phrasings will provide some guidance into proper usage. Not all Web sites use proper English, so be cautious; prefer reputable sites and high-quality scientific journals as your models in style.

ADVICE FROM AN ESL SCIENTIST

Zhiyong Meng, Research Professor, Peking University

To achieve high-quality scientific writing in English, it is important to do well in four aspects: 1) Be clear about the main point of the manuscript. 2) Organize reasonable and persuasive evidence in an easy-to-follow way. 3) Be logical and self-consistent. 4) Use correct and commonly used English expressions. The first three aspects are more in the realm of scientific research and critical thinking, so I will focus on the fourth. In my opinion, there are several ways to improve scientific writing in English.

First, read well-written journal articles by native English speakers, make notes of useful expressions, and review them from time to time. I have a document of useful expressions in different categories such as transitions, figure descriptions, numbers, equations, comparisons, and conclusions. Writing new text becomes much easier by having a word bank to look up useful expressions.

Second, use that word bank frequently through small writing projects such as summarizing after reading an article, writing weekly work notes, and communicating with peers in formal English. After drafting a piece of work, careful and multiple revisions can improve the quality by examining if a statement can be expressed in a better way or with fewer words by checking the word bank or previous literature.

Third, learn from mistakes. Ask someone else, especially a native English speaker, to review your writing. Remember your weaknesses, and avoid them in the future. The biggest difficulties I had in writing were using articles (*a*, *an*, and *the*) and overusing the same words and phrases. A native English speaker can express an idea using only one short sentence whereas I have to pile on several sentences or even a paragraph. I learned a lot from the tracked changes or handwritten comments made to my documents by my advisor and my colleagues when they reviewed my writing. Insist upon these from your coauthors.

All in all, no easy way exists to improve scientific writing. Besides the above three tips, persistence is important. It is not realistic to become a good English writer overnight. In my experience, the best way to improve scientific writing in English is to read the literature, practice your writing, and learn from mistakes. In time, you may find writing in English is not that hard any more.

16.3 USING THE LITERATURE AS YOUR WRITING COACH

You can teach someone the skills to be a world-class diver: the mechanics, the right twists and turns, the spring off the board. If you show her the moves through example, then let her try it, she will start to learn. But doing so requires hours in the pool and on the diving platform with the coach. Ultimately, after having mastered those skills, she can develop her own style and push the limits of diving.

Being a better scientific writer is the same. You can read this book and learn the mechanics of how to assemble the manuscript, what to say, and how to say it. You can study grammar and be near-perfect. To make the most rapid progress, however, you need the right skills in hand, excellent examples of how

CAREER-SPANNING ADVICE FOR ESL AUTHORS

- Do a daily writing exercise (Section 16.3).
- Maintain a word and phrase bank.
- Maintain an article file, and read the articles frequently.
- Read scientific articles.
  - Read with a critical eye.
  - Improve your language skills.
  - Expand your knowledge.
- Research questions about proper usage in scientific articles or via a Web browser.
- Maintain a list of your personal writing weaknesses, and refer to them often.
- Find a native English–speaking colleague to serve as mentor, editor, or coauthor.
- Read Chapter 31 on how to improve your skills.

it is done, and support from colleagues to help you learn. Most of all, you need practice. And this practice best occurs before writing even starts.

Like the diver tries to imitate her coach, and then develops her own style later, imitating your role models is the route for you to develop your own style. One of the best ways to improve is to learn from good examples by reading the literature. Consider the practice part of your training not only as a scientist, but also as a student of the English language. Reading good examples of literature closely related to your research allows you to develop a sense of the terminology of the discipline, how it is used, and the grammar to say it properly. An approach recommended by Montgomery (2003, pp. 163–165) is to identify 15–25 recent articles in your discipline. The articles need not be written by well-known or highly cited researchers, but they should be written in a style you admire, are easy to understand, and are well organized. Get recommendations from your supervisors and colleagues about the best articles to select.

Once you have a hardcopy file of these articles, read and reread them regularly. Appreciate how the author laid out the problem for the audience in the introduction, explaining the purpose of the research and the methods in detail. Admire how the author develops the literature synthesis, not just as a list of accomplishments, but as an integrated, critical review of past work. Identify especially striking passages that you want to emulate.

As you continue to read, listen to the words in your head, or even read them aloud. Feel the flow of the words and how the author uses the language to express the science. Copy a few paragraphs by hand to see how the sentences are constructed and how they flow from one to the other. Study the writing style of the author. What words does the author choose that resonate with you? What sentences does the author use that communicate clearly to you in as few words as possible?

As you get more comfortable with the language, add sentences of your own into the author's text, trying to emulate the style. The sentences can be made up or derived from your own research. After writing and editing to your satisfaction, get a native English–speaking colleague to look at your exercises for any additional assistance.

If the writing of your manuscript stalls, reread the articles for inspiration and ideas. Saving your own writing exercises in a notebook and referring to them may also give you ideas. Do this exercise regularly, once per day or so. Being a better writer of scientific English will happen if you expose yourself a little every day. Repetition is crucial.

16.4 TRANSLATING YOUR NATIVE LANGUAGE OR WRITING IN ENGLISH?

After having rehearsed your writing skills, the time has come for you to begin writing (Chapters 5–7). One common question asked by ESL authors is whether they should write their documents in their native language first and then translate them into English for formal publication.

If you are an ESL author who is comfortable thinking in English or have been in an English-speaking environment for years, write your manuscript in English. If you do not know how to say some word, phrase, or sentence, write it in your native language and then ask for help from others later. This approach allows you to continue working on the manuscript without getting stalled by the language.

ESL authors who are not as comfortable with English should work toward that goal but write in the language that best facilitates delivering the content. If needed, the document can be translated into English. Because enough differences exist between languages, translating a document from a foreign language directly into scientific English is generally not straightforward. Direct translations of documents from foreign languages cannot reproduce the structure of the sentences in English, or even perhaps the order inside a paragraph. Scientific words may not have an equivalent in the foreign language. Therefore, translations, unless done by someone knowledgeable in the language, and possibly even the science, can be quite time consuming. Rather than translate a document directly, working with an editor or coauthor with more English experience can be a more productive way to write your manuscript.

Ultimately, the more practice you have using and thinking directly in English, the more quickly you will become fluent in scientific English. Although writing a single paper in English directly may take longer and be more frustrating, writing two papers (one in your native language and one in English) will almost certainly take a longer time.

16.5 SEEKING HELP

Despite ESL authors knowing their weaknesses and wanting to improve, opportunities to seek and obtain help are often unavailable at their institutions. Some native English speakers think that working with ESL authors requires too much time, especially for postdoctoral fellows who may be around for only a year or two. Volunteer peer reviewers and editors at journals are increasingly rejecting papers that do not meet guidelines for clear writing without attempting to glean kernels of scientific truth from them. The increase in such submissions to journals may have created a bias against ESL authors in the peer-review process.

These challenges are especially difficult for early career scientists. Pagel et al. (2002, p. 114) summarized their survey of faculty and postdoctoral fellows at an academic medical center on the topic of scientific writing challenges for ESL authors as follows:

> We are led to the conclusion that ESL scientists have been dealt a Faustian bargain. At the time the bargain is made, the senior faculty know they want smart, hard workers that they do not have to spend too much time with, and the ESL fellows and junior faculty want an opportunity to study and work in the United States. The Faustian nature of the bargain goes unrecognized until the ESL fellow or junior faculty member learns that the growth and recognition that come from writing are not part of the bargain.

Thus, ESL authors must be even more proactive than native English speakers in developing these essential skills for career advancement. Before selecting a graduate school program or new job, find a supportive environment that has produced productive and respected ESL scientists in the past. Discuss with your potential supervisor or professor what your expectations are about being mentored in scientific communication skills. Search out Web sites and

---

A CHECKLIST FOR ESL AUTHORS BEFORE MANUSCRIPT SUBMISSION

- ☐ Enlist colleagues' help early in the writing and research process.
- ☐ Double-check the items in Table 16.1.
- ☐ Check your list of weaknesses for their occurrences in the manuscript.
- ☐ Run spell and grammar checkers on the manuscript.
- ☐ Visit the university writing center for help.
- ☐ Have the manuscript proofread by a native English speaker.
- ☐ Enlist a professional editing service, if needed.

other learning opportunities to improve your skills. Universities often have free writing centers that can help any writer, not only ESL authors.

Having supportive colleagues (including peers and fellow students) is important to having a successful and satisfying career, even beyond writing. Identify people, especially native English speakers, who can help you, and take the time to do so. Collaborating with them helps gain their confidence and support. Ask them for advice on your writing and presentations. Get their input on your writing *early*, and do not take advantage of them by expecting them to fix all your grammatical errors throughout the entire manuscript unless they are happy to do so. Instead, ask for general advice about the science throughout the manuscript, and specific questions about the grammar, even if only on one or two sections of the paper. Take their suggestions and rework the rest of the manuscript based on their advice.

Furthermore, do not be surprised if native speakers have difficulty helping you with grammar issues. Native speakers of any language learn what is right and wrong at an early age, but often cannot explain the rules of grammar to others as adults because they just know "what sounds right." Sometimes ESL authors have problems in knowing whether native English–speaking colleagues' recommendations are based on an authentic need for changes to produce clarity or merely on differences in personal writing style. Receive clarification on their comments, if they have not explained them specifically to you.

16.6 COLLABORATING AND COAUTHORING WITH ESL AUTHORS

In my experiences working with ESL authors, most want help with the microscale aspects of the manuscript (i.e., groups 4–5 in Table 16.1), often not considering the larger scales first (i.e., groups 1–3). If you are a native English speaker working with an ESL author, clearly distinguish the revisions the author thinks are necessary (microscale in Fig. 7.1) with those that would best serve the paper (synoptic scale). Do not deny the author the discussion of microscale aspects that he or she is seeking, but redirect the emphasis saying that improving the large-scale aspects of the paper needs to be done first to better serve the grammar. Then, later sessions can focus on improving the microscale aspects of the paper.

Grammar is the zero law of the communication process: necessary but very far from sufficient. —Scott L. Montgomery (2003, p. 22)

Use a specific paragraph to relate the small-scale issues to the larger-scale purpose of the paper. Then, go down the writing/editing funnel to the sentence level and word level. If transition is a problem, ask how paragraphs are designed to link together. Work through only part of the manuscript, allowing the author to learn from your comments, make these revisions, then apply your comments to the rest of the text.

PROFESSIONAL MANUSCRIPT EDITING SERVICES

Mary Golden, Chief Editorial Assistant, Monthly Weather Review, *American Meteorological Society; Technical Editor; and ESL Coach*

When writing a manuscript, you will reach a point at which you have made the science and the writing as strong as you can. Even native English speakers rely on others to help them improve their writing—it is quite difficult to proofread one's own work. The better the manuscript is that you submit for publication, the more likely you will succeed. Although a journal's editors will contribute their expertise, do not submit a "draft"—send your best effort.

Begin by sharing your finished manuscript with your coauthors and carefully consider their suggestions. Whether your institution requires you to run it by native English–speaking colleagues or not, doing so may help you improve it. Nevertheless, due to other time commitments, they may not give you recommendations detailed enough that, if followed, will make your paper concise, your presentation engaging, and your conclusions persuasive. (Part of the problem is that they may be poor writers themselves!)

But what if you do not have access to or receive adequate assistance from such colleagues? First, check with your institution to see if it can provide you with the services of a professional editor on staff or on contract. If not, you may need to hire a technical editor yourself and pay for it from your grant, research group resources, or your personal funds. (It is a good idea to always include a line item for technical editing services in your grant applications.) People in many fields invest in their careers by routinely hiring professional editors because they know that their reputations depend upon good writing. Selecting a skilled editor who is knowledgeable in your field can improve your chances of receiving favorable reviews and publishing your research quickly.

Where do you find such an editor? To accommodate the growing number of ESL authors who wish to publish their scientific studies in English-language journals and need help in reducing grammatical errors and writing clearly in a second language, a number of Internet-based companies and freelance editors provide professional manuscript-editing services to scientific authors. Many of these freelancers also work for scientific journals, and some contract employees of editing companies are graduate students in the sciences.

How do you choose the company or editor who is right for you? The first thing to do is to ask your colleagues for referrals to professional editors they have worked with and trust. Some publishers, such as the AMS, list the contact information for editing services on their Web sites, though they do not endorse any particular one. By viewing the Web sites of several professional editors, you can compare their qualifications and develop a list of questions tailored to your particular paper. Before entering into a contract with a company or an individual, get satisfactory answers to the following questions:

1. How much experience does your editor have working with scientific manuscripts? With atmospheric science manuscripts (or manuscripts in your field)? How many and what percentage of the papers edited were published, and where and when? Are before-and-after samples of work edited by that particular editor available for you to examine? Is contact

In addition, you may need to lower the author's expectations by treating this first session as just one step toward producing a revised manuscript—many authors expect a perfect draft to be the end result of just one meeting. You know that writing takes time and many revisions. Impress this upon them.

information available so you can check references? What percentage of clients constitutes repeat business? If the service will be provided by a company, is that company registered with an organization to verify that it follows good business practices and complies with all applicable laws? What will be the legal relationship between you and the editor? Is a copy of the editing contract available for you to review and revise before work begins?

2. How will you work together? Will you have the option of choosing a light edit (basically, proofreading punctuation, spelling, and grammar) or a heavy edit (correcting sentence structure, transitions, and flow)? Will the editor query you if it appears that a suggested change might affect the meaning of a sentence or section? Will the editor not only make changes but add explanatory notes that tell you why? Will the editor identify patterns of errors to aid you in overcoming them? Will you be able to discuss your paper with your editor on the telephone or will everything be handled via e-mail only? Where will the work be done? How will it be transmitted between you and the editor? How quickly will you receive your edited paper? What happens if the editor fails to meet your deadline? What recourse do you have if you are not happy with the work? How will any dispute be resolved?

3. If the service will be provided by a company, what is the name of the specific editor who will edit your manuscript and what are that individual's qualifications? Will the editor communicate directly with you or only through the company? Will you have the same editor throughout the editing process? Is the editor a native English speaker? Depending upon the complexities of your manuscript and the types of errors you tend to make in your writing, you may want an editor with a Ph.D. in science or an editor whose strength is primarily in writing. Not all editors need to have a Ph.D. or be credentialed grammarians to be effective, but they should have a sense of what is good science and excellent writing. Some graduate students may not be adept as editors, even if they attend the best universities; others may be quite thorough. Ask for a free sample edit of a page or so from your manuscript to help you see what to expect and how well you and the editor may work together as a team.

4. Will the editor or company guarantee the confidentiality of your manuscript? What safeguards exist?

5. What is the fee, how is it paid, and are there any additional bank fees? What services will you get? Some companies or individuals may quote a price that is only a fraction of what other editors propose; however, neither a high price nor a low price is any guarantee of quality. For instance, does the fee cover all work from submission of the original paper to final decision by the journal? Or does it cover only one edit? Will you have the opportunity to review the proposed changes to the original submission and send it back for another pass by the editor before you submit it? If the journal requests a revision, will another fee be required for editing it? Does the fee include editing your responses to reviewers? Will you receive a money-back guarantee of 100% satisfaction? How will that be determined?

Technical editing services can do quite a bit to help your manuscript. Although no ethical editors will ever promise that your paper will be accepted by a journal, some will stay with you throughout the peer-review process. If you choose wisely, your editor can become an essential member of your team for future publications. If your affiliation or research grant will pay for it, so much the better.

For native English speakers who may be asked to serve as coauthors of manuscripts written by ESL scientists, recall from Chapter 14 that if your name is listed, you are responsible for contributing to the success of that paper, even if it means fixing the grammar yourself. If you do not have the time to put into the manuscript, or do not want to do so, remove yourself from the author list.

Finally, I close this section with this thought. If every native English–speaking author identified one promising ESL author to mentor and worked closely with him or her on improving communication skills, international science would benefit. The number of international collaborations would increase, cultures would be exchanged, and the quality of the science would improve.

17

PAGE PROOFS, PUBLICATION, AND LIFE THEREAFTER

Getting a manuscript accepted for publication is not the end of the process. Sometime after acceptance, page proofs will arrive. Page proofs are the final step before publication, and the last chance to make any changes, albeit generally small ones. Once the page proofs have been scrutinized and the proposed changes sent to the publisher, sit back and await publication. After your article appears in print, alert your friends and colleagues. Post the article on the Web; issue a press release. Then, celebrate.

You have addressed the reviewers' concerns, and the editor has accepted your paper for publication. Congratulations! Now, you can revise your curriculum vitae to list this manuscript as "in press" rather than "submitted." (If the final title has changed since you last updated your curriculum vitae, remember to change it there, too.) Relax, and savor a job well done. Take a vacation, and reconnect with your family. You deserve it. Or, let the momentum of finishing an old project carry you into a new one. Do not think that your job is done, however. The final step in seeing your manuscript through to publication is correcting page proofs.

17.1 PAGE PROOFS

Several weeks to several months after the paper is accepted in its final form, the publisher will e-mail to the author *page proofs* (formerly called *galley proofs*). The page proofs are what the article will look like when finally published, the result of being handled by copy editors, technical editors, and layout specialists to make the manuscript consistent with the journal format. Authors receive the page proofs for one purpose: to ensure that the article contains no errors, either originating from the accepted version of the manuscript or introduced during the editing and layout process. Any suggested revisions need to be communicated to the publisher. The page proofs are the author's final check on the article.

As with the final version of the manuscript, always perform near-final edits of the page proofs on single-sided paper. Perform these edits when fresh and undistracted. If the journal has the option, request the marked-up, or red-lined, version of your original manuscript where the editors have identified their changes to the manuscript. Check the page proof against the changes indicated in the marked-up manuscript. Perhaps enlist a friend to efficiently and thoroughly check page proofs. One person reads aloud the accepted version of the manuscript while the other checks the proofs. Alternatively, give the proofs to someone unfamiliar with your article—they can more easily catch some types of errors. Several rounds through the page proofs with several people (especially the coauthors!) may be necessary to catch all the errors. If you make a large number of revisions, request a second round of page proofs as confirmation your revisions were correctly made.

Besides ensuring the accuracy of the text, figures, and tables, here is a list of other details to check in the page proofs:

> Figures are well-laid out and are close to their citation in the text.
> Figure captions are complete and accurate.
> Section, subsection, figure, and table numbers are sequential and correct.
> Header and page numbers are sequential and consistent.
> Symbols and equations are correctly typeset and accurate.
> Footnotes and references are accurate.
> Corresponding author address and current affiliations are updated, if necessary.

Another task at the page-proof stage is responding to any queries from the editors. Such queries commonly include requests to provide more information on references and verifying that changes in wording made by the editors did not change your intended meaning.

Keep a copy of your comments to compare to the final published version. Return proofs (on paper) by trackable mail or by e-mail. Remember that the journal makes the final decisions on your proposed changes, so, for changes that are substantial, unusual, or controversial, provide explicit justification in a cover letter for their necessity. If needed, cite previous papers published in the journal that followed similar approaches. These types of arguments can help support your position.

After submitting the page proofs, your job is finally done. Publication usually follows in a few weeks to a few months, depending on the journal.

Proofreading is nothing like reading for pleasure. Your eyes must stop on every word, indeed on every syllable of multisyllabic words where errors are beautifully camouflaged from the hasty reader. . . . Personally, I would rather lend someone money than read his proofs, so intensely boring do I find the task. But it's a job that must be done, and done not merely well but perfectly; so after the appropriate groans, mutters, oaths, and procrastinations, give the task the fullest measure of your concentration.
—Richard Curtis (1996, p. 198)

REVIEWING AND CORRECTING PAGE PROOFS

Michael Friedman, Journals Production Manager,
American Meteorological Society

Follow these rules for revising page proofs and sending comments back to the publisher:

1. Do not rewrite significant portions of the paper. As tempting as it may be to do so, remember that your paper was accepted by the editor and should only be edited to fix minor errors, correct mistakes introduced by copy editing or typesetting, or clarify language that could be confusing to readers. If too much rewriting is done, publication delays will occur, extra fees may be assessed, and the paper may even be returned to the reviewers to ensure the corrections do not substantively change the paper.

2. Understand and follow the publisher's instructions for marking and returning the annotated proofs. Usually, the most efficient way to make your corrections is by annotating the PDF file with the commenting and markup tools in Acrobat. These tools are available even in the free Acrobat Reader. Because correcting math may be a challenge, however, format your revisions with Word, MathType, or a similar program, and attach this external file at the appropriate location in the proof. If your corrections are simple and few in number, simply sending an e-mail or text file to the publisher describing the changes is most efficient. Please be sure to identify the locations of the edits in the paper precisely, and be clear when describing what needs to be changed.

3. Communicate your corrections clearly and concisely. Your strategy for identifying the corrections should include having your accepted version handy for comparison to the proof. If the publisher allows it, you may be able to request a red-lined file that shows all the copy edits that were made. Focus on the most critical sections of the paper to make sure any press errors or copy editing changes do not alter your intended meaning, and remember that some formatting changes may have been made to agree to the publisher's press style. Also, review key figures or tables, including noting their proximity to when they are first cited—sometimes typesetting requirements can cause a figure to be placed far from its discussion in the text.

Finally, remember that you and the publisher have the same goal—to get your paper published as quickly and cleanly as possible. It is a cooperative effort, and being able to communicate clearly and positively will enhance the efficiency in seeing your work published.

17.2 PUBLICATION

The time waiting for your article to appear in print for the rest of the world to appreciate can be frustrating, especially for your first few articles. Unfortunately, your article may be published without you even hearing about it. Unless you receive the journal in the mail or an e-mail with the latest table of contents posted online, you may not even be notified that the article has appeared online or in print.

When your article finally appears, take time to celebrate. One research group at the Finnish Meteorological Institute has a small celebration for an author when his or her article appears in print.

17.3 MARKETING YOUR PUBLICATION

Our scientific publications are like our children. We raised them, fed them, nurtured them, sent them on their way into the world, and we hope that they will be a credit to their parents. Nevertheless, like a proud but nervous empty-nester, we still want to give our children that little extra edge in the real world—the extra few spending dollars in college, the hand-me-down Cadillac, and the family vacation home on Cape Cod. Our manuscripts could use a little help, too.

Science is part marketing, and publications require a bit of marketing. Not everyone wants to talk about this aspect, believing that the merit of the article should carry itself. Unfortunately, many meritorious articles published each year go largely unappreciated. How can you help yours be more appreciated?

Once you publish your article, its fate is mostly out of your control. Nevertheless, there are ways to market your article to increase its availability, exposure, and impact. Send reprints (paper copies or PDFs) and e-mail notifications to colleagues, especially those who are acknowledged and cited in the article. People generally enjoy the personalized touch of getting signed reprints in the mail, and, when they do, they are more likely to read your article.

If the author agreement with the journal is such that you can legally do so, place the article on your Web site or in a repository (e.g., arXiv, university repository) as an effective means of increasing free access to the article. Articles that have been self-archived have two to six times more citations than non-self-archived articles (Harnad and Brody 2004).

Go to conferences and present the work, both before publication and shortly after publication. Give talks at your home institution, universities, laboratories, local chapters of the AMS, etc. However, do not oversell or give the same presentation to the same audience repeatedly (e.g., giving nearly the same talk at the Weather Analysis and Forecasting Conference three times in a row). However, it is probably okay to present your results at the European Radar Conference and the AMS Radar Conference in subsequent years because the two audiences will be largely different.

If appropriate, alert the communications or public affairs office of your institution to see if a press release is warranted. Articles that address current events (e.g., local tornado outbreak), hot topics in the news (e.g., global warming and hurricanes), or quirky science questions (e.g., Does it rain more on the weekend?) may be ripe for media attention. More on communicating with the public through the media appears in Chapter 30.

If your university, laboratory, or professional society has a paper-of-the-year competition, consider submitting your best work for consideration. For example, the AMS has the Father James B. Macelwane Annual Award for the outstanding paper written by an undergraduate student.

These steps to increase the visibility of your article may help ensure the longevity of your scientific ideas from among the thousands of others published each year in the atmospheric sciences. Most articles are only cited a handful of times, if at all, and 50% of the published articles never get cited, not even by the authors who wrote the article (Garfield 2005). Specifically, for geoscience articles published since 1998, the Institute for Scientific Information Web of Knowledge reports that the average article is cited only eight times, and half of the articles have received three or fewer citations. Given these statistics, disillusionment about scientific publishing is understandable. As Alley (1996, p. 253) says, do not expect to receive satisfaction from others. The satisfaction has to come from inside you—you had the idea, you conducted the research, you wrote the manuscript, and you shepherded it through peer review. Given that atmospheric-science journals reject an average of 37% of submitted manuscripts, your published article is a testament to your abilities.

METHODS AND APPROACHES TO WRITING FOR THE ATMOSPHERIC SCIENCES

18

Case studies, climatologies, model experiments, and forecasting methods are common types of research projects in atmospheric science. Each one, though, has its own set of informal rules that governs the design and execution of the research and determines whether the work is publishable. This chapter exposes some of these previously unpublished informal rules, providing guidelines for various types of research projects.

Chemistry has laboratory experiments. Field biology has population surveys. Psychologists have human experiments. As with these fields, the atmospheric sciences have several common types of studies. As long as the wind blows and the rain falls, case studies will be popular for describing poorly forecast, devastating, or unusual weather events. Yet, guidance on how to write these types of papers, along with alerts to possible reviewers' concerns, is generally not available. This chapter provides some do's and don'ts for writing good articles in these specific formats.

This guidance, however, is not a formula to a successful manuscript. The success of your manuscript begins with the question asked (Section 2.1). A carefully presented argument to a poorly posed question will have trouble in the review process regardless of how well your manuscript conforms to these rules.

Furthermore, not all articles can be categorized into such distinct categories as case studies, climatologies, etc. Sometimes, research may need a unique approach for presentation, and, in that case, the path is your own. Nevertheless, you may benefit from some of the elements discussed in this chapter. I start this chapter with three general points about research methods (classification schemes, automation, and thresholds), before discussing a sampling of representative, not exhaustive, research approaches for atmospheric science.

18.1 CLASSIFICATION SCHEMES

The human mind looks for organization from among the random, and classification schemes are one means of creating that organization. Through classification, we can begin to understand the processes that affect the observed variation. For example, classification of convective systems into different morphologies such as supercells and squall lines fueled research starting in the 1980s to understand the interplay between the morphology and the dynamics of convective systems.

Because the atmosphere is a continuum, and continua are challenging to categorize into well-defined boxes, how you deal with cases within the continuum may determine the success of your classification scheme. For example, construction of a classification scheme should result in each item appearing in one and only category. Furthermore, unless you define a mutually exclusive classification scheme (e.g., squall lines with tornadoes vs squall lines without tornadoes), your classification scheme should have an unclassified bin (Doswell 1991). Pigeonholing different cases into a rigid classification system invites the next case that occurs to lie outside your classification scheme.

More than one way exists to perform a classification—the result depends on the goals of your research, the interests of your audience, the available data, and the tools. For example, to develop a classification scheme for convection in a relatively data-poor area, the large-scale flow from a global analysis and forecast system might be one way to begin to classify these events, as such analysis will be useful in providing long lead times for potential prediction of events. Different large-scale flow regimes may favor or inhibit convection over the target domain, allowing a classification scheme based on the occurrence of convection. Alternatively, the morphology of the storms from satellite or radar will be apparent once the storms form, so, classification schemes for convective-storm morphology require a different set of data and methods.

18.2 AUTOMATED VERSUS MANUAL TECHNIQUES

Several of the types of papers in this chapter involve sorting through a large dataset to create a reduced dataset with a certain set of properties (e.g., tornadoes within Indiana, cirrus clouds with cloud tops colder than −40°C, cyclones with strong warm fronts over the eastern North Pacific Ocean). Creating this reduced dataset may involve an automated technique or a manual technique.

I am consciously avoiding the terms *objective* for "automated" and *subjective* for "manual" for two reasons. The first reason is that even so-called objective techniques can be quite subjective. For example, specific quantitative thresholds may be essential elements of an automated technique (e.g., short-wave troughs with absolute vorticity exceeding 5×10^{-5} s^{-1}). How should such thresholds best be determined? Is there a natural break in the population at this threshold? Is there a substantial difference between absolute vorticity maxima $4.9 \times 10^{-5}\,s^{-1}$ and $5.1 \times 10^{-5}\,s^{-1}$? What about errant absolute vorticity maxima unrelated to the phenomenon of interest that may be associated with vorticity streamers downwind of orography? If these unrelated features are not what is being classified, then additional criteria will be needed to eliminate these cases from your reduced dataset. These criteria invariably involve subjective decision making.

The second reason is that the words "objective" and "subjective" carry a connotation that subjective techniques are inferior. Scientists are supposed to be unbiased, so why would they choose a subjective technique? The use of these words has cast an undeserved negative cloud over manual techniques. A handmade chair sells for more than one produced in a factory, so why should manual methods be denigrated in our science?

Manual methods have certain characteristics that make them appealing for constructing a dataset. First, the data do not have to be in digital form for a manual approach, which can be particularly useful for identifying features on weather maps either in paper form or on microfilm. Second, if the criteria for defining your event are not amenable to definition in a rigorous algorithm (e.g., whether a given structure is present in radar imagery), then manual approaches will be superior. Third, manual methods are most appropriate when you have a "small" number of events. Your definition of "small" depends on how patient you are in going through the data, how many cases you expect to collect, and how complicated your classification scheme is. For example, if your classification scheme is determining whether it is warmer ahead of or behind a front and you have a complete set of analyzed surface maps, then a large number of days with potential events could be processed relatively quickly. On the other hand, if you must first analyze all the surface maps for frontal locations, determine whether it is warmer ahead or behind the fronts, use representative soundings to examine the stability difference in the air on either side of the fronts, and relate this to the number of severe weather reports, then much more time will be needed to yield a comparably sized dataset.

A common charge levied against manual approaches is the possible introduction of bias in the selection process. For example, the person doing the manual approach may select more marginal cases to enlarge the dataset. An additional problem is a subtle shift in the definition of the event as more events are examined during the manual selection process. One way to avoid time trends in manual approaches is to go back to the beginning of the data after completing the analysis to see if your perspective has evolved. Multiple passes through the data are often worthwhile to ensure every event is consistently identified. Alternatively, having two or three people perform the same analysis can also ensure constructing a manual classification scheme with as few errors as possible. Taking such steps will strengthen your ability to justify the construction of your dataset.

Do not underestimate the importance of these steps in defining the dataset clearly and consistently from the start. Because the rest of the paper hinges on a well-chosen dataset, you want to move forward confidently and without regret. Were a reviewer to find a flaw in your methods, you would have to re-create the dataset with new criteria. Care in designing and constructing the methods from the beginning can help avoid wasted effort.

Although these approaches minimize bias in manual techniques, many people often do not appreciate that automated systems have bias in them as well. Specifically, unless every event selected *and omitted* by an automated technique is checked manually, then being assured that all the possible cases have been collected and unusual events have been excluded is difficult. Often, the robustness and usefulness of the automated scheme is not known until it is tested on the data and the results are analyzed. As such, the design and implementation of an automated scheme may involve several iterations to develop the best scheme.

18.3 PICKING THRESHOLDS

For some studies, you may want to break up your dataset into a number of categories. For example, if you have created a list of downslope windstorms in Boulder, Colorado, what is the best way to segregate out the most intense cases? There are several approaches one could choose.

The first way would be to rank the events from most intense to least intense, then pick the top 25% and the bottom 25%, for example. (Other percentiles could also be reasonably selected.) Because this approach allows the distribution of the data to segregate the data, arguing that your method is biased is difficult. A slightly different alternative would be to rank the events, then let the number of cases you are comfortable working with determine what the threshold will be. For example, if 800 cases in ten years meet a certain threshold and your method is a time-consuming manual approach, is analyzing such a dataset realistic? A different approach in that situation would be to pick the 50 most intense cases, the top 5% of cases, or another reasonable number to work with through some other means, and no arbitrary thresholds have to be defined.

A second approach would be to choose a threshold with a physically meaningful connection (e.g., temperatures below 0°C, cloud tops colder than −40°C when homogeneous nucleation of ice is generally expected to occur). For example, the criterion used by the U.S. National Weather Service for severe hail (¾ in., or 1.9 cm) came from "the smallest size of hailstones that could cause significant damage to an airplane flying at speeds between 200 and 300 mph [89 and 134 m s^{-1}]" (Galway 1989; Lewis 1996, p. 267).

A third approach would be to plot the distribution of the data, and then exploit natural breaks in the data. The choice of threshold can also be optimized using the Relative Operating Characteristic (ROC) curves (Wilks 2006, Section 7.4.6).

The choice of threshold is important because reviewers will likely question your rationale. Therefore, convincing the readers that your threshold is the best or the most reasonable choice usually results in your manuscript having a smoother ride through the review process. Testing the sensitivity of your results to a variety of thresholds is one way to address such concerns before peer review.

18.4 RESEARCH APPROACHES FOR ATMOSPHERIC SCIENCE

This section presents a few of the types of research approaches that are common in the atmospheric sciences, and possible issues when performing and writing them. Although hardly comprehensive, I hope this section provides some insight to help improve the research and writing of your own project.

18.4.1 Case studies: Observations and models

Case studies are descriptions of particular weather events using standard observations from the Global Telecommunications System (GTS) or special data from field research programs. Since the advent of mesoscale models as tools for synoptic meteorologists, the ability of more people to run mesoscale models, and the ready access to operational model output online, case studies using model output, either alone or in combination with observational data, have grown in popularity. Case studies are often a staple of undergraduate- and graduate-level class projects.

The advantage of a case study is that the processes responsible for a particular event can be described in detail. Special data collected during field programs can help elucidate structures of weather systems that were hitherto unknown. Even careful examination of the GTS observations can yield insight into the atmosphere and benefit forecasting. Output from a faithful simulation of an event can be used as a surrogate for diagnosis. As such, case studies can be powerful scientific tools.

Case studies are also the most deceptively easy article to write—how difficult is it for a meteorologist to describe what happened? Yet, writing a concise, relevant, and informative exposition of a case appears to elude many authors. Typical problems plaguing case studies include the following.

No point to the case study. Often case studies end up being descriptions of the case, believed to be sufficient to qualify as a published manuscript. Documentation alone is not the same as explanation or understanding. Case studies should have a purpose that goes beyond documenting a weather event, even for student projects. Professors will be more interested in reading your project (and giving it a higher grade) if you find something unique to say about the case. Likewise, authors who take the obvious case and spin it from a different perspective are more likely to succeed in getting it published. Case studies of phenomena seen elsewhere, but documented for the first time somewhere else (e.g., the morphology of lake-effect snowstorms over small lakes in northern Europe), may be acceptable if a thorough, but not superficial, comparison is performed. Alternatively, put your work into historical and modern context. Why is it worth studying? What does it tell us about past events or the present state of forecasting?

Poor justification for the uniqueness or commonness of the case. Case studies can be useful to demonstrate how typical weather patterns come together or they can demonstrate unusual or climatologically infrequent events. Does the author develop a climatology to illustrate how common or uncommon an event is? Do other documented cases exist? If the event is uncommon, how uncommon is it (e.g., anomalies, percentiles, records at individual stations)? Such information provides context to the reader. Forecasters will read and interpret a case study of a flash flood differently if the flood is a yearly occurrence or if it is a 100-year flood. The representativeness of the case is one of the common issues raised by reviewers, so it behooves authors to have at least thought about this, if not to have explicitly written something into the text.

Too many figures versus too few figures is in the eye of the beholder. What matters in the end is that the selected figures work hard for the paper. Showing figures without weaving them into a story is like "show and tell" from grade school and accomplishes nothing.
—Lance Bosart, The University at Albany/State University of New York

Too many figures. A common mistake of case studies is to present all the constant pressure charts for each time of the case study. Are the 850-, 700-, 500-, and 300-hPa maps every 12 hours all needed? Such level of documentation might be appropriate for a technical memorandum, but not for most journal articles, theses, or class projects. To include too many charts might bore, and thus lose, the reader. Instead, can you describe the case with a series of three 6-h surface charts and a 500-hPa chart at the time of the event? Think about the essential figures needed to tell the story and argue your point based on a minimum number of figures. Avoid tangents—a few may bring some color to the case, but include too many and the audience may lose focus.

Too little interpretation. Some authors write as if the event is so obvious from the figures that the audience can understand what is going on themselves. Wrong! Effective scientific communication guides readers through the story you tell—leaving it up to the readers to figure it out invites them to be frustrated and abandon your paper. Write text that explains rather than simply tells. Annotate complicated maps for the benefit of the audience.

Too much speculation. Case studies often suffer from too much speculation supported by too little evidence. The existence of a moist low-level jet stream alone is not sufficient to claim responsibility for the resulting convective storms. Ideally, you would want to *prove* that storms would not have developed without this feature. Unfortunately, that could require quite a bit of work, if it were even possible.

Poor organization. When presenting a case, make the organization clear to the audience. Presenting observations before model results generally makes the most sense. Present data before speculation. Present observed quantities (e.g., wind, temperature) before derived quantities (e.g., frontogenesis, deformation).

Not using the right tool. Make sure you use the right tool for the right job. One example is frontal analysis. As discussed on page 355, fronts are defined by thermal, not moisture, discontinuities. Therefore, using θ_e or θ_w for frontal analysis is not appropriate.

Attributing a weather event to a single factor. Atmospheric processes are often the result of multiple processes acting together. To claim that a weather phenomenon is "responsible for," "plays a primary role in," or "causes" would be to oversimplify the actual mechanisms. For example, deep moist convection requires three ingredients: lift, instability, and moisture. Therefore, claiming that a single factor such as the unstable environment "causes" the resulting convection would be to ignore the other two ingredients, which must be present as well.

Some papers provide two or more case studies to show that the first case is representative of other cases. Unless there is a legitimate reason to compare the cases with each other, a paper with two or more cases is usually tedious. In particular, I find myself getting the main point of the paper from the first case, then skimming pretty quickly through the later cases. If you are considering such a paper, make a compelling argument for why multiple cases are needed.

A model study can be an effective way to enhance or support the results of an observational analysis, especially in the same paper. Although case studies with a modeling component may be plagued by the same problems listed above, modeling studies have the following additional challenges to producing interesting and publication-quality manuscripts.

How much model verification is needed? The simulation does not need to be perfect to study the physical processes, but the model should capture the essential features being explained. Take care that the model captures the physical processes effectively (e.g., that you are not trying to understand processes that are parameterized or inadequately resolved in the model).

The model should do more than just reproduce the case. Just showing model output that reproduces some phenomenon in the atmosphere is rarely a reason for a publication. Some research articles are driven primarily by the simulation at the expense of the scientific content. What is the scientific question being addressed? Diagnosis of the relevant physical processes is needed.

One caveat about any modeling study is that the model may get the right answer for the wrong reason. Specifically, the model might produce a realistic-looking evolution, but the physical processes operating in the model may not reflect reality. For example, the model may produce a reasonably well-forecast precipitation field, but how the model partitions the hydrometeors into different microphysical categories may not be well reproduced at all. Unfortunately, this caveat is rarely stated.

18.4.2   Model sensitivity studies

Model sensitivity studies are characterized by a suite of simulations where the model physics or initial conditions are altered to evaluate their relative importance to the simulations. Sometimes such sensitivity studies can be greatly enlightening. Other times, authors, answering no particularly deep scientific question, just flip switches on and off because it is easy to do. For example, in the early days of mesoscale modeling of extratropical cyclones (late 1980s and early 1990s), the role of latent heat release in their intensity was just beginning to be understood, so turning the latent heat of condensation to zero was a worthwhile experiment to consider. After numerous sensitivity experiments, the role of moisture in extratropical cyclones is better understood, so such simple studies are not particularly worthwhile anymore. Interesting studies remain to be performed, but the methodologies generally need to be more advanced than simply turning latent heat release on and off.

A further complication, especially with moisture, is that the timing of when surface fluxes are turned on or off may affect the outcome of the result (Kuo et al. 1991). For example, turning the surface latent heat fluxes off at the time the cyclone is already mature may make little difference to the strength of the system because the moisture needed for the storm has already been picked up and its latent heat will eventually be released. In contrast, turning off the fluxes 48 h or more before the storm starts may have a much bigger impact on the development of the storm.

A related aspect happens when multiple processes in the model are acting, perhaps together. For example, flash-flooding events in southern Europe are commonly associated with moistening of the inflow air over the Mediterranean Sea and topographical ascent. A control simulation with both moisture and topography present can be run, as can a run without moisture, a run without

WHY FORECASTERS SHOULD PUBLISH

Jim Johnson, Retired Forecaster, National Weather Service

Sadly, operational forecasters are often reluctant to document their meteorological experiences. There are a number of excellent reasons why they should do so.

No one is closer to day-to-day atmospheric anomalies. Every forecaster has seen many cases where conceptual and computer models of the atmosphere failed miserably. A case study of those experiences from the forecaster's viewpoint, when published, allows research meteorologists to refine those models, thereby reducing their failure rate. Thus, forecasters potentially benefit from even a simple write-up of these events. Publication makes such case studies available to the research community.

There is a misconception on the part of operational forecasters that research meteorologists dwell in the high ether of the science producing Great Truths of Meteorology on a daily basis. In fact, research is mainly dog work trying to piece together bits of a dismayingly stingy collection of half facts and unrelated events. The Great Truths, what few are ever found, generally come from assembling a multitude of tiny observational discoveries that can eventually be molded together into a working hypothesis. Those tiny observational discoveries, more often than not, come from a collection of forecasters' published case studies found in collections of journal articles!

Moreover, forecasters *are* researchers. Their job requires constant researching of the available data for familiar features. In so doing, forecasters often *see* unfamiliar features that later turn out to be significant in the evolution of the atmosphere! A few ideas jotted down at the end of a forecast shift can lead eventually to better understanding of these unfamiliar features and their impact upon the current atmospheric problem. In this way, easy documentation is available, making eventual publication of a possibly significant atmospheric phenomenon fairly simple.

Perhaps the greatest value in publishing by the forecaster, however, lies in the knowledge gained in the process. It is a fact in the world of applied sciences that no one knows all there is to know about any particular topic. The peer-review process of formal publication can be an arduous task for the operational forecaster or it can be a journey to exciting discoveries. Literature research for the purpose of publication exposes the operational forecaster to a great store of knowledge that was not obtainable in a baccalaureate or even masters degree program. Rather than drudgery or "paying your dues" to the system, enjoy the research process for the many new things that you will learn.

Finally, any pain caused by poring through the literature for supporting material and then assembling the findings in a coherent manuscript can usually be greatly mitigated by teaming up with a professional researcher on your project. Most research meteorologists are only too happy to have operational forecasters who are in the trenches daily come to them with proposals and ideas. Doing this can greatly ease the overall work of producing the manuscript and also accelerate the publication and peer-review process.

---

topography, and a run without both. Unfortunately, nonlinear interactions will occur between moisture and topography. The factor separation technique (e.g., Stein and Alpert 1993; Krichak and Alpert 2002) accounts for the nonlinear interactions between phenomena. In principle, factor separation can be used for any number of processes, although the physical meaning of such an approach becomes more difficult to understand with more than three factors.

The best way to present model sensitivity studies where multiple model experiments are performed is to decide whether the relevant physical processes being tested are essential to the story. Before turning switches on and off, design a framework to organize and discuss the physical reasons for these experiments. Be aware that nonlinear interactions may affect your ability to unambiguously ascribe physical significance to the results and that multiple physical processes, not just a single process, are usually responsible for any weather phenomenon.

18.4.3 Climatologies

As discussed in Section 18.4.1, one of the potential weaknesses of a case study is its representativeness. Representativeness can be addressed by performing a climatology of the event. The strength of a climatology is that the characteristics and distributions of a large number of events can be presented. For example, if a case study showed that cold-air damming lasted three days, a climatology might show that three days is a typical length for a cold-air damming episode, although events could range from one to seven days. Other characteristics of weather events that can be explored in the context of a climatology are listed in Table 18.1.

How many events do you need to collect in order to claim some measure of representativeness for your climatology? If you have 100 events over Ohio, that is different from 100 events over the entire United States. When in doubt, it is best to just be honest and say "A five-year climatology of...." People may argue about whether five years is enough, but they cannot say that you are misrepresenting the data.

18.4.4 Synoptic composites

After creating a climatology of an event, creating a composite evolution may be useful to characterize the flow preceding the event, at the time of the event,

---

Table 18.1 Characteristics that can be explored in climatologies

Spatial distribution
Distribution in time (e.g., annual numbers of events)
Intensity
Annual cycle
Diurnal distribution
Initiation and ending times
Duration of event
Associated weather events
Synoptic regimes

---

and after the event. One of the key points is to define the onset time very clearly so that all cases are uniformly defined. These are called lagged composites. For events whose duration may vary, lagged composites can be constructed at the onset and the ending times of the event.

All events in the composite should be examined to ensure that the features in the composite are present in each of its members. If this criterion is not met, then the validity of the synoptic composite can be legitimately questioned. Furthermore, each of the bins should have a large enough number of members such that the composite is not overly influenced by too few events. The composite can be presented as the mean field and anomalies from the mean. Anomalies should be checked for statistical significance from climatology or some other type of mean.

18.4.5 Forecast methods

By some measures, forecasting is the ultimate goal of what we do as atmospheric scientists. If we can understand a phenomenon better, we can hope to forecast it. Although papers appear in the literature describing different forecast measures, some lack general utility for one or more of several reasons.

Ingredients-based approach not employed. When considering a phenomenon in the atmosphere, all the factors that affect the production of that phenomenon should be included in a forecast scheme. Documentation of these factors results in what is called the ingredients-based approach. Ingredients-based approaches were first discussed by McNulty (1978), and further elucidated by Johns and Doswell (1992), Doswell et al. (1996), and Schultz et al. (2002). For example, the three essential ingredients for deep, moist convection are lift, instability, and moisture. The absence of enough of any of these three ingredients means deep, moist convection will not occur. Forecast methods should employ aspects of all ingredients, if known (the ingredients for some weather phenomena have not been determined or articulated).

Using ingredients-based thinking strengthens the presentation of your paper. Because an ingredient is necessary for the occurrence of an event, an ingredients-based methodology prevents omission of important ingredients, thus providing a focused analysis. Many papers whose authors do not employ ingredients in their thinking are scattered—showing a bunch of different parameters because authors have seen other papers use them—rather than incorporating the relevant physical processes. An ingredients-based approach is a unifying theme for forecast studies and allows a proper connection between forecast variables and physical processes.

Improper construction of forecast parameters. Some papers derive methods or diagnostic variables (parameters or indices) to help improve forecasting. For example, many attempts to obtain better diagnostic variables to

WRITING FORECAST VERIFICATION STUDIES

Tom Hamill, Research Meteorologist, National Oceanic and Atmospheric Administration (NOAA), Earth System Research Laboratory

Forecast verification is the process of assessing the quality of forecasts. A verification study may examine a particular aspect of the forecast ("Do forecast thunderstorms resemble observed thunderstorms in intensity, coverage, and spatial propagation in this model?") or it may be more general ("Is model A demonstrably better in precipitation, wind, and temperature forecast skill than model B?").

In many respects, good verification studies resemble any other good journal articles. However, there are a few aspects of verification studies that may differentiate them from other types of articles. First, forecast verification studies are commonly used to assess the performance of a model that is highly dimensional, perhaps rainfall forecasts at a set of grid points. Consequently, the characteristics of the forecast can and should be diagnosed in several ways (e.g., Murphy 1991). Is the deterministic forecast both accurate and unbiased? Is the probabilistic forecast both reliable and sharp? Does the performance of the model change with season, with location, with synoptic situation? Because one forecast is rarely uniformly better than another, a good forecast verification study

examines enough aspects to help the reader assess the relative merits of the two competing systems.

Ideally, any scientific study between systems A and B not only measures differences in system performance but also quantifies the statistical significance of such differences. Many past forecast verification studies have omitted these error bars; the high dimensionality and the large spatial and temporal correlations of forecast samples violate the implicit assumptions underlying many standard statistical tests (e.g., independent and identically distributed). However, with modern computers and numerical methods, bootstrap techniques (Efron and Tibshirani 1993) can be applied readily to verification studies. In the bootstrap, simple blocking techniques are used with correlated data (Hamill 1999). Application of bootstrap techniques may also teach the researcher an important lesson: the statistics may not support a grand conclusion without a large number of samples. Because of the computational and logistical expense, a study may attempt to support conclusions with only a few cases. The resultant error bars may reinforce what the researcher already knew: many cases spanning many synoptic situations may be needed to assure statistical significance. And as often as not, testing over a wider range of cases illuminates where more study is needed. For example, why did the new forecast model do well every season but spring? Good primers on forecast verification can be found in Jolliffe and Stephenson (2003) and Wilks (2006, chap. 7).

---

ascertain tornadogenesis from environmental parameters such as instability and wind shear have been published. Doswell and Schultz (2006) categorize different types of diagnostic variables based on their construction and utility to forecasting severe storms (although, in principle, these results hold for more than just severe storms). The creation of forecast parameters by the arbitrary multiplication of a number of individual diagnostic parameters is not recommended.

Inadequate verification. Forecast skill is determined by comparing the accuracy of the forecast scheme to the accuracy of some standard forecast method (e.g., climatology, persistence, model output statistics). Forecast schemes that show statistically significant skill by this comparison can be considered a useful forecast parameter. If you use skill scores, remind the readers which values represent a perfect forecast and a forecast with no skill.

Null cases not examined. If a scheme to forecast tornadoes relies on a signature in satellite imagery, looking only at cases where tornadoes form does not provide the ability to discriminate whether the forecast scheme works. Cases that produce tornadoes, as well as the cases that do not produce tornadoes, need to be examined. The simplest way to verify such forecasting schemes is to use a 2×2 contingency table (e.g., Wilks 2006, Section 7.2.1).

18.4.6 Other approaches

The goal of this section was to provide just a sampling of the types of possible research approaches, particularly the more common ones in meteorology. To articulate guidance for a larger number of studies is not the intent here. Instead, consider the type of research question needing addressed and see if any of the lessons in this chapter apply to your specific research project.

PARTICIPATING IN PEER REVIEW

EDITORS AND PEER REVIEW

19

Editors are a bit like the Wizard of Oz. Seemingly great and powerful, editors are simply ordinary men and women who volunteer to oversee the peer review of manuscripts. In this chapter, we learn how editors perform their jobs.

Fred Sanders, professor emeritus at the Massachusetts Institute of Technology, was a long-time editor of *Monthly Weather Review* (1986–1999). Being retired, but still active scientifically, Fred had the independence to work on his own research projects, the benefit of years of experience, and the free time to really savor his job as editor, all without the demands of having an affiliation (except for Sanders World Enterprises, the mock affiliation he would wear on his conference badges).

Receiving Fred's editorial decisions on your manuscript was a mixed blessing. In addition to reviews from two reviewers, he typically provided his own review of your manuscript, which was often more thorough and lengthy than the other two reviews combined. The time that Fred spent reading the

Frank and Ernest

© 2005 Thaves. Reprinted with permission. Newspaper dist. by NEA, Inc.

HOW EDITORS MAKE DECISIONS

C. David Whiteman, Research Professor, Department of Meteorology, University of Utah, and former editor, Journal of Applied Meteorology, *and Johanna Whiteman, former editorial assistant,* Journal of Applied Meteorology

Although the editor usually relies very heavily on the reviewers' recommendations, the editor must also ensure that the reviews are of good quality and that the summary recommendations are consistent with the reviewers' comments before making an initial decision on the suitability of the manuscript for publication. Because each reviewer has a different perspective on the quality of research and the effectiveness of writing, different expertise, and different external circumstances that affect the quality of reviews, the reviews and advice to the editor are sometimes widely divergent. One of the reviewers may catch a fatal error that the other reviewers have not seen, or one of the reviewers may come to an incorrect decision based on a misunderstanding or an error. In addition, a few reviewers will accept nearly all manuscripts regardless of quality, and a few will reject almost all manuscripts, though most reviewers are hesitant to reject a manuscript outright. Thus, the editor cannot simply average the reviewers' recommendations but must carefully consider the reviewers' comments and then come to an independent decision, keeping the goal of maintaining the quality of the journal in mind.

An example of peer-reviewer recommendations and editor's initial decisions on 45 manuscripts submitted to the *Journal of Applied Meteorology* is presented in Fig. 19.1. Here, the reviewers' recommendations are indicated with Xs, whereas the editor's initial decisions are indicated by ovals.

For this journal, the editor usually solicits three reviews. Occasionally, one of the reviewers may fail to get comments back to the editor before a decision has to be made (manuscripts 9, 10, 32, and 43). In these cases, the decision is based on the reviews received. If the reviews are quite divergent, the editor may seek an additional review, occasionally from an associate editor who has expertise in the subject matter. In other cases, more than three reviews are solicited when the manuscript is submitted (manuscripts 25 and 26).

Editor decisions usually follow very closely the recommendations made by the peer reviewers. Manuscript 8 is clearly an exception. It was rejected, despite supportive reviews, when it was determined that most of the material in the article had been previously published in other journals. Journals have strict policies regarding this issue. Sometimes the reviews are widely divergent, with one reviewer recommending "acceptance without modification" while another recommends rejection (manuscript 15). In such cases, the editor determines the appropriate course. Rarely will a reviewer recommend publication without modification (manuscripts 15 and 38). Manuscripts that receive suggestions for rejection or major modifications are often best rejected. In some of these cases, the editor will recommend that the manuscript be revised based on the reviewer comments and then resubmitted. Going forward on such manuscripts by requiring major changes can occasionally lead to "infinite loops" where the paper goes back and forth between authors and reviewers without resolution.

The final decision on a manuscript (accept or reject) is made by the editor on the basis of changes made to the manuscript by the authors in response to the initial reviewers' comments. The editor usually sends the revised manuscript back to reviewers who have recommended rejection or major changes so that they can review the changes. The reviewers then make a second (sometimes third, fourth, etc.) recommendation to the editor based on the authors' responses to their technical comments. Eventually, the paper is either accepted or rejected.

Initial decisions

MS #	accept	minor	major	reject	MS #	accept	minor	major	reject
200					231		(XX)	X	
201		(XXX)			232		XX	(X)	
202			X	(XX)	233		(XX)	X	
203		X	(X)	X	234		X	(X)	X
204		XXX	( )		235		(XX)		
205		(XXX)			236		(XXX)		
206			X	(XX)	237		X	(X)	X
207		(XXX)			238		XX		(X)
208		X	XX	( )	239		(XXX)		
209					240		X	(XX)	
210			(XX)		241	X	X	(X)	
211			X	(X)	242			XX	(X)
212		(XX)	X		243		(XXX)		
213			X	(XX)	244		X	(XX)	
214			XX	(X)	245		X	X	(X)
215			(XXX)		246				(XX)
216	X	(X)		X	247		(XX)	X	
217		X	(XX)		248				
218		X	(X)	X	249		X		(XX)
219		XX	(X)		250		XX		
220		XX	(X)		251		X		X
221		(XXX)			252				X
222		XX	( )	X	253				
223		X		(XX)	254				
224			X	(X)	255			X	
225			X	(XX)	256				
226		XX	(X)	X	257				
227					258				
228					259				
229		(XXX)	X		260				
230		X		(XX)	261				

manuscript and writing his own review provided him exceptional insight into your paper. Although some would argue that Fred's closeness to the manuscript prevented him from making an unbiased decision, others have argued that his decisions were among the most well informed because of his attentiveness to each manuscript.

Editors cannot be experts on every aspect of every manuscript submitted, which is why the assistance of reviewers is needed. Just like the output from a numerical weather prediction model provides guidance to the human forecaster, the reviewers only provide guidance to the editor. Reviewers make

Fig. 19.1 Initial decisions on manuscripts submitted to *Journal of Applied Meteorology* in one year, showing the reviewer recommendations (X) and editor decisions (ovals). (Figure courtesy of C. David and Johanna Whiteman.)

recommendations, and editors make decisions. Some reviewers are under the mistaken impression that they are the gatekeepers through which acceptable manuscripts must pass. In fact, the author has to please the editor, not the reviewers.

Editors are people, too. Most editors, unlike Fred Sanders, have to face the regular demands of our jobs, then oversee the peer-review process on manuscripts, all for no compensation and few, if any, fringe benefits. Sometimes editors have to make the difficult decision to reject a manuscript. Some of these manuscripts are easily rejected, their quality being so poor. Others are not as obvious, so the editor must carefully read and scrutinize the manuscript and the reviews, and carefully craft decision letters. Even some papers requiring major revision occupy much of the editors' time. If your manuscript is rejected, remember that editors are simply doing their jobs, which is to publish high-quality manuscripts as efficiently and as promptly as possible. Editors have obligations to the publisher, authors, and reviewers, and part of the job is to balance all of these obligations.

Specifically, these obligations require editors to judge each manuscript on its scientific content "without regard to race, gender, religious belief, ethnic origin, citizenship, or political philosophy of the author(s) (American Geophysical Union 2006)." The editor (as well as the reviewers) "should respect the intellectual independence of the authors," meaning that the reviewers should allow authors to present their manuscript in the form and style they wish, subject to a high standard of quality of science and presentation. The editor also has to avoid conflicts of interest, whether real or perceived. While a manuscript is in review, editors (and reviewers) are expected to treat the manuscript as confidential information, and any use of the material in the manuscript beyond the review process should be cleared with the author. Finally, if misconduct (Chapter 15) has occurred or errors in previously published papers are identified, then the editor should take responsibility for having the manuscript be withdrawn or retracted.

Thus, editors are invested with a lot of responsibility. Authors trust editors to handle their manuscript professionally, reviewers trust editors to value their input, and journals trust editors to make wise and fair decisions.

WRITING A REVIEW

20

Thousands of reviews are written each year by reviewers who receive no credit, for the benefit of authors whom the reviewers may not even know. The result of this review process is that authors receive guidance on improving their manuscripts and editors receive guidance on the suitability of manuscripts for publication. As such, reviewers are *the publication process. This chapter provides guidance to reviewers (or potential reviewers) on whether to agree to perform a review, how to critique a manuscript, and how to write a review.*

Whereas the last chapter focused on the editor's role in peer review, this chapter focuses on the role of the reviewer. Imagine you were one of the peer reviewers of your own manuscript. Faced with this external view of your manuscript, what would you think of it? What flaws in your science or your argument might be apparent?

> To help myself think about writing for the audience instead of myself and to make the manuscript more convincing, I envision that the readers are (1) my worst critics and (2) the best scientists in the field. I also imagine that the readers are the authors of manuscripts I'm citing. Thus, I'd better interpret their paper correctly and, if I'm critical or suggesting an alternative perspective, the writing better be tight, clear, and convincing. —Jim Steenburgh, University of Utah

Envisioning his toughest critics reading his paper brings out the best writing in Jim Steenburgh. In this chapter, you will learn how to critique someone else's manuscript and, in doing so, learn how to critique your own.

20.1 SHOULD YOU AGREE TO DO THE REVIEW?

Before the 1990s, you might have received a paper copy of a submitted manuscript in the mail without any prior notice, and you would be responsible for providing a review or sending the manuscript back to the editor unreviewed. If you were out of town participating in a field program, the manuscript might have lingered for weeks. Fortunately, times have changed, and most journals now send an e-mail asking if you would do the review, rather than assuming you would. Before you take on the role of reviewer, should you?

There are many reasons to accept a review. Reviewing a manuscript:

- Is good practice for writing and revising your own manuscripts;
- Means that you get to see new research before it is published;
- May force you to accelerate your learning on some topics;
- Will help improve the author's work and the quality of the published literature; and
- Is a way to give back to the atmospheric science community.

If you do not believe that you can perform the review by the requested deadline, say so to the editor. If you can perform the review, but with a modest extension of the deadline, then suggest that to the editor. Even if you have accepted the review and you subsequently find you cannot meet the deadline, please tell the editor. An editor would rather have you ask for an extension, decline a review, or return a manuscript unreviewed than face a situation where e-mails asking for an overdue review go unanswered or a promised

WHY EARLY CAREER SCIENTISTS SHOULD DO REVIEWS

I like to ask early career scientists to provide reviews. They often provide the most thorough and helpful reviews because they take time to think about the manuscript and provide constructive criticism to the authors. Some of the people I ask are flattered and politely decline because they do not consider themselves experienced enough to do the review. Here are five reasons to encourage early career scientists to become reviewers:

1. Reviewing papers is something that scientists should start early in their careers.
2. Editors would not have asked you to review a paper if they did not think you could do it.
3. Other reviewers will also be providing reviews, so your review is not the only one to be considered.
4. Reviewing manuscripts increases your exposure and visibility within the field.
5. Journals are always looking to enhance their list of potential reviewers. If you are interested in serving as a peer reviewer for a journal, but have not been asked, you might e-mail the editor and express your interest.

review never comes. Too many reviewers earn the ire of editors by not owning up to their responsibilities.

Prospective reviewers are selected for their expertise on the topics covered by the manuscript. Although a paper may address more than one topic (say, statistics of climate model output), reviewers may have expertise in only one topic. Thus, reviewers should indicate to the editor that their expertise only pertains to one aspect of the paper and that their review cannot comment on other aspects with which they are not familiar.

Consider declining the review if you do not have the expertise to perform the review or if you are unable to give an unbiased review. Such situations may occur if you have a conflict of interest with the authors (e.g., financial, supervisory); you agree or disagree strongly with the authors, methods, or conclusions; or you have a personal relationship with one of the authors.

20.2 OBLIGATIONS OF REVIEWERS

The peer-review system is based on volunteerism and the honor system. That the process works as well as it does is a testament to the integrity of the majority of scientists who participate in the process. Before accepting your job as reviewer, be aware of your responsibilities:

They say it's easy to be critical, or negative, or destructive, but it isn't really. To stick to serious, negative, unconstructive criticism takes a lot of thought and effort. —Dwight MacDonald, American writer and editor

- If you submit N papers a year, expect to perform at least $2N–3N$ reviews a year, more reviews for authors with more experience or specialized expertise.
- If the manuscript is of high quality, respect the authors' right to present the manuscript in the style they choose.
- Never criticize the author personally in your review.
- Manuscripts should be treated as confidential documents. Do not show or discuss them with others. Specific questions to others are all right, but disclose that information to the editor. (This information is required for some journals, such as those published by the American Geophysical Union.) Obtain the consent of the authors to present unpublished information to others.

20.3 HOW TO APPROACH A REVIEW

Writing a review can be a bit overwhelming the first time. However, this fear subsides as you write more reviews. To provide some guidelines, here are the steps I employ when approaching a review:

1. Print out the manuscript single-sided. If possible, separate the figures from the text. If the manuscript is single-sided, comparing figures and text on

HOW TO READ AND CRITIQUE
A SCIENTIFIC PAPER

Pamela Heinselman, Research Meteorologist,
National Oceanic and Atmospheric Administration
(NOAA)/National Severe Storms Laboratory

Whether reading a paper for personal interest or as a reviewer for a journal, a critical reading strategy is needed. A key to critical reading is looking for ways of thinking rather than looking solely for information. In addition to reading a paper to obtain information, think about how the information is presented, how evidence is used and interpreted, and how the text reaches its conclusions. One way to engage in critical thinking is to ask yourself questions as you read through the paper. The template below is a basic guide to thoughtful reading that will help you develop or fine-tune your critical reading skills.

Template for Reading Critically

Title
- Does the title clearly and correctly represent the research?

Abstract
- Does the abstract summarize what the paper covers?

Introduction
- What is the purpose of the paper? What are the hypotheses?
- What previous research forms the basis of the study?
- Who is the audience?

Background
- What concepts (e.g., terminology, theory, conceptual models) are needed to understand the paper?
- How are these concepts used to organize and interpret the data?
- What are the boundaries of this particular work? Can you identify: What is known? What the new research adds? What remains unknown? What assumptions are made, and why do you agree or disagree with them?

Data and methods
- What data are used? Are they described in adequate detail?

different pages will be easier. In my experience, reading the manuscript onscreen is not nearly as effective.

2. Read the manuscript once for pleasure. Take some notes, or do not take any notes at all. Just try to understand the manuscript and get a feel for its quality. (If you fail to understand the manuscript, the problem may lie with the author and not you!)

3. Consider the questions from the Template for Reading Critically (sidebar on pages 232–233).

4. Allow yourself a deeper appreciation of the manuscript, its science, and its applications.

5. Read the manuscript a second time, writing the review as you go along.

6. You may wish to read the manuscript a third time to confirm your impressions and to make sure you have identified all your concerns.

7. Proofread your review. Reword instances where you used language that may offend. Sandwich especially tough criticism between positive state-

- Are the data appropriate for addressing the hypotheses?
- What methods are used? Are they described in adequate detail?
- Are the methods appropriate for addressing the hypotheses?

Results
- Is the purpose of the paper addressed?
- Are all posed hypotheses addressed?
- Are the data interpreted correctly?
- What kinds of evidence are used (e.g., statistical, analytical, anecdotal)?
- Do the findings offer evidence to support the hypotheses? Why or why not?
- Could the evidence be interpreted differently?
- Are results discussed in light of previous research?
- Do the figures and tables support claims made in the text? Are they readable?

Conclusions
- Are conclusions clearly stated?
- Are conclusions overstated?

- Do the conclusions provide a complete picture of the study?
- What are the limitations and assumptions of the study?
- How do they affect support for the research results?
- What might an alternative explanation for the evidence be?
- Is the research a significant or a small contribution to science? How so?
- How does this work change or expand our understanding, if at all?

Future work
- What research questions remain unanswered?
- Do any new questions come to mind after reading the paper?
- Are there any new experiments that could be done to give further support to the paper?

References
- Are the references appropriate to the research?
- Are any relevant articles missing?

ments about what you liked about the paper and suggestions of how to improve. Such layering of criticism is called the feedback sandwich or the hamburger method of constructive criticism.

Sometimes, the manuscript may be difficult to read because the manuscript needs more proofreading (e.g., figure numbers in text do not match figure numbers, figures are poor quality, poor-quality English language and grammar). If you feel that such difficulties prevent you from understanding the paper, you are free to return the manuscript to the journal unreviewed or recommend rejection.

One of the most common criticisms levied by reviewers is that the manuscript is too long. If you are reviewing a manuscript that you believe is too long, offer specific areas where the manuscript could be shortened, figures could be removed, sections could be eliminated, etc.

Should you as a reviewer double-check all the derivations? Perspectives differ on this question. Some editors are adamant that all the derivations be confirmed by the reviewers otherwise errors in the derivation or gaps in describing the steps of the derivation might not be revealed. Others say that to verify another author's perhaps complicated derivation might be too time consuming.

20.4 MAKING THE DECISION: REVISE OR REJECT?

A reviewer considers two broad aspects of the manuscript. First is the quality of the scientific content (defined as the idea, execution, choice of data and methods, results, interpretation, etc.). The second is the quality of the presentation (e.g., organization of manuscript, neatness, effective figures, grammar, spelling, format consistent with style guide). As discussed previously in connection with Fig. 2.1 on page 13, these two aspects can be displayed on a graph with quality of science on one axis and quality of presentation on the other (Fig. 20.1).

This graph shows that papers with high-quality science are likely to be rejected if presented poorly (lower-right corner). Similarly, the lowest-quality science cannot overcome rejection, even if presented well (upper-left corner). Thus, authors need to submit manuscripts of high-quality science that is presented well to maximize their success in publication.

If, in your opinion as the reviewer, the manuscript does not meet the standards of the journal or a revised version of the manuscript will be unlikely to meet the standards, then recommend rejection. Other considerations include whether the proposed revisions can be done within the recommended time, or if the author is likely or able to complete the revisions at all. Some

Fig. 20.1 Outcomes from the review process as a function of quality of science and quality of presentation. Even high-quality science that is presented poorly is likely to be rejected (Fig. 2.1 updated).

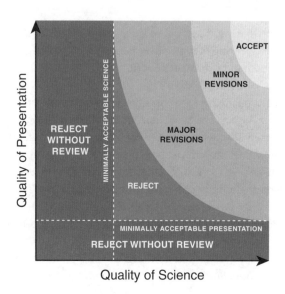

reviewers think that recommending rejection can send a serious message to the authors, a very different message than they could send by recommending major revisions.

Many reviewers think that their criticisms may be overly harsh, especially after recommending major revisions or rejection. In general, I would not worry about this. The review is your assessment of the quality of the manuscript, and there are no right or wrong answers. As long as you keep your criticism scientific and not personal, be as critical as you can defend in your review.

Finally, you may encounter the unfortunate occasion where you discover plagiarism, omission of authors from the author list, or other scientific misconduct (Chapter 15). Alert the editor to the misconduct. The editor will handle it from there.

20.5 WRITING THE REVIEW

After having considered these aspects of the paper, how is the review to be written up? The generic structure of a review has the following components (Table 20.1). Not all components may appear in each review, however.

The first part of the review consists of the preamble. The review should be titled with the name of the manuscript, the authors, the manuscript number assigned by the journal, and your reviewer number or letter, if known. Some reviewers provide a summary of the paper with its principal contribution to science and whether the manuscript is relevant to this particular journal. If not, reviewers might suggest other journals where the manuscript would be more appropriate. Say what the positives and negatives are of the manuscript, paving the way to justify your recommendation. Your recommendation should appear within the preamble of the review.

Different journals may have slightly different recommendations. The basic list includes:

- **Accept.** The manuscript can be published in its current form with no revisions.
- **Return for minor revisions.** The manuscript will likely be acceptable for publication pending minor revisions. These tend to be relatively small concerns, relating to clarity, figures, or grammar. Usually, such revised manuscripts are not returned to the reviewer for a second round of revisions because the editor decides whether the author has adequately responded to the reviewer's concerns.
- **Return for major revisions.** The manuscript will likely be acceptable for publication pending major revisions. Major revisions typically include serious concerns with the quality of presentation and the quality of science. The

Table 20.1 Contents of a review

Preamble
"Review of *Title* by *Authors*–Manuscript Number X by Reviewer Y"
Significance of the manuscript to science and the journal
Summary (positives, negatives, reason for recommendation)
Recommendation (e.g., accept, revisions required, reject, transfer)

Numbered concerns
Fatal flaws (especially if recommending rejection)
1.
2.
. . .
Major comments
1.
2.
. . .
Minor comments
1.
2.
. . .
Typos and misspellings
1.
2.
. . .

reviewer is usually accorded the opportunity to provide a second review. Should the author fail to adequately address the reviewer's concerns, the editor can reject the manuscript for publication. Some journals only have "revisions required," with no distinction between minor and major revisions.

- **Reject.** Rejection can occur when the manuscript lacks adequate organization, originality, or scientific competence. This manuscript either cannot be corrected with any revisions, the revisions cannot be performed within a specified time frame, or a revised manuscript would be so substantially different from the original that it would constitute a new manuscript. Manuscripts can also be rejected solely because the grammar or manuscript preparation is inadequate. Finally, manuscripts can be rejected for scientific misconduct (e.g., plagiarism, redundant publication). Some journals have a softer level of rejection called "revise and resubmit" or "reject and revise," recognizing that the manuscript may be publishable pending revisions, but that these revisions are so substantial that the authors should take their time and not be under any deadlines for resubmission.

● **Transfer to another journal.** When a manuscript may be more appropriate for another journal, the reviewer may recommend a transfer.

Following the preamble are numbered lists of the reviewer's specific concerns with the manuscript. A good starting place for these concerns is the Template for Reading Critically presented in the sidebar on pages 232–233. The reviewer should rank the concerns from the most serious to the least serious (as the progression from fatal flaws → major comments → minor comments → typos in Table 20.1 implies). Rather than listing each occurrence of a certain problem separately, grouping comments into a smaller number of manageable chunks will help convey the message more forcefully and make it easier for the author to make revisions. Within each group (e.g., major comments, minor comments), list the comments in the order they appear in the manuscript. If a concern is repeated throughout the manuscript, but listing all occurrences would be time consuming, you can make your comments on a hard copy of the manuscript and send them to the editor to be forwarded to the author (an annotated manuscript). Or, you may indicate the type of comment that appears, and say in the review, "these minor comments should be considered representative rather than comprehensive."

When identifying a concern and recommending a change, the reviewer may be worried that the author will be threatened by or not appreciate the recommendation. Reviewers should describe their concerns constructively, not judging or criticizing the author. Employ some of the recommendations in Table 20.2. Use the feedback sandwich (page 233): authors are more likely to listen to your meaty negative criticisms if they are sandwiched in between warm positive supportive bread. If you make a criticism, be specific by providing the evidence, the reason for the change, and the suggested revision. Authors respond more favorably to suggested changes when they know why the change is being requested or are shown a suggested revision. For example, just saying the paper needs to be shorter does not help the authors know how to improve the paper. Pointing out which paragraphs of the manuscript can be trimmed is much more helpful. In addition, positive examples from the authors' own writings can help make your point. For instance, if an author does not describe a few figures in enough detail in the text, your review could point to other examples that were described well in the present or even other manuscripts. By showing good examples, the reasons for your criticisms may become more obvious to the authors.

Rather than writing comments such as "awkward" or "unclear," provide more specific comments such as "If I understand what you wrote, you meant . . ." or "Do you mean to imply that . . . ?" By refocusing comments into questions, you demonstrate your willingness to understand the text rather than criticize it.

Table 20.2 How to recommend substantial changes to a manuscript

Sandwich criticism between more positive surrounding comments.
Explain why the change is recommended.
Provide supporting evidence for the change, including citations to literature.
Suggest possible experiments or additional calculations.
Use positive examples from the author's writing to motivate revisions.
Offer a suggested revision.
Appeal to the readers or audience of the manuscript.
Indicate why the change would benefit the manuscript.

How specific and thorough should you be when writing minor comments? The answer depends on the reviewer. How patient are you? How much time do you have to write an extremely thorough review? Some reviewers relish careful and thorough point-by-point reviews on every typo. Others just tell the author to clean up the manuscript, citing representative examples.

Remember that being a reviewer means respecting the intellectual independence of the authors. Discriminate between recommendations that are required for improved readability versus those where reasonable people can disagree (e.g., split infinitives). Said differently, Alley (2000) posits a scale of "errors that would unsettle many readers" to "errors that would distract only a few readers." Be cautious when asserting too much stylistic control over the manuscript; recognize that the English language is not constant in time (e.g., the verbose style of technical writing is becoming outdated) and in space (e.g., different journals require different syntax).

However you choose to write your review, number your comments sequentially and identified by location within the paper, particularly on points for which you want a response. For example, "1. Page 4, line 13: Delete `it has been noted that.'" Trying to respond to a paragraph of rambling comments by a reviewer is difficult. Other expressions that are commonly used in identifying locations within the text: "3 lines from the bottom of the page," "third paragraph, line 5," and "page 5, line 4, and elsewhere throughout the manuscript" (to avoid citing numerous similar changes that need to be made).

Finally, review the paper in front of you, not the one that you wished the authors had written! We have all been disappointed by manuscripts that promised to deliver one thing (or we wanted to deliver one thing) and delivered something else. Use these opportunities to ask the authors if they had considered your ideas, but do not mandate it. Eager authors may be inspired to do the work, but the editor may not require it. Alternatively, consider this an opportunity to add a new scientific colleague and work with them to develop your ideas, perhaps as a follow-up to the manuscript being reviewed.

20.6 TO BE OR NOT TO BE ANONYMOUS

Much has been written about the strengths and weaknesses of peer review. Currently, anonymity is one of the key tenets of the peer-review process. The argument goes that, without anonymity, reviewers are not free to criticize papers being submitted without fear of reprisal. Some journals have variations on the anonymous reviewer. For example, some journals have tried double-blind peer review, where the author is also (supposedly) unknown to the reviewers. Stensrud and Brooks (2005) discuss their experimentation with the double-blind review process at *Weather and Forecasting*. Another variation is at *National Weather Digest*, where the reviewers are encouraged to reveal their names in the review. Yet another variation is *Atmospheric Chemistry and Physics* where nonanonymous reviewers, in addition to the anonymous formal reviewers, can comment on submitted manuscripts in a public forum. The *Electronic Journal of Severe Storms Meteorology* makes all the reviewers and their substantiative comments public. Even *Nature* (2006) tried a form of open peer review.

Despite all its problems and the fixes that have been experimented with, anonymous peer review still stands as the principal means by which manuscripts are judged appropriate or inappropriate for publication. One might invoke Winston Churchill's witticism to describe peer review as well: "democracy is the worst form of government except all the others that have been tried."

Although in most instances reviewers remain anonymous and good reasons exist for doing so, some situations may arise where revealing your name is beneficial. For peer-reviewed articles, when should you reveal your name as a reviewer?

- To expedite closure on a paper by encouraging discussions outside of the peer-review process between the author and the reviewer.
- If the author is a colleague and you know that your review would be welcome or could jumpstart further rewarding discussion.
- If the review will expose you as the reviewer through citing your own research or sharing specialized knowledge in the review.
- To get credit for your contributions toward improving the article.

Some advocate that anonymous reviews are self-serving and recommend that all reviews be nonanonymous. Whether you decide to relinquish your anonymity or retain it is largely your personal preference.

20.7 PROVIDING COMMENTS TO OTHERS

Although much of the time you will be masked behind the cloak of anonymity, you may be asked personally to review a manuscript for a colleague or you may provide comments on a manuscript for which you are coauthor. For example, some laboratories or research groups expect manuscripts being submitted to journals to have passed either an internal or external review. If you are perceived to be an excellent writer, you may be asked frequently for your advice. What is the best way to review your colleagues' work? What pitfalls might you face?

First, find out two things: the time frame within which the author would like comments and what level of editing is expected from you. Knowing what level of editing is expected helps you gauge how much effort to put into the manuscript. Is he or she anxious to submit the manuscript by the next week (implying that only minor revisions can be expected), or can you take more time (ensuring that you can provide a more thorough review)? If the authors show you an early draft requesting your general impressions, then do not spend much time on sentence-level editing.

The first read of the manuscript can be an important time for you to assess the state of the manuscript and determine if the authors are realistic in their assessment of how close the manuscript is to submission. Recognizing what stage of revisions within the writing/editing funnel is needed to improve the manuscript (e.g., paragraph level, sentence level) will determine how much time you need to give comments. As discussed previously, revisions should be carried out at the largest scales first. Usually when I first meet with an author requesting an informal review, I start with the biggest issues first. Only after several rounds of revisions do we fine-tune the text on the sentence and word levels. Approaching the time of submission, we make only small-scale revisions.

Once you have made your comments, several questions remain.

Should I make my edits on paper or edit online? Editing on paper is best when you are working with a colleague and want the colleague to physically transfer your changes from paper to the electronic document. Such an approach is useful for working with students or less experienced coauthors who would benefit from your mentoring. In contrast, editing electronically is most efficient when a large number of small edits need to be made to a document. This approach works best late in the editing process or when deadlines are fast approaching. If using Microsoft Word, you can use "track changes" to indicate your suggested changes.

Should you resolve issues in person or by e-mail? Most people will respond better to constructive criticism when you can sit down and walk them through your changes rather than dumping on their desk a manuscript bleeding from corrections. The meeting allows you to gauge whether the author agrees with your suggested revisions, and perhaps why your suggestions should not be adopted. Not everyone will respond to criticism in the same way, which is one of the principal challenges to collaboration.

If multiple people are commenting on a manuscript, is it better to edit in parallel or in series? When more than one person is working on the manuscript, the author may have difficulty managing the different revisions. Parallel editing (sending the manuscript to everyone at once) minimizes the wait for comments, although resolving conflicting comments from the reviews and juggling different versions of the manuscript may be more challenging for the author. Serial editing (sending the manuscript to reviewers in sequence) lays out a more clear chain of command, perhaps from those most involved in the research to those just providing informal reviews. When edits will be made directly to the manuscript, serial editing is more efficient.

RESPONDING TO REVIEWS

21

The reviewers have provided their input, and the editor has made the decision to return your manuscript to you for revisions. How do you make revisions and respond to the reviewers to maximize your chances of publication? What do you do if your paper is rejected? What can you do if you are dissatisfied with the review process? This chapter deals with these and other questions.

Manuscripts sent out for peer review almost always are returned to the author with revisions requested by the editor and reviewers. Thus, authors should be prepared to receive comments. If these suggested revisions are particularly detailed, thorough, or harsh, such comments may sting initially. Remember those comments will improve the manuscript into something that you can be proud of years in the future. Criticism is rarely so severe that it cannot be addressed, resulting in a much improved manuscript.

If you are troubled by the reviews, set them aside for a week, so as not to focus on the negative. Do not do anything you might regret. Vent your anger to your friends, colleagues, or hamster, but do not respond immediately and angrily to the editor. Eventually, your better self will begin to realize that changes can be made to improve the manuscript.

21.1 MAKING REVISIONS AND WRITING THE RESPONSE

Comments from reviewers usually fall into one of four categories:

1. Comments that you agree with and are easily or sensibly addressed through revisions.

THE SEVEN STAGES OF EMOTIONS DURING MANUSCRIPT REVISION

1. **Shock** at receiving negative reviews.
2. **Anger** at reviewers.
3. **Fear** of not getting the manuscript published.
4. **Bewilderment** at some of the comments.
5. **Frustration** at making revisions and responding to the reviews.
6. **Acceptance** of having to do the work to improve the manuscript.
7. **Happiness** with a revised and improved manuscript.

2. Comments that may be more difficult to perform, but raise important concerns that need to be addressed to improve the manuscript.
3. Comments that are a matter of style or taste between different people and are easily rebutted in your response to the reviews or incorporated into a revised manuscript.
4. Comments that misinterpret your manuscript or are incorrect because reviewers are fallible or may not have understood your manuscript. These comments occasionally happen, but also consider the possibility that you might not have carefully explained yourself in the manuscript and have confused the reviewer.

As you start to make revisions to the manuscript, two strategies can be adopted. The first is to address the easy and minor comments first. Doing so provides a quick sense of accomplishment and prepares you to address the tougher comments. The second strategy is to address the major comments or the comments that require the largest changes to the manuscript first. This strategy is more consistent with the writing/editing funnel approach, which argues for making the biggest revisions to the manuscript first.

However you address the revisions, you need to determine whether the revisions recommended by the reviewers should be incorporated into the revised manuscript, and how difficult the revisions would be. You should also consider whether such revisions would affect other parts of the manuscript or whether the revisions would introduce any inconsistencies with other parts of the manuscript. Even if you disagree with the reviewers, consider what weaknesses in the paper gave them those impressions. Would this view be widely held in the community? If so, then responding positively to each of their comments improves the readability of the paper for your audience and greatly increases the probability of it being published.

When writing your response, you may wish to ask whether your editor has a preference for how your response should be structured. One approach is to quote all of the reviewer comments in your response, then address them point by point. Such an approach can ensure that you have addressed all the comments and makes it easier for the editor to see how you have responded to the reviewers.

The editor is your colleague in trying to get your manuscript published, not your adversary. The editor is generally willing to provide advice or specific instructions about how to address reviewer's concerns. Ask the editor for clarification from the reviewer when necessary.

When you can guess who the reviewer is, proceed through the peer review as though you do not suspect. Some reviewers may not like being "discovered." Or, you might even be wrong. For example, Peter Csavinszky, a physics professor at the University of Maine, would adopt the style of others in writing his review, sometimes assuming a Chinese, Bulgarian, or Spanish accent; or Csavinszky might list references that should be included, highlighting those that are not his own, calling them "pioneering" or "seminal," to throw the authors off the trail (Brownstein 1999).

When returning your revised manuscript and the responses to the reviewers, also submit a cover letter to the editor. Cover letters can explain in general terms what revisions were performed and why. Cover letters can also

ASK THE EXPERTS

STRATEGIES FOR RESPONDING TO REVIEWS
Roger Samelson, Professor, College of Oceanic and Atmospheric Sciences, Oregon State University, and former editor, Journal of Physical Oceanography, *and former editorial board member,* Journal of Nonlinear Science

1. Identify each conceptually independent reviewer comment in some straightforward way: quote the first few words, refer to numbers if given, etc.
2. For each such comment, list the corresponding revisions made to the manuscript, by section, page number, etc., as specifically as possible. If practical, quote added or edited passages in the response. Avoid general statements, such as "Section X was rewritten to address the reviewer's comments," that give little specific information as to what changes were made and to which paragraphs and sentences.
3. If no revision was made in response to a comment, say that and explain why. Recognize that if a detailed response to a reviewer comment is necessary, the inclusion of at least some portion of the response in the revised manuscript is frequently merited, even if it is a rebuttal. Most of the questions that occur to reviewers will occur to other readers.
4. In general, focus on clearly identifying what revisions were made, or requested but not made, and explaining why they were or were not made. Avoid responses that are not clearly tied to specific changes in the text or figures or that do not specifically rebut certain suggested changes.
5. Include an introductory statement that briefly outlines the main elements of the response, especially if major changes were made or suggested but not made. This statement can be helpful simply because it allows the editor or reviewer to estimate quickly how much time the review will take to complete.

be useful for discussing issues with the editor that are not to be shared with the reviewers.

When revising, never take for granted that an initially favorable decision leads to a final favorable decision (i.e., publication). In 2006, 89 (13%) of the 685 manuscripts that were rejected from AMS journals were initially sent back to the authors for revisions. These rejections occurred because of one or more of the following reasons:

- ▶ Authors failed to address reviewers' concerns adequately.
- ▶ The reviewers and author may not have been converging toward a resolution as quickly as the editor would have liked.
- ▶ Authors said that they made revisions, but careful inspection of the original and revised manuscripts indicated that few substantive revisions were made.

Let me expand on this last point a bit. Submitting your manuscript to the journal is a privilege, not a right. It is a privilege that can be revoked by the editor or publisher of the journal. You are imposing upon an editor and several reviewers, all volunteers, to improve your manuscript for publication. Even the most critical reviews offer advice that can make your manuscript better. *To ignore their efforts and resubmit the manuscript with only minor changes is a blatant disregard for the time of others.* Most editors do not tolerate such behavior and your manuscript will be rejected. A similar infraction occurs when authors are found to be shopping around rejected manuscripts between journals. Because atmospheric science is a relatively small discipline (compared to physics or chemistry, e.g.) and your area of specialty may be even smaller still, chances are that some of the same people that knew about your original manuscript at the first journal will see it again at the next one. Not making major revisions to a rejected manuscript, whether or not it was submitted to the same journal, is simply unacceptable.

Thus, the colloquial expression "accept with major revisions" should rather be called "return for major revisions." Consequently, until the editor has said that the manuscript is accepted for publication, the author should not assume this outcome.

21.2 RESPONDING TO SPECIFIC COMMENTS

When responding to reviews, you may encounter some comments that do not warrant implementation in the manuscript as revisions. This section provides a list of those situations.

The reviewer asks you to perform additional analysis that would be tangential to your manuscript. If your manuscript has a clearly defined purpose in its introduction, this rebuttal is easily justified. Respond to the reviewer that you agree, but that such additional work is "beyond the scope of the present manuscript." If the proposed additional analysis is a useful suggestion, this might be the topic of your next manuscript!

The reviewer apparently missed one of your points. Sometimes the reviewers may identify a potential concern with the manuscript, which you felt has already been addressed. In this case, consider whether you have made your point as clearly as you could have. Perhaps rewording your point or restating it using different words one or two more times throughout the abstract and paper can help clarify your argument to a confused reviewer.

The reviewer gives you little substance to respond to. In your response, focus on the substance in your paper, doing your best to rebut the reviewer in a professional manner. Defend your ideas and show the failure of the reviewer's arguments. Focused and substantive responses usually will win over the editor in these situations. If possible, try to lump your concerns together in a prefacing statement, such as "I believe that major comments 1 and 2 raised by the reviewer were already addressed in the manuscript on pp. 4, 6, and 12 in the following excerpts. . . ." Using this approach, you can easily dismiss what might look like a long and rambling review with a few strong arguments right up front in your response. Then, dismantle the argument item by item, providing specific citations to the overlooked material. For the parts where you may not understand the reviewers' comments, do not know what the question is, or cannot follow the relevance of the argument, say so diplomatically.

The reviewer is hostile. In the rare situation where you must deal with a contentious reviewer, revisions may be time consuming and annoying, but you have to play the peer-review game, and unfortunately, sometimes you have difficult opponents. Hope that the editor sees through such reviewer antics and dismisses them. If you feel you need to say something, politely address your concerns about the reviewer's tone in the cover letter to the editor. Always maintain the moral high ground.

21.3 DIVERGENT REVIEWS

Sometimes your manuscript may elicit divergent opinions from reviewers. For example, the three reviews may recommend minor revisions, major revisions, and rejection. Editors receiving such reviews face a difficult decision. Why might your manuscript receive such disparate reviews?

- The manuscript may be so innovative that its significance is not appreciated by some of the reviewers. Manuscripts like this are truly rare, however.
- The manuscript may be controversial, and the reviewers may be split among different camps.
- One or more of the reviewers failed to give an adequate assessment of the manuscript because of inexperience, carelessness, laziness, expedience, or bias. For example, one reviewer may have recommended minor revisions on a manuscript that requires much more work, as noted by other reviewers.
- The most likely scenario is that the author has submitted an average or above-average manuscript, but has failed to adequately state its purpose and utility. When this happens, reviewers can provide widely varying recommendations. Some reviewers may see the manuscript only needing minor revisions, but more discerning reviewers will see the underlying problems. One reviewer may want one improvement, whereas another reviewer wants the opposite. With such manuscripts, the author needs to more clearly present the material inside the manuscript and defend it.

In all of these four situations, conscientious editors with insight and a strong will can sort through the conflicting assessments and make a reasonable decision.

21.4 DEALING WITH REJECTION

In the unfortunate event of your manuscript being rejected for publication, what can you do? First, carefully read the editor's decision letter and the reviews. The editor may indicate what revisions are needed so that a substantially revised manuscript could be acceptable for publication (e.g., "we would welcome a revised manuscript"). Alternatively, the editor may indicate that the manuscript should not be resubmitted to this particular journal. Sometimes, editors may reject papers that they wish to see published because they think that the authors need an exceptionally long time to adequately make revisions. An e-mail or phone call to the editor may be appropriate for further clarification. In no way should authors berate editors about their decisions. Editors rarely respond well to such behavior.

Rejection is sometimes a blessing in disguise—it frees the author to make improvements without having to meet a publication's deadlines, and the author is advantaged by the reviews and editor's perspectives. —Mary Golden, Chief Editorial Assistant, Monthly Weather Review

Manuscripts are rejected for many reasons. The most common reason is that the science is of poor quality because the manuscript has made some fatal flaws or has failed to make a contribution to the science. Sometimes the manuscript is out of the mainstream of what is acceptable research. Perhaps the manuscript is written by people outside the discipline or with a specific agenda. Another leading cause of rejection is that the manuscript is poorly written. Perhaps the manuscript is disorganized, or the English grammar is so poor that reviewers cannot understand the science. Sometimes, the manuscript is not appropriate for the journal to which it was submitted.

Occasionally, even good science may alarm some reviewers, and they will reject a paper. If this is you, do not be disheartened. Consider the recommendations of the reviewers and editor seriously, revise the paper, decide on a strategy (same journal versus different journal), and resubmit the revised manuscript if you think the work is good enough. If you resubmit to the same journal, state in your cover letter that this manuscript was previously rejected and include what revisions you have made to address the reviewers' concerns.

This last point is important. Even if it is not required, I encourage authors to write a response to the original reviewers and send that response to the editor when they resubmit to the same journal, as if they were resubmitting following major revisions. There are several advantages of this approach:

- Because writing the manuscript helps clarify your argument, writing a formal response benefits the clarity of the revisions to the manuscript.
- The authors get their say against hostile reviewers.
- The response allows the editor to weigh the relative merits of the reviewer's comments versus the author's comments.

Based on how effectively the editor thinks the author rebutted the reviewers, the editor can choose to send the revised manuscript back to the original

PERSISTENCE AND PRECEDENCE

On 31 May 1988, Prof. Peter Grünberg's newly submitted manuscript was received by the prestigious American Physical Society journal *Physical Review Letters*. After review, the paper was rejected. Undaunted and believing that his research was definitely publishable, Grünberg revised the manuscript and submitted it in December to a different journal published by the American Physical Society, *Physical Review B*. The paper was eventually accepted and published in March 1989 (Binasch et al. 1989), but Grünberg insisted that the original submission date to *Physical Review Letters* be the one listed on the published article.

Peter Grünberg won the 2007 Nobel Prize in Physics for his paper in *Physical Review Letters* describing Giant Magnetoresistance, or GMR, the effect that enabled a huge increase in the storage capacity of hard-disk drives in computers, digital cameras, and digital music players. He shared the award with Albert Fert, who had independently discovered the same effect, but had submitted his paper to *Physical Review Letters* in August 1988, where it was published in November 1988 (Baibich et al. 1988). With the earlier submission date on his manuscript, Grünberg was able to patent GMR, despite Fert's paper being published first. Consequently, Grünberg's patent has earned him millions of euros in royalties from commercial applications of GMR.

reviewers (with the response) or choose entirely different reviewers who are unfamiliar with the history of the manuscript.

It is rare, but the review process can go badly. Reviewers and authors may get hostile, and the editor may do little to resolve the dispute and keep the dialog constructive. A manuscript may be rejected without adequate justification. If you think you have been wronged during the review process and the editor is unresponsive, the author should go to the chief editor, the publications commissioner, or the publisher. Be warned, however, that taking matters to this level should only be reserved for the most egregious errors in the review process. Alternatively, you may wish to turn your back on that journal, at least until that editor is gone, and try someplace else that might be more receptive.

Amid all these issues with the review process, remember that editors represent the journals. Their job is to ensure that only appropriate and high-quality manuscripts are published, ensuring the reputation and success of the journal. Editors serve the authors, the reviewers, and the publishers. Balancing these competing interests sometimes leads to difficult decisions with few easy resolutions.

PREPARING AND DELIVERING
SCIENTIFIC PRESENTATIONS

HOW SCIENTIFIC MEETINGS WORK

22

Interactions with others at scientific meetings can fuel new research ideas, provide feedback on preliminary research results, and test-drive the science before the manuscript is finished. Meetings can also be used to market your work, look for jobs, reconnect with friends, and travel to new places. Unfortunately, meetings are also places to sit in overly air-conditioned conference rooms, be bored by others who presented the same material at the same workshop two years ago, and get behind on your work back home. How do meetings get organized? How do you choose the best meetings at which to communicate your science? Which are right for your career? This chapter discusses these and other aspects of scientific meetings.

Scientific meetings are expensive to attend (about $1500 or more, depending on location), each attendee requires a large carbon footprint, and most meetings consist of many poor-quality and uninteresting presentations. Despite these problems, scientific meetings (e.g., conferences, workshops, symposia) are the primary way by which new scientific results are communicated in person between scientists. Meetings can be useful for networking and meeting face-to-face with people you would not have had the opportunity to meet otherwise, especially friends and colleagues that you have not seen in several years. Moreover, you may even spend productive time with colleagues from your home institution because you finally have a chance to get away from the daily responsibilities of work back at the office and participate in social activities together.

Large conferences usually consist of sessions during the week where 10–30-minute oral presentations are given by participants. Usually the number of submissions exceeds the 40–60 slots that such a format can accommodate, so poster sessions are also held in large halls where participants stand in front of a poster of their research and talk with passersby. If the meeting is even larger (200 people or more), parallel sessions may be held where two or more presentations may be occurring at the same time in smaller rooms. Meetings may also have panel discussions, group discussions, break-out sessions, or other activities. In addition, many meetings include social events such as icebreakers with hors d'oeuvres, trips to a local historical site, and banquet dinners.

Although this chapter and the rest of this part of the book discuss presentations from the perspective of scientific meetings, the principles discussed are relevant to more than just that. Your presentations in the hour-long weekly department seminar series, your Ph.D. defense, and even an impromptu research group presentation all will benefit from the information within.

22.1 HOW MEETINGS ARE ORGANIZED

Conferences are typically organized by some kind of sponsoring organization, such as a professional society or a government agency, and run by a scientific advisory group. In the AMS, these groups are called the Scientific and Technological Activities Commission (STAC) committees, and their members are drawn from across the AMS to represent certain fields (e.g., atmospheric chemistry, weather analysis and forecasting, mountain meteorology). These advisory groups are composed of up to a dozen people, of whom just a few may actually do work for the conference. These groups plan the meeting and determine the theme, format, and program. The sponsoring organization usually provides staff to arrange the logistics of the conference: everything from renting the conference facilities to ensuring name tags for attendees. Some conferences, especially the smaller workshops, may be run by individuals. In these cases, the organizers or their affiliations often handle the logistics.

A *call for papers* usually announces the meeting to the scientific community. The call for papers is a description of the theme of the conference, topics upon which the meeting may be focused, and instructions on submitting an abstract for a potential presentation at the conference. Knowing who is organizing the meeting and the style of the meeting can help you plan whether to attend the conference, whether to submit your research, or what research topic to submit. Some meetings may accept a large percentage of talks from early career scientists, whereas other meetings may favor more senior scientists. You can find out more about the type of conference from people who have attended previous meetings or from the conference organizers. You may ask the conference organizers whether your proposed topic will fit into the meeting's theme. The style of the meeting may also give you an indication of whether you are likely to receive an oral or poster presentation. If you are aiming for a poster to get the maximum feedback, you may wish to know the format of the poster session and how long your poster will be displayed.

22.2 PICKING THE RIGHT MEETINGS TO ATTEND

The costs of attending meetings should be properly weighed against the costs of staying behind, and these costs go beyond the monetary cost to your affiliation or research grant to attend a conference. Considering the absence from the office with e-mail and other work piling up, the time away from your family, and the cost to the environment, a reasonable question to ask is whether conferences are worth it. If they are worth it, what are the best ones for you to attend?

Meetings, especially for early career scientists, are places to get exposure for you and your research. Those attending conferences are usually viewed as being active members of a scientific community, and promotions and job opportunities may go to those who present. A well-presented research project may spark someone in the audience to offer you a job or a research collaboration.

Meetings come in all shapes and sizes, from two people to thousands. The smallest meetings may arise from a personal invitation to visit another laboratory or university to give a seminar and have one-on-one interactions with a colleague or research group. Such intimate meetings are often the most effective forum for collaboration. Workshops of 20–30 people may be focused on tackling a specific problem or organizing a field program. The smallest conferences (50–100 people) with their focused topics, number of people, and shorter length provide the most effective and interesting potential for scientific interaction among the conference-style meetings.

Meetings the size of several hundred people usually offer a greater diversity of topics, at the expense of being personal. Earlier in your career, these larger meetings may potentially expose you to more people and their research, but establishing yourself may be harder unless you know organizers personally. The largest meetings are composed of many different conferences and symposia (such as the annual meetings of the American Geophysical Union, AMS, and European Geosciences Union), and you have the opportunity to be exposed to a scientific discipline you might not have seen before. On the other hand, so many sessions are running simultaneously that attending all the sessions you want to see may be a challenge.

Although attending such conferences can be an isolating, lonely experience if few of your close colleagues attend, sometimes venturing out from the standard conferences in your field can be valuable (e.g., the sidebar on

I just can't write a decent paper that hasn't at some stage been a talk. If I haven't prepared to stand up and point to the key graphs and points, in a cultivated order, and talk an audience through the sense of it all, then that sense never quite comes into being. The paper remains a dead jumble no matter how long I bash around editing it. I used to think the key was audience feedback. But actually the important audience is my writing brain. I have to explain my science to my writing brain out loud. Once the writing brain has been duly informed, it takes over. —Brian Mapes, University of Miami

COMMUNICATING WITH DECISION-MAKERS AND OTHER STAKEHOLDERS

Eve Gruntfest, Director, Social Science Woven into Meteorology, University of Oklahoma, and Professor Emeritus, Geography and Environmental Studies, University of Colorado at Colorado Springs

Communication outside of the meteorological profession is important for meteorologists. Many governmental and nongovernmental agencies benefit from collaborations with meteorologists, but too frequently the meteorologists and decision-makers come from different cultures that use different languages.

Meteorologists are fond of their technical terms—the scientific language helps advance the science and also gives the meteorologists a group identity. As is true in other scientific communities, connecting via acronyms and niche terms helps distinguish one scientific subspecialty from another.

Nevertheless, professional development and communication with decision-makers and stakeholders requires learning a new language. When meteorologists want their research results to have impacts outside of meteorology—in the world of policy makers, other scientific communities, or with public groups—they need to speak a different language. They need to temporarily unlearn the meteorological language to connect effectively. It is not easy to shift gears, and the process can be frustrating, but taking the time to understand how your words are heard by others can be tremendously rewarding. The payback comes in the interaction between you, as a scientist, and the larger community that can use, learn from, and inform your work.

The Weather and Society*Integrated Studies movement (WAS*IS; Demuth et al. 2007) is building bridges both between the cultures within meteorology and between meteorologists and a wide range of stakeholders. As of August 2009, 200 official WAS*ISers have participated in multiday workshops to introduce social-science concepts and tools to meteorologists. One of the main topics covered in the workshop is communication without acronyms. In the United States, Hurricane Katrina appears to have mobilized early-career meteorologists toward more societal impacts work. This disaster showed that even well-forecasted events can have devastating societal impacts.

How can you develop your communication skills to speak more generally to decision-makers and other stakeholders? In addition to changing wording and speaking patterns to be more understandable to a wider community, participate in meetings and conferences outside your usual professional contexts. For example, meteorologists who participate in floodplain management meetings, community watershed meetings, local National Weather Service workshops, or the National Hydrologic Warning Council conference see how weather research is recognized and what problems different professional and public groups consider most pressing. You also can find new collaborators who are approaching questions similar to yours, but in different ways.

By attending these meetings, you also can get a more comprehensive picture of where your work fits. For example, the WAS*ISers were surprised to learn which types of weather information were most useful to stormwater managers and urban floodplain managers during flood events. On a fieldtrip to the flood site, meteorologists realized how hydrological processes complement precipitation and what sorts of information were most useful (or not useful at all) to city officials and managers during and following the flood event.

Making the effort to translate your scientific findings or research questions to reach broader societal and scientific communities can enrich your work in many ways. It is rewarding to see your work be applied as you intended it to be, and also, public interactions and engaging in the decision-making process can provide a richer set of considerations for how you frame your future questions and results.

page 256). Exposing your work to a different audience can open yourself up to interdisciplinary and multidisciplinary research that you might never have been involved in otherwise.

Who the other attendees of a conference are may contribute to your decision to go. Talk to people in your field to find out which conferences are the ones most worthy of your attendance. Find one or two conferences where most of your discipline goes every 1–3 years, and try to attend these regularly to stay linked with your community. Meetings with lots of invited talks tend to be more highly attended because, presumably, the speakers were selected partially because they are excellent speakers. As your career develops, you will get invited to more meetings and be expected to go to more meetings. Remember that attending conferences takes time away from doing the work that gets you to conferences in the first place, so choose judiciously.

22.3 HOW TO BE A GOOD AUDIENCE MEMBER

When I look back on some of the most enjoyable meetings I have attended, their success was not only because of the high quality of the speakers, but also how engaged the audience was. Compelling audience discussion after a mediocre presentation can make the time spent seem almost worthwhile. Whether at a conference, your department's weekly seminar series, or a Ph.D. defense, "Be engaged in the talk," says Prof. Daniel Jacob of the Harvard Atmospheric Chemistry Modeling Group. "You're not watching TV; you're at work. Sit up front. Concentrate. Don't phase out." Mental engagement also will offset any tendency to drift off to sleep. Standing up in the back of the meeting room can prevent innocent visits from the sandman.

Show good manners. Should you need to leave early for another appointment or are expecting an important call on your cell phone, sit in the back or at least where you can leave gracefully and without causing too great a fuss. Let the speaker know that you will be leaving early. Put your phone on silent or vibrate, or better yet, turn it off.

The audience should be attentive and professional to the speaker. Avoid checking e-mail, grading papers, or skipping talks to check out the tourist sights until after the session ends. Nod in agreement with important points to show your support.

When a presentation ends with no questions being asked, my impression is that the audience was not actively engaged during the talk. Do not think of the talk as passive, quiet time. Use the time during the talk to repeat or summarize the speaker's material. Note-taking helps engage your critical thinking skills about what is important and worth remembering from the presentation. Take notes in your conference notebook, not only about what you learned, but also what questions you have for the speaker. By the end of the talk, you might have written down a number of interesting questions from which to draw the best one or two for the speaker.

Be respectful when asking questions. Do not ask questions to be aggressive or show off. Instead, ask questions to be helpful, amplify a point made by the author, provide a better explanation for the audience, or to take the audience on a discussion somewhere else, to contribute to positive discussions. If the question-and-answer session runs long, visit with the speaker afterward and ask your question. Speakers generally appreciate answering questions privately—it shows that the audience was interested in their presentation.

Before closing this chapter, I want to make a very clear statement about attending conferences. If you go, take advantage of the conference to the fullest extent. Every presenter at the conference deserves the same attendance and attention that you hope for during your presentation, even if the speaker is the last one of the conference. Most people have trouble sitting through four or five straight days of talks. Nevertheless, wait until after the afternoon sessions are over to hit the beach, or better yet, schedule a few vacation days *after* the meeting ends. As Orville (1999) argued, "scientists being supported by public funds to attend scientific conferences have a responsibility to attend and to contribute to the entire meeting. . . . Poor participation in scientific conferences is really a rip-off of public funds."

Finally, conferences are an opportunity to interact with more than just your friends, so do not be shy about meeting other people whose talks or posters got you thinking about your research differently. You may never know whose talk leads to a fruitful collaboration—you never know what poster may inspire your next big idea.

THE ABSTRACT AND EXTENDED ABSTRACT

23

Having decided to attend a certain conference, your ticket to receiving a presentation at that conference is the abstract. If your abstract is accepted into the conference, you may be asked to produce an extended abstract. What are the strategies in writing and submitting successful abstracts? Under what circumstances should you submit an extended abstract? This chapter discusses these and other questions related to the abstract submission process.

Because conference organizers determine the conference program from the abstracts that have been submitted, the abstract is the key to getting into the conference program. Therefore, a well-written abstract, especially if your name is not well known to the conference organizers, is essential.

Before submitting an abstract, consider the following:

1. Because abstracts are due many months before the conference, most abstracts are works in progress, written before some, if not all, of the conclusions have been reached. Make sure at least some research is ready to present by the time of the conference. As long as there are some interesting results to present, the work does not need to be completed at the time of abstract submission. Do not submit an abstract on research that has not even been started on the hopes of having the deadline of the conference presentation provide the inspiration to do the work. Such abstracts are called *fabstracts* and are generally unethical, particularly if you are requesting an oral presentation.

2. Research that has already been submitted elsewhere for formal publication or is even in press can be submitted (assuming that the work has not been published by the time the abstract is submitted). Such presentations can

be an opportunity to get feedback on submitted manuscripts or to market a forthcoming article.

3. Ensure you will have resources to attend. Do not fill up space in the conference program in place of someone who is definitely going to attend. If you make a habit of submitting abstracts and not showing up, conference committees will wise up and not include you in future conferences.

4. Unfortunately, some organizations will only fund people to attend conferences who have something to present. Such issues can provide challenging situations for scientists who are submitting abstracts and hoping to be in the conference program.

There are likely to be three audiences for the abstract, so your single abstract may have to satisfy multiple audiences, each with their own purpose for reading the abstract. The first group is the conference organizers. They are interested in reading the abstract to determine, first, if you will be accepted into the conference, and second, if your presentation warrants an oral or poster presentation.

The second group to see your abstract is the conference-goers. They may see the abstract in the conference volume or online. This group reads your abstract to determine whether they are interested in attending your talk. Thus, your abstract is an advertisement for your research.

The third group is the online audience. Many conference abstracts are bound in a volume or posted online for perhaps years after the conference ends. As such, your abstract documents your research results at a snapshot in time.

Balancing these three audiences with a single abstract can be challenging. In the next section, we explore how to satisfy the first group, conference organizers, but touch upon how to also inform and attract the second and third groups.

As a convenor, I work to serve the 400 listeners, not to rub the egos of the 16–50 people who submitted abstracts. I want to build the session into an aggregrate, finding talks that complement each other. This is not a competition—it is not so that the six best get a bow (like in a cat exhibition). —Elena Saltikoff, Finnish Meteorological Institute

23.1 CHARACTERISTICS OF A CONFERENCE ABSTRACT

To determine what conference organizers are looking for in an abstract, we need to go inside their heads. Organizers want an interesting, diverse, and entertaining meeting devoted to the advertised topic. Organizers tend to pick interesting research topics, good speakers, and speakers from a variety of different affiliations (so that any particular group does not receive too much attention). Therefore, if you wish to get invited to the conference, you need to address three things, in order of easiest to most difficult:

1. Your abstract must be relevant to the conference theme.

2. Your abstract must be of interest to the conference audience.
3. Your abstract or your prior reputation must convince the committee that you are capable of giving a high-quality presentation.

Consequently, the best abstracts have the following attributes:

▶ **New.** An abstract that presents new results is more likely to get accepted than the same material you presented at last year's conference.

▶ **Interesting.** The abstract should discuss the unresolved aspects of the research topic, express your contribution, and suggest that you have an exciting presentation to deliver. Because conference abstracts are not peer reviewed and the desire of most conference committees is to have a lively meeting with lots of debate, conference abstracts can be more brazen and controversial than the typical article abstract.

▶ **Focused.** Given the relatively short length of most conference presentations, you generally cannot present an entire manuscript, let alone your entire dissertation. Pick one or two points that you want to make in the abstract and spend time motivating, developing, and demonstrating these points in your presentation.

▶ **Informative.** Vague abstracts with little scientific substance, particularly results, are not attractive to the conference committee; the committee cannot risk taking the chance that your presentation will be devoid of content. Be specific about your results in the abstract. However, if you reveal all your results and conclusions, you take away the motivation for some in the audience to see your presentation, especially those that are only marginally interested in your topic. Therefore, you may prefer a more coy approach by not disclosing all your results in the abstract.

▶ **Understandable.** Because the audience will generally be much broader (e.g., more varied education, more diversity in research interests) at most conferences than the audience of your journal article, more background information, fewer technical details, and more motivation for why your work is important need to be presented in the conference abstract.

▶ **Enticingly titled.** The title should accurately describe the work, but also entice people to see it. The title should encapsulate the spirit of the presentation.

▶ **Well written.** The abstract should make compelling arguments, be well organized, and be free of grammatical errors and typos.

A look at this above list of characteristics of the abstract of a scientific conference suggests many similarities with abstracts for journal articles (Section 4.4), but do not confuse the two! Their purposes are different.

23.2 WRITING AND SUBMITTING A CONFERENCE ABSTRACT

Before writing the abstract, read the instructions. Your abstract should comply with all requirements put forth by the conference (e.g., submitted before deadline, no references in abstract, contact information of corresponding author). If there is a length restriction, the abstract should approach, but never exceed, the maximum length. Too short or too long are both bad.

With the characteristics of the abstract defined and the requirements known, writing the abstract is the next step. As with an abstract for a journal article, the conference abstract should contain information about the purpose and motivation of the study, data, methods, results, and some conclusions. Similarly, a conference abstract needs to convince the program committee that you have done the work you say you did and your research is not at a premature stage. Avoid the future tense. Make clear what material is going to be presented at the conference by the content of the abstract. Avoid vague statements like "results will be discussed" without saying what at least some of those results are.

Most importantly, the abstract must be well written. Start writing the abstract several days before the deadline. High-quality writing rarely happens at the deadline. Proofread the abstract, then exchange it with other colleagues going to the conference to get their feedback before submission. The abstracts may be bound into a volume, stored online, or put on a CD-ROM for conference attendees, so they become permanent public documents as well. You want to present your work publicly in the best possible light. Thus, submitting a high-quality abstract goes beyond just getting accepted to the conference.

Finally, never write the abstract from scratch directly into the online submission form. If the computer goes down, Internet connection is broken, or power is lost, you can lose your abstract. Although it may be possible for the conference committee to accept a late submission personally, never ever count on it. Instead, write the abstract in your favorite word processing program, and, when complete, paste the abstract into the online submission form.

23.3 ORAL OR POSTER PRESENTATION?

When you submit the abstract to the conference, you may be asked whether you prefer an oral or poster presentation. Most conferences receive too many abstracts to give everyone an oral presentation, yet most program committees are adverse to turning away people because attendance is money. Therefore, most of the abstracts at conferences with more than a few hundred attendees will be given posters, whether they have asked for one or not.

Decisions about who gets oral presentations versus poster presentations are usually made by the program committee. The program committee may have to sort through hundreds of abstracts to determine which ones would work best as oral presentations. Not every research project or even published paper deserves an oral presentation at a conference because the abstract may not meet the criteria in Section 23.1. Note that I apply no value judgment in saying that oral presentations are better than poster presentations. I have seen many worthless talks, and many excellent poster presentations.

Many conferences will not allow more than one oral presentation per author. Therefore, be judicious in your requests for oral presentations. If you are submitting more than one abstract, prioritize your submissions and do not request oral presentations for all your abstracts.

Request a poster if your topic is complicated material that would take too much time to explain or if your topic is esoteric or of interest only to a small number of people. Do not waste the time of 300 people at a conference if your message is only relevant to five specialists. You may also consider requesting a poster if you are attending a conference where you may not be comfortable with the native language. Finally, if you have a strong fear of speaking in public, you may request a poster presentation, where the opportunity to communicate one-on-one in front of your poster may be more comfortable for you.

23.4 THE EXTENDED ABSTRACT

Extended abstracts, preprints, proceedings, and postprints (all referred to as extended abstracts for simplicity) are documents published before or after the conference, documenting the work presented. They are typically a few pages (1–4) long, but with the advent of digital files stored online or on a CD, these page limits may be replaced by a maximum file size. Usually these extended abstracts are not peer reviewed and therefore should not be listed on your curriculum vita in the same category as peer-reviewed scientific literature. Because the extended abstract is not a formal publication, there is no prohibition against presenting the work at multiple conferences with different audiences to get different feedback.

Extended abstracts have multiple functions. For some who do not aspire to formally publish their research, they represent the end product of research. For others, they represent a stage along the way to formal publication. For these people, extended abstracts are valuable to help formulate their thoughts, serve as rough drafts of formal publications, get ideas out to others in an informal setting, or leave behind some documentation of the research if the work is never completed.

Because submitting extended abstracts for some conferences is optional, your decision of whether to submit an extended abstract may come down to one of the above issues. If you think your time is best spent working on the formal publication that will follow in a few months, perhaps documenting your work in the form of the extended abstract is not necessary. On the other hand, the extended abstract can serve as the motivation to draft your formal publication.

When constructing an extended abstract, the conference organizers will often provide a template to follow. If not, find a colleague who can provide the raw file from an earlier conference. Mimic this file, as it is easier to borrow the format and style than to create your own working template. You may also wish to see previous conference volumes, if available, to see the style, format, length, and tone of the extended abstracts.

Extended abstracts should contain the elements of a scientific paper, without the abstract. Previous literature can also be reduced or eliminated, as the extended abstract should focus on your work. If your talk presents a surprise ending, you may omit it from your extended abstract. However, keep in mind that the conference volume will not have documented your results, which will inhibit later investigators from discovering what you did if you do not eventually publish that research more formally. As with the abstract, reading "results will be presented in the talk" in someone else's extended abstract can be frustrating. Authors taking such an approach should be aware of the risks of alienating your readers or potential audience members.

ACCESSIBLE ORAL PRESENTATIONS

24

Most of the time at conferences is spent listening to others talk. Given the overwhelming challenge for the audience to sit through, absorb, and remember material from all these presentations, a speaker must be memorable to connect with the audience. As the previous chapters focused on scientific writing, this chapter starts by distinguishing written communication from oral communication. Oral communication requires a new set of skills to speak most effectively with the audience, organize presentations, and deliver presentations professionally and with style, the topics of the rest of this chapter.

Science in the early nineteenth century used to be the rock shows of today. People would pay to see the latest scientific, and pseudoscientific, presentations. Standing-room crowds of hundreds would cheer wildly for the presenter. Lectures were often accompanied by wonderful new machinery, explosions, and astounding physical experiments to entertain the audience, who were held in rapt attention by speakers with powerful speaking voices, physical energy, and personal rapport.

Think of the last scientific lecture you have attended. How similar was it to this spectacle? Can you even remember what the speaker's principal conclusions were? What has happened to us over the last 200 years?

24.1 HOW WRITING DIFFERS FROM SPEAKING AND WHAT THAT MEANS FOR YOUR PRESENTATION

A successful scientific presentation has many of the same characteristics as a successful scientific manuscript. Just as the science in an article should be worthy of publication, a speaker should be presenting material worth listening to. Like the author of an article, a speaker should also determine who the audience is and how best to reach them. Finally, similar to the purpose of a manuscript, the purpose of a talk should clearly frame its content. Despite these similarities, however, the differences are crucial to understanding how we create and deliver effective presentations:

- ⊙ **Presentations must be focused.** Generally, writing requires the author to elaborate on details and provide all the evidence supporting the conclusions. Speaking requires keeping the audience focused, which usually entails limiting the number and depth of details in the presentation.
- ⊙ **Presentations have more flexibility.** Articles have fixed layouts and delivery methods (print or online). In contrast, presentations are more flexible, with more ways to emphasize material, either through the speaker's delivery or multimedia content. And, including color figures does not cost extra.
- ⊙ **Presentations are received by a captive audience.** Articles are read by the audience at their convenience and at their own pace, sometimes again and again until the results are finally understood. Presentations are delivered to a mostly captive audience in a room at a fixed time and place. The audience members are beholden to the pace of the speaker, and they get a one-time viewing.
- ⊙ **Presentations involve feedback between the audience and the speaker.** Who reads your journal article is largely unknown, so the author has to write for an unspecified audience who cannot ask questions or provide feedback to improve the article. In contrast, the audience faces the speaker, and their feedback can be received in real time—facial expressions, approving nods, questions, notetaking, yawns, reading e-mail, talking on the phone, and booing—all are indications of the level of audience participation.
- ⊙ **Presentations can be provocative.** Whereas scientific journal articles are peer reviewed and permanent, speaking is not. Presenters can be more informal, provocative, and controversial.
- ⊙ **Presentations can contain fresh content.** Once published, the content in the article is fixed. Unlike writing where the content is months or years old, the content of a talk can be only minutes old. Presentations can be updated for variety, spontaneity, different audiences, or different occasions.

For these reasons, presentations are not spoken summaries of a paper! Effective speakers know how to take advantage of the above differences to create a focused, flexible, attention-commanding, interactive, provocative, and fresh presentation. Before you jump in and start composing your PowerPoint presentation, consider the following seven admonitions in this chapter, which, when ignored, plague many poor presentations.

24.2 FOCUS YOUR MESSAGE

Speakers need pithy presentations because time, especially for conference presentations, is quite limited. If focusing your message and concision were important lessons in Part I on writing, they become essential for presentations.

Consider the following. Reading a journal article can take several hours. Reading that same article aloud to a seminar audience would take much longer. Thus, scientific presentations by necessity cannot have the same level of detail and complexity as the scientific articles upon which they are based.

Because you cannot go as deep into material in a talk as you can in a manuscript, keeping the audience focused will require you to eliminate details that might be essential for duplicating your results, but otherwise would be a distraction to most of the audience. Such an approach may sound dishonest, but the audience can ask specific questions about the approach in the question-and-answer session, talk to you afterward, refer to the extended abstract, or wait for the published manuscript.

Most speakers overestimate the importance of their material to the audience, thinking that they can condense their manuscript into a ten-minute talk. Many manuscripts make several points, and not all those points can be adequately justified to the audience in an oral presentation. What one or two things do you want your audience to remember? A rule of thumb from Charles Doswell is that five minutes are generally required to deliver one point of substance. If you are giving a 10-minute conference presentation, that is two points, maximum. Having decided on these key points, build the talk around them.

Avoid trying to present too much content, which either ends up rushing the presentation or forcing you to speak too long. Scrutinize the necessity of

I maintain the focus of the audience on my talk by my visualization techniques, by varying my voice, by moving around, by using hand gestures, by inserting one-liners (which actually are devices used to gauge attention, as well as entertainment), and by involving the audience (making them give me feedback, asking them to guess, to vote on alternatives I give them).
—*Robert Fovell, University of California Los Angeles*

LEARNING A LESSON–THE HARD WAY

I was once asked to give a talk about snow to a local ski club. I took my hour-long scientific presentation on snow microphysics and reworked it for a nonscientific audience. The result was a Titanic disaster, and when I realized I was sinking, all I could do was rearrange the deck chairs. By the time my talk was over, the audience had mentally checked out early and had become amused by this nerdy scientist talking over their heads.

Why did I fail? I failed to consider the audience, a group of nonspecialists who did not want to know about the details of the Bergeron–Findeisen effect (surprising, I know). I failed to focus my message— "The vertical profile of the atmosphere determines the type of snow that falls and how good the skiing will be." I failed to build a new talk from scratch around this point, only employing graphics from my prior scientific talk in hand. And, importantly, I failed to add some levity to the talk, given that this was a club that liked to have fun.

every slide. Does it add to the content of the talk, or are you keeping it because you cannot bear to remove such a nice-looking slide?

24.3 KNOW WHY YOU ARE GIVING THE TALK

As with writing a manuscript, having a clear definition for the goals of your talk *before* composing the talk is essential. Why are you giving this talk? Why were you selected and not someone else? Was it invited? What is the topic? Are you trying to persuade forecasters to adopt your methods? Or, were you looking for your laboratory to pay for your trip to this Hawaiian conference?

What is the reason for the talk? You may want to inform, persuade, confront, inspire, educate, or some combination of these. Different types of talks require different approaches. Whatever the purpose, speak to the occasion. If the occasion is to honor a venerated scientist, make sure you say something about how the honoree inspired you. If you were invited to train forecasters, make sure you give them usable information to be better forecasters. When strong action is required to shake up or inspire an organization, Prof. Cliff Mass of the University of Washington, a master of such presentations, advocates, "Never go ad hominem or personal." Instead, he argues that a stronger, more insistent, and serious tone is needed than with a traditional scientific presentation.

For meetings, find out more about your position in the schedule. If you have been invited to give a presentation and are opening the session, your remarks can be more introductory and forward-looking, with a tip of the hat to the people that follow you. If you are the last speaker in a series, try wrapping up the comments and building connections among the various speakers that preceded you. You may even wish to contact the other speakers before the meeting to ensure that everyone's messages are complementary rather than redundant.

Are the slides for you or the audience?" —Terri Sjodin, youtube.com

24.4 ADDRESS YOUR AUDIENCE

Ask yourself, "What does this audience want from me? And, why is it important?" Then, figure out how to connect with them. Respect your audience. Do not show contempt or disregard for them by not understanding their needs. They took the time listen to you, so make sure their time was well spent. Remember that you are trying to impress them.

As with papers, your audience will determine your presentation style and content. If you are presenting to an audience of nonspecialists, you need to alter the standard scientific presentation you would give to your colleagues or peers. What background information do you need to present? What jargon do you need to define or eliminate?

Do not overshoot or undershoot your audience. At the NOAA/National Severe Storms Laboratory, we would have the occasional visit from an administrator from NOAA headquarters who would tell us about all the great NOAA initiatives that were going on, often things that we already knew were happening because they were *our* projects. In another situation, we had a speaker come to the lab and give a presentation about climate change, talking to us as if we were high-school students. If you do not know your audience, do your best to find out from the conference organizer or the sponsor of your visit before you arrive.

If you suspect that the audience might have a strong negative reaction to your presentation, avoid biasing them against your material too early in the talk. Present noncontroversial material early, then systematically reveal the discrepancies with the current thinking until they have no choice but to agree with your overwhelming evidence.

24.5 DELIVER THE CONTENT AT THE RIGHT BAUD RATE

The human brain has a high capacity to process information. Unfortunately, to connect the speaker's brain to the brains of the audience members requires communication through a narrow channel that delivers information both verbally and visually. The baud rate of this information channel needs to be carefully managed. If the speaker transmits more information than the narrow information channel can accept, the audience will not receive the information. On the other hand, a speaker with a small baud rate underutilizes the information channel, and the audience members' brains idle and start to daydream.

Planning an effective talk must take into account the volume and rate of information that the channel can support, which will be a function of the education level of the audience, the level of material being presented, the content and quality of the presentation and presenter, how fast the presenter speaks and displays information, and how distracted the audience is. Furthermore, your audience will be applying filters related to their own backgrounds, experiences, and values. Some of your messages may be understood quite clearly, others may not be. Incorporating that knowledge into the design of your presentations will ensure more reliability in transmitting information.

It is far better to be understood by your audience—even if you convey less information than you hoped—than to convey everything you intended and be incomprehensible. —Stephen Benka (2008)

24.6 CREATE A SYNERGY BETWEEN YOUR WORDS AND YOUR VISUALS

The mind processes information through all the senses. For presentations, these are generally verbal and visual. To take maximum advantage of the brain's processing capability, the speech and slides need to complement each other rather than contradict each other or be redundant. The brain cannot process information when it arrives both written and verbally at the same time, resulting in the following failures:

- Slides with too much text spread the audience's attention between reading and listening, so that they do neither well.
- If the text is read verbatim off the slide, then the speaker is redundant because the audience can read the slides faster than they can be said aloud.
- Having few connections (or even inconsistencies) between the speaker's words and the material on the slides confuses the audience, reducing comprehension.

Thus, the spoken word and the visual cues on the slides must be synchronized. The best approach is to favor relevant photos and graphics over text on the slides, do not read the slides verbatim, and speak articulately about the material on the slide.

24.7 UNDERSTAND THE DISTRACTIONS TO YOUR AUDIENCE

Your goals as a speaker are to connect with the audience, hold their attention on your topic, and help them remember it. How effectively you can do that is determined by the presentation quality, the presenter quality, and the audience quality. You can imagine that the best presentation given by the most energetic lecturer would still fail to connect if the audience were distracted, uninterested, or asleep. Although some of the factors that lessen the audience's ability to pay attention may be out of your control, others are entirely within your control (Table 24.1).

As you speak, watch your audience. Get a sense of everyone in the room, not just a few individuals. Some people will fall asleep no matter what, so do not judge your performance too harshly based on them. Does your audience seem attentive? Do they look confused? As the speaker, you must take control.

Table 24.1 Factors that lessen an audience member's ability to pay attention

Boring speaker
Topic not of interest
Before coffee break or meal
After a meal
Late in the afternoon
Laptop, e-mail, or cell phone
Illness
Personal issues and distractions

Back up and reiterate your point using a different approach. Ask a question of the audience to wake them up, to get them to actively participate in your presentation, and to get feedback on how well your message is being received.

24.8 ADDRESS EVERYONE WITHIN A DIVERSE AUDIENCE

Most speakers must balance two competing effects. While oral presentations require a focused message, most audiences usually have a diverse background. The speaker must therefore balance a lot of detail (narrow, but deep) with a wide perspective of the research (broad, but shallow). This lack of consideration of the depth and breadth of the presentation plagues many conference presenters who focus, for example, on describing intricacies of the data collection methods or the simulations. Unfortunately, these details may appeal to only a few people in the audience, while the majority of the audience is bored, left unappreciating the potentially interesting reasons for the study or the implications to the larger research community.

Communicating with your audience in these situations will require you to broaden your material, to make it more interesting to more people. Do not worry about speaking too long at a general level for the specialists. Most people probably would rather spend their time in a well-presented but general talk than a poorly presented but specific talk.

Broadening is not your only possible strategy when speaking to a heterogeneous audience. Speak to their diversity throughout the talk (Fig. 24.1). Start out by discussing the topic in a way that everyone can understand. As the talk progresses, dive down to depths at various points, reaching more specialized portions of the audience. At the end of a topic and especially at the end of the talk, come back out to the big picture. Connect what was just learned back to the whole audience, so even nontechnical audience members know the implications of what just happened, even if they did not understand the specifics. Repeat this cycle for as many times as you need to.

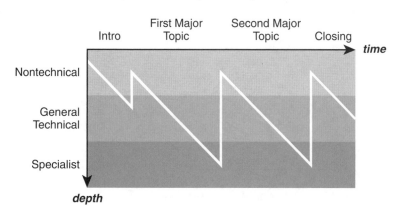

Fig. 24.1 Timeline showing the presenter reaching multiple audiences by beginning at the surface of a topic for nontechnical audience members, diving into a subject for the specialists, and then surfacing to gather the entire audience again. (Caption and figure adapted from Fig. 2-3 in Alley 2003.)

The vertical axis of Fig. 24.1 could also represent different disciplines. For example, in a talk about the societal impacts of flash flooding, you may be speaking to an audience of meteorologists, hydrologists, and social scientists. As your presentation follows the curve of Fig. 24.1, touch upon topics that relate most closely to the meteorologists, then the hydrologists, then the social scientists. Such a cycle does not need to be repeated or be in a specific order, but by making at least some portions of your talk relevant to each segment of your audience, you can deliver a talk that your whole audience will appreciate.

CONSTRUCTING EFFECTIVE ORAL PRESENTATIONS

An electronic presentation is the visual companion to the verbal component of a scientific presentation. As such, a poorly constructed electronic presentation can cripple an otherwise excellent speaker. Speakers need to carefully consider the design, construction, and delivery of the slides, topics to be covered in this chapter.

Bashing PowerPoint seems to be all the rage. Whether through phrases ("Death by PowerPoint," "PowerPoint Phluff"), magazine articles ("PowerPoint is Evil"), music videos ["Power(Point) Ballad"], or books (*Why Most PowerPoint Presentations Suck*), people love to spew their frustration with bad presentations at Microsoft's software.

Is it fair to criticize electronic presentation software (which, here, includes Keynote, Beamer, LaTeX, Impress, etc., what Edward Tufte calls slideware)? Not really. Blaming slideware for people enduring bad presentations is equivalent to blaming all table legs for stubbed toes.

Indeed, we have all been subjected to boring slideware presentations, but previously we were subjected to roughly the same percentage of bad overhead transparencies or bad chalkboard experiences. Thus, we understand: it's not the tool, it's how you use it. Regardless of the medium, a majority of speakers demonstrate a lack of forethought, preparation, and expertise in constructing their presentation.

This chapter looks at how to plan and create clear and effective slides to support your oral presentation. In Section 7.4, the writing/editing funnel described the means to approach writing and editing from the largest scale to the smallest scale. Although not an exact analogy, a similar funnel could be envisioned for presentations. The largest scale would be the storyboard for the presentation, the overall organization, and flow. The next largest scale would

NOT ALL SITUATIONS REQUIRE POWERPOINT

Some companies have banned electronic presentations in their meetings because, in the time people spend creating their graphics, connecting the laptop to the projector, and making the presentation, they could simply have just talked. Slideware has made it easier for us to communicate scientific results in an organized, clear, and colorful way, but it is not the only way. Some situations may be suited for simple oratory, instead.

Imagine if Abraham Lincoln had given the Gettysburg address by PowerPoint. Peter Norvig did. Over 1.6 million people have viewed it, the *Wall Street Journal*, *Lancet*, and *The Guardian* have cited it. The title of one slide was labeled, "Review of Key Objectives & Critical Success Factors" and contained the sub-bullets: "New birth of freedom" and "Gov't of/for/by the people." In the real Gettysburg Address, Lincoln said, "The world will little note nor long remember what we say here." If he used PowerPoint instead, maybe the world *never would have* remembered.

The speaker before Lincoln spoke for two hours. In two minutes, Lincoln was able to honor the fallen soldiers, dedicate the cemetery, and resolve to give the fledgling United States a new birth of freedom, all in the same number of words of a typical abstract of a scientific journal.

Before you sit down to your computer, decide if you really need PowerPoint, or whether you could obtain more *power* by making your *point* some other way.

be the layout of the individual slides, followed by graphics, text, and the fine points of the aesthetics. Let's begin at the largest scale.

25.1 STORYBOARD YOUR PRESENTATION

The term *storyboard* comes from the entertainment industry and is the planning by which the scenes of a movie are illustrated on separate sheets of paper and arranged to display the entire shooting sequence. Done this way, scenes can be rearranged and revised to develop the plan for filming. A director skipping storyboarding would have to shoot the movie without knowing where the scenes would be filmed or in what order, let alone details such as the camera position.

Although most directors would not imagine filming scenes of a movie without storyboards, most speakers prepare electronic presentations without even the thought of creating a storyboard. How often do we receive an invitation to speak and our first step is to launch slideware and begin composing without even planning the presentation? Instead, we ought to slow down, think about the presentation, and sketch out the storyboard without even turning on our computers.

The time spent storyboarding pays off later. Usually, too much time is wasted working and reworking slides before the message is even honed. Storyboarding also eliminates the need to repetitively switch between views of the whole slide show (to display and organize the overall structure of the presentation) and the views of the individual slides (to edit individual slides). This unnecessary switching is tiresome, especially in a presentation file with a large number of slides. Thus, efficiency demands creating the slides only after storyboarding is completed.

Most importantly, storyboarding forces you to focus on the theme and content of your presentation (the difficult part) rather than on the style and visuals (the easy and fun part). Your storyboard can be as specific or as general as you prefer, but the process of articulating on paper the structure to your presentation is easier than doing it in the slideware.

Create the storyboard with explicit drawings of each slide. Lay sticky notes or index cards for each slide on a wall or on a conference table. Having each slide displayed separately allows the organization to be worked and reworked to tell the story in the most sensible order. Focus on the message, the content, and the order of the slides to tell the story. Details such as the specific graphics, background color, and style can be refined later. Simply put, storyboards do not require a lot of detail on each slide because they should mimic the final presentation—simple and relevant.

25.2 STARTING TO CONSTRUCT YOUR PRESENTATION

After the storyboard is complete, open up your slideware and start creating on the computer. Start with your storyboard and write notes for what you want to say about each slide. (PowerPoint has the speaker notes area for this purpose.) Each slide should have one important point, and that one point should be made extremely clear by the slide. If no discernible point exists for the slide, delete it from your presentation.

DILBERT: © Scott Adams, dist. by United Feature Syndicate, Inc.

Do not put the obvious on your slide. An example would be to say that hail is an important forecasting problem at a severe storms conference. Instead, provide statistics, the number of events, the economic losses, or other facts that demonstrate to the audience why it is an important forecast problem.

The rest of this section addresses the beginning and the ending of the talk, often the most difficult to construct and arguably the most important.

25.2.1 First few slides

Most presentations start with a title slide that presents the title of the talk, author, coauthors, and affiliations. This slide is usually on display while the speaker is introduced and while the speaker gives some introductory remarks. This slide can be quite creative or quite simple. Most often, it will not be on display for very long.

What comes next in many talks is the "Outline of my talk" slide. For a 10–15-minute conference presentation, the outline usually consists of some-

Fig. 25.1 (a) An outline slide in a talk about fronts in the Intermountain West of the United States. Such a slide is unnecessary. (b) In contrast, a strong motivating slide is a much better approach.

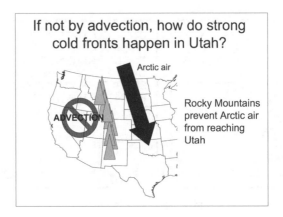

OUTLINE OF MY TALK

- Introduction
- Data
- Methods
- Results
- Discussion
- Conclusions

If not by advection, how do strong cold fronts happen in Utah?

Arctic air

ADVECTION

Rocky Mountains prevent Arctic air from reaching Utah

thing obvious to even a naïve audience member: an outline of how any scientific manuscript would be presented (Fig. 25.1a). Introducing your talk this way wastes time and loses the interest of the audience immediately. A better approach is to present the motivation, purpose, or goals of the presentation. Alternatively, following the title slide, present some shocking evidence that contradicts traditional thinking. By using the slide in Fig. 25.1b, the speaker wanted to motivate the question of how strong cold fronts occurred in Utah. By challenging the belief that such fronts might not be caused by the advection of Arctic air through the western United States, the speaker posed a question to get the audience thinking. Remember, with oral presentations, you have the ability *and the right* to be provocative.

Although literature reviews are effective in manuscripts, avoid presenting your literature synthesis in slide form. This admonition does not mean to omit all references, but putting previous work into context in the same way as you do in a manuscript is generally not needed. If your talk is going to resolve issues that have been debated in the literature, then painting the landscape of your presentation with a thoughtful, but focused, discussion of the previous literature is appropriate. Wherever possible, try to frame the debate around concepts rather than a recitation of "paper X did this, paper Y did that."

25.2.2 Last few slides

Your last slide should be *one* well-considered and briefly worded conclusion slide. Do not make extremely general statements that are obvious to anyone paying at least some attention during your talk ("The model was capable of reproducing the Madden–Julian Oscillation."). Select real results that summarize your talk in a few key points, which should already have been done during the storyboarding.

Although all slides should be simple, concise, and clear, the conclusion slide is especially important to emphasize the take-home message to the audience. A different style, but still quite an effective approach, to the conclusion slide is to present a conceptual model or graphical schematic. With Fig. 25.2, the speaker had tied all her results together with simple schematics or figures that were repeated from the talk.

Because the conclusion slide contains the summary of your presentation, leave it on display for as long as possible to allow the audience to fully absorb the message. Avoid presenting more after the conclusion slide, which will frustrate the audience who had hoped your talk was over. Do not close with a "Questions?" slide, "Thank you" slide, or the list of references from your talk. Moreover, be careful with "Future work" slides, unless you have some exciting prospects. Slides that include calls for more data or more case studies often do not value the audience's time and are an anticlimatic way to close.

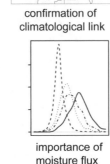

Summary

PV-streamer and
heavy precipitation

confirmation of
climatological link

For more details see:
Martius et al. 2006
Int. J. Climatol.

seasonal variation
of the link

importance of
moisture flux

Fig. 25.2 An effective schematic conclusion slide from "A climatological analysis of the link between breaking synoptic-scale Rossby waves and heavy precipitation events in the Alps." (Figure courtesy of Olivia Martius.)

25.3 DESIGN ATTRACTIVE SLIDES

The general design, layout, and color scheme of slides alert the reader to the type of presentation you will give. The slides complement or may influence the audience's opinion of the speaker. Well-constructed and attractive slides convey professionalism and credibility, whereas poorly designed or sloppy slides convey the opposite.

Maintaining a consistent look to your slides (e.g., background, font, color, transition between slides) indicates that the slides are part of a consistent message from the speaker to the audience. Too much variety becomes visually taxing to the audience. The advantage of maintaining a consistent look to the slides is apparent when the background or other style component is purposefully changed in the middle of the presentation. This dramatic change places a different emphasis on this material, allowing the speaker to indicate a change in direction to the talk.

Most organizations offer you a background slide to brand their image into the audience or as a watermark declaring the owner of the intellectual property. If gaudy, such backgrounds do a disservice to the presentation. Although small logos of your affiliation on the title and conclusion slides may be appropriate, such graphics on every slide are visually distracting. Another item of reducible clutter that often appears on each slide is the footer containing the speaker's name and date of presentation.

Avoid standard backgrounds packaged with your presentation software—most are overused or poor quality. Sometimes individuals create their own backgrounds, but they may have the same disadvantages as the standard backgrounds. Some companies sell professionally designed backgrounds online; some of their simpler backgrounds may be useful. Avoid photos as backgrounds. If the photo is important, show it as a separate slide or as an insert, but remember to display the photo at the original aspect ratio, do not stretch the image to fit the available space.

All components of the presentation should be clearly visible from the back of the venue at which you will be speaking. One approach to assess the legibility of your slides is to run the slide show on your monitor and stand several meters away. Such tests can alert you to potential problems with your graphics. Use nonserif fonts (e.g., Helvetica or Arial), which are easier to read than serif fonts (e.g., Times).

Use light colors on simple dark backgrounds. Dark green, blue, or purple with white and yellow letters are an effective combination. Light-colored backgrounds, especially white, are not ideal for the following reasons:

- Red lasers (especially if the laser light is weak) may not show up well on white backgrounds.
- Slides with white backgrounds lose contrast if the room is not dark enough.
- Color to the slides is refreshing and not as tiring on viewers during long presentations.

Graphs, however, are generally more legible with dark colors on white backgrounds.

Do not use yellows, light greens, or light blues on a white background, no matter how good it looks on your computer. These colors simply do not show up when projected. Avoid contrasting color combinations: no red lettering on blue backgrounds.

*When we see or hear a change, we expect it to mean something, so every visible or auditory change should convey information. This idea runs counter to the habits of many Power-Point users, who include decorations or interesting (but essentially random) visual changes, thinking it makes the talk more attractive. But if words, shapes or effects don't convey information, they distract.
—Stephen Kosslyn (2007)*

25.4 HEADLINES ARE BETTER THAN TITLES

The default in many slideware applications is to produce a centered title with a relatively large font (e.g., 44 point). Usually, these titles are too short to be meaningful or are entirely obvious from what the speaker is saying (e.g., Introduction, Data, Conclusions). Instead of a title, provide a headline in a slightly smaller font. Just as a headline of a newspaper article is a phrase or sentence that summarizes the article, a headline on a slide summarizes the slide or conveys the most important information. For example, rather than a slide called

"Results," the headline might read "Zonal wind variations, not heating, cause Kelvin wave amplification." Other points to consider about headlines:

- The headline forces you to define the main point of slide.
- A sentence headline orients the audience and speaker to the topic of the slide.
- Headline titles replace relatively nondescript titles, allowing more material to be removed from bullet points and reducing the number of words on the slide.

HOW TO CONNECT WITH YOUR AUDIENCE

After having determined who your audience is and what they want, what techniques can you use to connect with them? Think back to the last seminar you attended. What aspect do you first recall? Was it something shocking the presenter did? Was it an interesting story or fact? Was it a joke that the speaker told? These personal connections, what Alley (2003) calls "flavor," can make the audience remember the talk.

Address the concerns of your audience. Forecasters generally want to know about techniques to improve forecasting, whereas researchers generally want to understand the science better. Therefore, if your topic is of interest to both audiences, have two versions available, depending on who your audience is. A third version may be needed when both are present. Talking to conference organizers or to people at the conference before your talk will give you some idea of your audience's concerns, which you can work into your talk.

Provide some personal insight. Put your presentation in the context of your own story. What motivated you to look into this problem?

Tell stories. Knox and Croft (1997) elaborate on the importance of storytelling in the classroom, but the same is true for a presentation. A historical narrative about how the jet stream was discovered or the societal impacts of a tornado outbreak help the audience remember the material better.

Use props. Dusan Zrnić, a radar engineer at the NOAA/National Severe Storms Laboratory, uses props in his seminars. To show an example of anomalous propagation of radar waves due to the refraction of the radar beam by a higher-density medium, he puts a meter-stick representing the radar beam into a beaker of water, showing the bending of the light rays. In another example, he uses a key ring, rope, and his body to illustrate the three-body scattering effect associated with hail (Fig. 25.3).

Fig. 25.3 Dusan Zrnić displays three-body scattering using a piece of rope (radar beam), two fingers (water-coated hail stones), a key ring (the ground), and his mouth (radar transmitter and receiver). (Photo by Jelena Andrić.)

- The audience can always read sentence headlines that they might not have heard.
- A sentence headline shows a perspective on a topic that a title phrase generally cannot.
- Headlines are useful if you lend your talk to others.

Because headlines are different from titles, they need to be properly placed on the slide so that audience reads the headline first. The sentence headline should begin in the upper-left corner of the slide, be 28–40-point font, be left

Use analogies, facts, or observations. Even we scientists can be so jaded by seeing the inevitable stretching of the vertical coordinate in cross sections that we need to be reminded that, if the earth were an apple, the thickness of the skin would be comparable to the thickness of the troposphere. Or, a radar can detect a single bee 10 km away from the radar. Not everyone may appreciate the Twomey (1974) effect, but nearly everyone can be engaged by observations of clouds from commercial airplanes. Furthermore, did you know that visibility can vary by as much as a factor of 10 for a given value of liquid-equivalent snowfall rate? Judicious use of metaphors also can be a powerful tool for engaging audiences.

Deliver a surprise. Gimmicks can make the audience remember your talk. At a conference, Prof. Peter Hobbs of the University of Washington played a narrated animated video that presented his group's conceptual model of extratropical cyclones in the central United States. At a Cyclone Workshop, John Nielsen-Gammon's presentation on potential vorticity concluded with him donning a hat bearing the words "PV Boys."

Ask the audience questions. Remember to talk *with* the audience not just *at* them. Engage the audience with questions. During a presentation on a climatology of drizzle in the United States and Canada, then University of Virginia student Addison Sears-Collins surveyed the audience about where they thought the most drizzly place was with the following

slide (Fig. 25.4). Questions also gauge the audience's knowledge of the topic. If you can be flexible, the answers to questions can be valuable for you to trim out material the audience already knows, or elaborate on topics the audience may not know well enough.

On the other hand, if you ask a question to the audience to gauge knowledge, be prepared for the worst possible answer. I was at a conference where the first thing the speaker did was to ask the audience, "How many people know about *knowledge management*?" When only one person in the audience raised her hand, the speaker said, "I have a problem." My immediate thought was, "I guess he didn't manage the knowledge well enough."

Challenge the audience. Audiences want to come to a talk and be inspired. Challenge them to new heights in your talk, then give them the knowledge or the skills they need to reach those heights.

What is the Most Active Place for Drizzle in North America?
a) Hoquiam, Washington
b) Cannon Beach, Oregon
c) Norman, Oklahoma
d) St. Paul Island, Alaska

Fig. 25.4 Survey the audience in your talk. (The answer to this question is St. Paul Island, Alaska, which receives 403 hours of drizzle a year [Sears-Collins et al. 2006].)

Table 25.1 Examples of headline titles for electronic presentations

---

Microwave precipitation algorithms underestimate rain rates in shallow convection.
Eddy energies scale with the mean available potential energy.
Our parameterization is not sensitive to the sea-salt flux from the ocean surface.
The daytime convective boundary layer decreases the wave drag.

---

justified rather than centered, be absolutely no longer than two lines, be colored differently than the rest of the text on the slide, and be written in active voice. Table 25.1 illustrates some examples of potential headlines.

25.5 DELETE UNNECESSARY WORDS

Most slides are cluttered by too many words. Some sources recommend no more than eight lines per slide, others no more than six lines per slide, yet some even recommend no more than six *words* per slide. I do not believe that a single recommendation can be uniformly applied to all situations. Instead, consider your audience and the purpose of the slide when deciding on its text content. For nontechnical audiences, favor fewer words than for more specialized audiences. Resist the temptation to place too many words on the slide as a crutch for yourself. Instead, remember what you want to say by using handwritten note cards or the speaker notes function in many slideware packages, or, best yet, through repeated rehearsals.

Some situations may be helped by putting more words on a slide, however. If you are speaking English in a foreign country, or you are speaking in a language that is not your native tongue, the audience will follow you more easily the more complete the phrases and sentences are on your slide.

The indiscriminate use of bullets on most slides should also be questioned. Could items that would otherwise be bulletted be simply indented or better yet eliminated? An even greater question is whether bullet points even adequately describe the relationship between the items. Bullets imply that all items are of equal value, which may not best represent the relationship between the points on your slide.

If you do use bulleted lists, keep them short, generally under four items, as the audience cannot remember much more. Long lists are useful to overwhelm the audience with strong evidence for your point. Leave empty space to prevent adjacent lines from blurring into each other. Aim to keep each bulleted item or headline on a single line, or at most on two lines. Make items in the list parallel (Section 9.4). Use well-constructed phrases rather than sentences, and skip nonessential punctuation. List items in a sensible order (e.g., chronologic, by importance).

USE HUMOR, ALBEIT CAREFULLY

Despite the oft-repeated adage to start your presentation with a joke, there may be good reasons to be more cautious:

- Although you want to show the audience you are comfortable and they are going to have a good time, some people's humor is too dry, sarcastic, or slapstick to appeal to everyone in the audience, and some cultures may be confused or offended by sarcasm.
- The beginning of a talk, when the audience is assessing the type of speaker you are, is a bad time to tell a joke. Additionally, you are not warmed up and may be slightly nervous, further ensuring a bad delivery.
- Some professional situations are not appropriate for any humor at all, especially racy or other off-color humor.

Instead of forcing humor, engage your natural humor. Humor gives an audience a needed mental rest from your talk, which is why it is useful to work humor in later in the talk. Use humor, but sparingly and effectively, and only if it is relevant to your talk. Do not put entertainment over substance. Not only will the audience question your credibility, but even professional comedians know that the serious material in between the punchlines is necessary for the audience to appreciate the humor.

Visual humor usually goes over poorly, especially if a cartoon has multiple panels or lots of text. The time to delivery of the humor can take too long, and you are never sure when all the audience have read the text and gotten the joke so that you can move on. The silence and the sparse chuckling can be too uncomfortable. Single-frame comics without words and with an obvious joke provide immediate impact.

Humor works best in a packed room where laughter is contagious and sounds loud. In a huge auditorium populated by 30 people, even a hearty laugh has the potential to sound muted. If the audience fails to catch your first attempt at humor, better to avoid it through the rest of the talk. Not everyone is suited to make a crowd laugh.

Avoid gratuitous equations, as they slow down the pace of the talk and make your presentation more difficult to understand, even for mathematically oriented listeners. Presenting equations, especially derivations, usually requires too much time and demands too much patience of the audience. Present your ideas in words or graphics, wherever possible. When equations are definitely needed, define the variables and use annotations to explain the physical interpretation of the equation.

AVOID LONG STRINGS OF CAPITAL LETTERS. THEY ARE MORE DIFFICULT TO READ THAN LOWERCASE LETTERS, AND THEY TAKE UP MORE SPACE. Use left justify only, not both left and right justify; the unequal spacing that looks professional on the printed page makes reading short blocks of text on a slide more difficult. To create emphasis, words can be accentuated with color, italics, or upper case. Fonts should be 18-point font or larger.

25.6 INCLUDE RELEVANT AND CLEAR GRAPHICS

Some books on electronic presentations advocate that each slide should have one image or graphic. Because graphics are more visually stimulating than words, such graphics can amplify your point and potentially increase audience retention. As with most pieces of advice, it can be taken to the extreme. Do not embed a photo simply because it was a pretty picture or you felt that one was required. If you use photos, make sure that they relate to the content of the slides. A weak connection between the slide and the photo leaves the audience confused.

ASK THE EXPERTS

CREATING A MEMORABLE PRESENTATION

Svetlana Bachmann, Senior Research Engineer, Lockheed Martin Maritime Systems and Sensors

A professional electronic presentation can awaken even a lethargic audience, spark the curiosity of spectators, and open a surge of questions and suggestions. The following guidelines will help you achieve clear and entertaining presentations.

Try to avoid abbreviations, or provide a repetitive look-up option or legend. Often, I provide the meaning directly near the acronyms, using a smaller font. When such a direct approach clutters the slide, I create a placeholder (i.e., region for inserting text) for the descriptions (Fig. 25.5). Such a placeholder can be located in any portion of the slide, but should be clearly visible and readable. This placeholder can be left blank on the slides without acronyms or used for cartoons and notes.

Use color to draw attention. In my presentations, I sometimes use illumination. Suppose a slide has a white background and is partitioned in four sections. I place four gray-shaded rectangles with some degree of transparency over the sections, completely covering the slide. To draw attention, I remove one rectangle at a time—the white background makes the section pop

up, just like if it were illuminated with sunlight. Darkening portions of the slide to illuminate a particular section is also a powerful tool to draw attention.

Repeat important statements to highlight key points. Strategically placed statements can be used to keep your audience focused as you transition between slides. I often replicate the key item from a previous slide to create a smooth flow between the slides and to provide extra exposure for the key item.

Use amusing graphics or cartoons in your presentation. Humor can lighten the atmosphere, and the audience response tends to refocus any attention that may have drifted. Hint: exaggerate to engage viewers' imagination—for monetary advantage, show images of extreme luxury items; for an ingenious solution, show a maze with one obvious path out.

Include simple animations and graphics. Your visuals can suggest to your audience to be surprised, perplexed, satisfied, or unhappy. Hint: a cartoon scratching his head and shrugging his shoulders will indicate a difficult problem, whereas a cartoon jumping up and down and clapping his hands will signify a desired outcome.

Viewers who are exposed to a plot for the first time always need help understanding the meaning of this plot. You will definitely have to spend time explaining the axes, the shapes, and the meanings. An appropriate cartoon can save you the time and the

For graphs, the tendency is to use the same publication graphics you created for the manuscript. Sometimes, publication graphics do not make the best presentation graphics. What reads well on a journal page may not read well on a projector to an audience of 500. Take special care with axis titles, axis value labels, and other such items that may need to be enlarged to be readable.

Put as much descriptive (captionlike) material on the slide as possible. Doing so will prevent you from having to spend precious time during your presentation to explain the figure to the audience. Remember that the audience can read faster than you speak, and most in a scientific audience know

explanations. Once, when presenting a novel concept in clutter filtering for weather radar, I placed a plot of unfiltered Doppler spectra on one side of the slide and gathered cartoons of a radar, cloud, and building on the other side of the slide. I used colors to relate different portions in the plot with the appropriate cartoons. I replaced the unfiltered curve on the plot with the filtered one and used an animated bulldozer to push the cartoon building off the picture. Obviously, we do not bulldoze buildings to detect weather, but the joke was understood and the plots did not need an explanation.

Although slideware has many tricks for creating animations, keep animations simple and few—too many visual effects might prevent viewers from focusing on your topic.

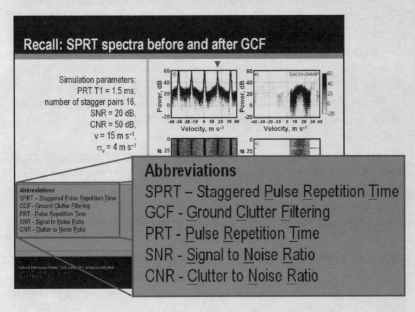

Fig. 25.5 A placeholder for abbreviations. (Courtesy of Svetlana Bachmann.)

how to interpret a graph if they know what they are looking at. In such cases where you are presenting an uncommon style of graphic, you may have to take some time to explain how to interpret the graph to the audience.

Annotated graphics also have the benefit of reminding the audience about details on the slides. The audience becomes momentarily distracted by the new visuals when slides change from one to the other and they may have missed hearing what the contours were. Or, your voice may not have carried well. Or they may have been daydreaming. Having that information handy in the title of the graph or in the margin allows them to continue to follow you.

Acknowledge the source of graphics you downloaded online, even if only a small acknowledgment in the bottom corner of the slide. Because of the prevalence of sharing graphics online, be careful to track down the original source and respect copyright. Digital images copied from Web sites (especially logos) might pixelate and look unprofessional if the graphic is stretched too much. Either use the image at the designated size or obtain a higher-resolution logo from the organization.

Should you use clip art or photos? Opinions vary about this. Slick stock photos with professional models may send the wrong message in a scientific presentation. Clip art is more generic, albeit less professional looking; photos of real scientists in real situations or photos of abstract concepts are generally more appropriate. Photos of real people help the audience relate to the topic on an emotional level. For example, which conveys more emotion: a map of the rainfall distribution for a flash flood in Missouri or a photo of a flood victim with her head in her hands?

Be creative when presenting your graphics. For example, to compare two graphs, rather than have them side by side, could you blend the two of them by fades back and forth? Make a boring flowchart more interesting by having photos or other graphics pop up as you describe the different elements. Embedded animations and movies can enhance a presentation and pique the audience's interest.

Finally, avoid clutter on the slides. Graphic designers recommend no more than seven items on a slide (e.g., headline, three bullet points, main graphic, two anotations). Less-cluttered slides have a more powerful impact, so use empty space to keep the items on the slide well placed.

25.7 EXAMPLES OF HOW TO IMPROVE SLIDES

To demonstrate some of the problems identified with poor-quality slides in this chapter and how they can be remedied, consider the following four draft slides, all in need of much improvement (Fig. 25.6).

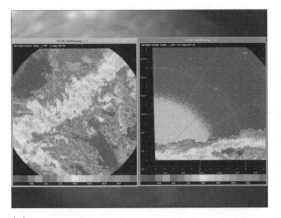

(a)

Conclusions

- A strong cold front moved through northern Utah on 14–15 February 2000.
  - Cold air arrived earlier aloft than at surface
  - Pressure trough preceded front at surface
  - DCAPE was large
  - Precipitation varied with elevation
  - Subcloud sublimation was important

(b)

Predictability

- Lilly (1972, 1990): From turbulence theory, and dimensional arguments:
  - Synoptic scales may have unlimited predictability
    - better initial state leads to better forecasts
  - Mesoscale may have limited predictability
    - better initial state yields little improvement
- BUT, Lilly (1990) predicted that cloud-scale models will provide more realistic simulations of cloud systems. (He was right!)

- Hence, the ultimate benefit of cloud-scale simulations may not be in forecasts of cloud-scale phenomenon. Rather, their main benefit may be in "removing damage due to inaccurate parameterizations"

(c)

The Melting Effect as a Factor in Precipitation-Type Forecasting

- Kain et al. (2000): December 2000 *Weather and Forecasting*
- Earlier work by Findeisen (1940), Wexler et al. (1954), Lumb (1961), Stewart (1984), Bosart and Sanders (1991)
- Frozen precipitation falling through an above-freezing layer melts and absorbs latent heat from the environment.
- If enough cooling occurs, melting precipitation can be inhibited and rain will change to snow.

(d)

a. The slide is obviously showing radar imagery, but we are not given any information about where or when this was. The scales are too small to read.

b. These "conclusions" are not really conclusions at all, just a list of observations of the cold front. A busy background distracts from the text. Because the five bulleted points give the impression of equal weight, the most important conclusion of the research ("Subcloud sublimation was important") is underemphasized.

c. The slide is too wordy. The last highlighted line is actually too dark. The title of the slide doesn't really say anything.

d. As in (b), the busy background makes the text difficult to read. "2000" is repeated twice. The list of references seems unnecessary.

Fig. 25.6 Slides needing much improvement.

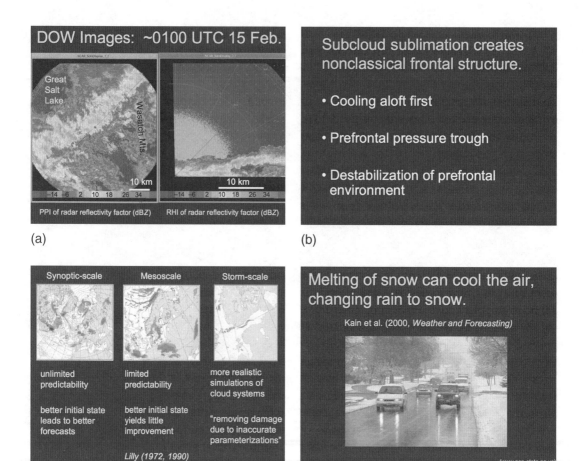

(a)

(b)

(c)

(d)

Fig. 25.7 Much improved slides.

The problems with these slides are remedied in Fig. 25.7.

a. Annotations now describe the graphs. Spatial scales and geographical annotation help identify the location. The small scales have been covered up, and new labels for the color-bar values have been created. The right image might be further cropped to highlight only the data, not the empty space.

b. The title for this conclusion slide is a headline, and all the bulleted observations support that conclusion.

c. Graphics and better organization to this slide provide a framework to see the differences in predictability described by Lilly.

d. The physical process becomes the headline, followed by a photo illustrating the consequences of rain changing to snow. The citation remains to emphasize the article upon which the talk is based.

CONSTRUCTING CROSS-PLATFORM-COMPATIBLE PRESENTATIONS

Just because two computers have the same software does not necessarily mean that you will be able to load an electronic presentation from one to the other and expect complete compatibility. At least once during every conference, someone's mathematical symbols are replaced by icons of a phone, exit sign, or finger-pointing hand. Other changes that may occur during file portage include the sizes and types of fonts, arrangements and sizes of text boxes, custom animations, and links to external files (e.g., movies, sounds). Such problems are more common when transferring presentations between different platforms (e.g., Mac, Linux, Windows), but also can happen when transferring presentations between computers of the same platforms. Reasons for such incompatibilities include different versions of the operating systems, different versions of the software, and different preference settings.

The following tips can maximize the chances of your presentation being properly ported when the big time comes:

1. Deliver your presentation from your own laptop. Doing so will avoid nearly all problems. (Occasional incompatibility between the computer and the projector can occur, however.)
2. Choose standard fonts (e.g., Arial, Times New Roman, Symbol) as often as possible.
3. Leave some margins in the placeholders instead of pushing the text to the edge. Some font resizing may occur after transfer.
4. Directly embed graphics.
5. Put all linked content for the presentation (e.g., movies, sound files) into a single folder with the presentation, *then* create the link to the external files within the presentation software. Move that folder as a unit between computers.
6. Create equations as graphic items, then embed.
7. If transferring from a Mac to Windows, avoid using Mac PICT images, narration (which is recorded in AIFF format), or Quicktime files. Do not use slide templates for pictures as they convert the graphics internally to PICT format; insert your graphics manually instead.
8. If transferring from Windows to Mac, avoid embedded objects (Word and Excel files or graphics).
9. Movies work best on both platforms when in AVI or MPEG-1 format.
10. Instead of creating animation within a single slide (e.g., bullet points appear sequentially), create a series of slides differing only in the animation. The difficulty with this approach is that any changes to one slide in the series will need to be edited on all slides in the series, and placement of the items on the slide is important so that the items do not jump around during the transition.
11. Rather than incorporating an animated gif of radar imagery, place single images on a series of slides, and manually step through each slide.
12. Convert your presentation from slideware format to a more portable file format such as PDF.
13. Finally, and most importantly, test your transferred file thoroughly before your presentation.

25.8   USE EFFECTIVE TRANSITIONS

Revealing information piece by piece can keep the audience focused, especially when presenting the whole slide might give away the punchline. Slideware also allows for each bulleted item to appear sequentially. Unless you have carefully designed your slides and are using a remote-control clicker, I recommend not revealing bulleted items sequentially, for the following reasons:

LESSONS FROM THIS CHAPTER

- Plan your presentation through storyboarding.
- Each slide should make one point.
- Keep slides simple.
- Do not state the obvious.
- Keep the audience focused.
- Save the details for the question-and-answer period or the manuscript.
- Close with one well-considered conclusion slide.
- Headlines are better than titles.
- Remove unnecessary words from the slides.
- Are those bullets necessary?
- Find ways to connect with the audience.
- Use meaningful graphics.
- Graphics should be legible to the audience.
- Favor simple transitions.

- Clicking takes time, as most speakers will pause during the click.
- Lots of clicks means that you are tied to the computer if not using a remote control, inhibiting you from walking around during your talk.
- If you need to reverse direction to reshow a slide, you will need to wait for all your sequential items to play out in reverse.

On the other hand, complicated graphics can be presented much more effectively by building the graphic piece by piece. Use transitions to walk your audience through complicated slides using arrows, lines, pictures, and animations. Prof. Robert Fovell of the University of California Los Angeles says:

> Complex graphics should build step by step. I present a conceptual model of a squall line like this. The first slide shows the cloud outline, used to explain the setting. Next adds precipitation. Next adds the cold pool. Next adds airflow arrows. Next adds convective cells, bright band, etc., until the model is complete. By this time, the model is busy but not inaccessible. It takes no more time than explaining a single very complex figure.

One of the most common devices that speakers use to add variety to their presentation is the transition between slides. Most transitions are too garish, slow, or inappropriate for most scientific presentations, so use a quick transition ("appear"), unless there is a specific reason for choosing a different transition ("dissolve," "flash," etc.). Do not set the transition on "random" unless you want your audience placing bets about what transition might appear next.

DELIVERING COMPELLING ORAL PRESENTATIONS

How do the best speakers seem so confident and comfortable? Their confidence and comfort arise from knowing their material and thoroughly preparing it. A natural speaking style and a bit of showmanship further engages the audience. Providing clear, concise, and accurate answers to audience questions also demonstrates command of their material. In this chapter, we look at how even an average slide presentation can become a compelling experience through effective delivery.

A seasoned politician, Al Gore did not need to read his words off the slides in *An Inconvenient Truth*. He used animated cartoons and performed dramatic stunts to emphasize his points. Importantly, and perhaps surprisingly, Gore's presentation was persuasive because of his delivery. He was comfortable with his material and *passionate* about it—a far cry from the stiffness he was accused of during his 2000 presidential campaign. Although we may not have his resources to produce such an experience, we can strive to emulate his delivery and to be one of those memorable speakers that engages the audience.

26.1 REHEARSE TO REDUCE ANXIETY

When I was a grad student at Albany, all students going to a conference would rehearse and critique their talks together, overseen by the faculty. Further revisions were approved by the advisor, but the group critique served several purposes: to rehearse the oral presentation, to improve the presentations, to increase confidence, and to build comraderie among the group.

Rehearsing your presentation is essential to giving your best presentation. Going over the slides in your head is useful but is not rehearsing: actually

FOUR WAYS OF DELIVERING PRESENTATIONS

Talking points. Speaking from talking points is how most scientific presentations are delivered. A well-rehearsed talk based on talking points conveys credibility and is the most comfortable way to listen to someone speak. Because the visuals cue the speaker for the next things to say, the challenge of delivering the presentation this way is to remember what to say and the order to say it in once the next slide appears. Slides can be creatively constructed to reveal information in a preferred order, or bullet points off to the side of a graphic can indicate the main points.

Reading. Reading a presentation is appropriate for a press conference or if reading a quotation within a presentation. Although there are obvious benefits to scripting the whole talk, audiences are turned off by a person reading text to them. Some authors will read text verbatim on the slide, in lieu of explicit talking points. Because the audience can read faster than the speaker can read the text aloud, slides should not be used as a teleprompter. If precision in your wording is important, I recommend scripting.

Scripting. Scripting is what TV news anchors do. They work off a script, but are so well rehearsed—part reading and part memorization—that it sounds natural. Extremely important speeches (e.g., invited presentations, inaugural addresses) can and probably should be entirely scripted. I have already argued that scripting the initial part of the talk can help ease the speaker into the meat of the talk smoothly. Small portions of a talking-points talk that are likely to trip up the speaker can be scripted on Post-it notes or note cards for reference by the speaker when needed; these approaches are mostly transparent to the audience.

Off-the-cuff. When speaking off-the-cuff, you run the risk of looking unprepared and it could potentially lead to disaster. Without visuals for support, the audience must focus on the speaker, further increasing pressure on the speaker. If you know you are going to speak, even if you just plan to ask a question of a speaker at a conference, why not prepare? Simply write down your question, or rehearse it in your head to ensure that you say it clearly and concisely.

stand up in a room with your slides and deliver the talk—either to yourself or others. Only by giving the presentation out loud can you be sure you can explain complicated topics comfortably, clearly, and within the time limits to an audience during the real presentation. Without a script or a fantastic memory, speakers often say things differently on stage than during rehearsal, so repeated rehearsals are essential for honing the language to perfection. More specifically, the shorter the talk is, the more practice is needed. Sixty seconds of flubbed material during a 12-minute talk is not the same as losing the same 60 seconds in a talk that lasts 50 minutes.

The more comfortable you are with the material, the less anxiety you will experience, the less likely you will forget something important you wanted to say, and the more likely you will say the words you intended. How many times should you rehearse? Many factors go into how much rehearsal is needed (Table 26.1), so a single recommendation is futile. Nevertheless, first-time presenters giving a ten-minute presentation at a major conference should rehearse their presentation at least five to ten times. Do not practice so much,

Table 26.1   Factors that increase the number of times rehearsal is needed

Shorter talk
Less experienced speaker
Particularly important talk
Never presented the talk before
Been a while since presenting the talk the last time
Presenting someone else's work
Difficult topic
Potentially hostile audience

however, that it creates anxiety. When you are not benefiting from additional rehearsals, you are ready. On the other hand, experienced presenters giving an hour-long lecture may not need much more than to just review their notes beforehand. If your talk is based on a paper or extended abstract, be sure to reread the manuscript before your talk. Often, we may forget details about the research that reading the published paper or extended abstract would refresh.

The value of scripting your talk for rehearsals can be demonstrated by the following experience. During rehearsal for her Ph.D. defense talk, my student would go through a few slides, clearly struggling with what to say, saying it differently each time. She was clearly frustrated. I told her to script out what she wanted to say about each slide, focusing on the order that it made the most sense to present the concepts before our next rehearsal. After doing this, she delivered the talk perfectly, not even referring to her talking points. The *process* of thinking about what she wanted to convey and writing it down was enough for her to remember the material.

26.2   PREPARE BEFORE THE PRESENTATION
Print out a copy of the talk and have it with you when you travel:

- If there is a catastrophic failure of the projector or computer, you can still talk off your notes.
- A printed version of the talk may reveal typos or minor changes that you missed onscreen.
- You can review the talk quietly before your presentation without the need for a computer.
- You can make notes on this paper, as reminders of things to say.
- You can take notes on the paper after the talk, in response to questions or comments about your presentation.

Table 26.2   Checklist before your presentation

☐ Arrive early.
☐ Pick a good seat close to the podium.
☐ Check in with the moderator.
☐ Upload the presentation or use your own laptop.
☐ Test that your uploaded presentation works.
☐ Turn off the screensaver and other intrusive applications.
☐ Does the projector display your presentation accurately?
☐ Have the presentation ready to start, if possible.
☐ Test the range and directionality of the remote-control clicker.
☐ Find the pointer, and learn how to use the laser.
☐ Have water, tea, handkerchief, or cough lozenges with you, if needed.
☐ Use the bathroom.

Arrive early to the place where you are speaking; there are many things you need to do before your presentation (Table 26.2). Scope out where you will be speaking from. Pick out the best seat—close to the podium, but not so close to the screen so that you cannot see the other talks well. Check in with the session moderator to make sure that they know you are present and your talk is uploaded to the computer. Find the bathroom and the emergency exit.

To the extent possible, make yourself comfortable in that environment. Control your speaking space. If there are cables on the ground that you are likely to trip over, move or secure them. Move the podium to where you prefer to speak from to see the audience. If there is outside noise coming from an open door, close it. If you want the lights on while you speak, request that they stay on. Do not speak to a totally dark room, as you do not want the audience falling asleep.

Once you get to the place from where you will be speaking, you may have the option of either uploading your talk to the computer already in place or using your own laptop. Both approaches are worth considering in more detail.

If you uploaded your talk, test it thoroughly. Formatting may change, animated sequences may need to be redone, and video and sound files external to the electronic presentation may not transfer, so these things should be double-checked before you speak. Audiences are quite disappointed when promised movie files do not work. Section 25.8 describes things you can do to increase the portability of your electronic files. Get familiar with the keys or mouse buttons that advance and reverse your slides.

If you are presenting from your laptop, be aware of how connecting the projector cable will affect your computer and how to project your laptop image onto the screen. Know which key strokes send the signal from your laptop to the projector. Test beforehand to ensure your computer or projector does not freeze up. Place the laptop in a place that does not force you to block the screen when you stand or pace around.

Check for graphics or files on the desktop that you may not want to show to the audience. Turn off the screen saver before you start talking. If it goes off, the audience will be distracted and wonder if you have been talking too long. Make sure you turn off other intrusive applications that may pop up suddenly during your presentation. I was at a conference where the speaker had finished his talk and was answering questions when a Skype window from his coauthor kept popping up. The coauthor was probably calling to find out how his colleague's talk was going!

Check for color accuracy between the monitor and the projector, and make adjustments, if needed. Sometimes the brightness or color settings on the projector may not match those of your laptop. For some reason, the colors on most projectors tend to appear lighter than on the laptop screen, so light-colored graphics may appear washed out.

Be familiar with how the laser pointer works (especially if it is integrated with the remote for the projector, and the power button and laser button are adjacent to each other). You will be surprised at how many people will be handed a laser pointer and will hunt, for what seems an inordinate amount of time, for the On button. Attach the microphone high on your body along the centerline, and notice how to turn it on. To keep a constant volume, avoid twisting your head too far away from your centerline.

Remote-control clickers that advance your presentation function like the prompting devices that weather forecasters use on television and can be great tools for giving presentations, freeing you up from always returning to the computer to advance your slides. Before the talk though, test the range and directionality of the clicker from places that you are likely to wander as you talk. If you spot potential problems, advancing the slides by hand is more convenient and less frustrating for you and the audience.

Have a water bottle or glass of water, if you need it. Do not be the speaker that takes a sip out of the bottle every other slide as a nervous habit. Because drinking will disrupt the flow of your talk, either do so during a pregnant pause or when being asked a question. Do not try to rush the sip, as you may end up inhaling liquid, coughing, and making a bigger scene than was intended.

Use the bathroom before the session starts. The urge to pee from the adrenaline rush, coffee, or both is usually strongest before going on, so do not wait until the last minute. If you forget, do not worry. It will usually be superceded by your focus on your presentation once you start talking. Also, a bathroom break is a good time to check if any of that wonderful pesto you had for lunch is in your teeth or on your white shirt.

Do not waste the audience's time with the technical side of starting your talk. Test your presentation before you take the podium, and have your talk ready to start, if possible. Aim to make a smooth transition once you are introduced.

26.3 DELIVER A STRONG OPENING

The start of the talk is your opportunity to connect with the audience. You need a strong opening to engage the audience. Otherwise, you risk losing them at the very beginning.

If the person who introduces you says your name and the title of your talk, do not repeat them to the audience, unless there is a point you want to make. If the format and length of your presentation is a bit flexible, invite the audience to ask questions during your talk. Doing so will help keep the audience a bit more alert and you can address their questions at the time they think of them.

The first slide (almost always the title slide) is usually showing while the speaker makes introductory remarks, thanks the host, tells an anecdote about the topic of the talk, etc. Do not spend too much time on this slide, however. The audience is looking to hear about your topic, and dwelling too much on any one slide early on makes them think that they are in for a slow, possibly painful, ride.

Your starting pace will determine your pace throughout the talk. Start comfortably quick; the quick pace will wake up the audience somewhat. If you start with a relaxed pace, include lots of tangents, and deviate from your slides, people will start daydreaming and you will lose them, as well as run over your allotted time.

Once you start talking, the audience is generally not focused. Your job is to get them to focus on *you* quickly. The longer they take to be attentive, the larger the risk of their missing some important introductory material. Within a few minutes, the audience will have settled down and begun paying attention.

Depending on the length of the talk, they may start to get anxious. For a conference presentation (about ten minutes long), they may start to get anxious a few minutes before your time is up. For an hour-long seminar or lecture, the audience has settled in for a longer haul, so after about 20 minutes, their attention span starts to drop. To maintain their interest, you will need to address this increasing attention deficit.

To handle the inevitable attention deficit, break up the flow of the talk with a brief video, a story, or some lighter material. Take questions to signify a change in topic. Disrupting the talk in other ways, such as using the drawing tools in the slideware, drawing on the chalkboard, having a discussion, or asking a question helps maintain the interest of the audience. You may also wish to blank the screen by either by typing "b" in PowerPoint or through the projector remote. Talking directly to the audience puts the emphasis on you and your words. If the session is more than an hour, give the audience a break of five minutes or more.

26.4 KEEP THE MOMENTUM GOING

Most presenters find that the butterflies in their stomach have likely flown away after several slides and they feel more comfortable. Keep the slides moving. Do not run off on a tangent or dwell too long on any individual slide. Leaving a slide up for, say, five minutes starts to tax the audience, who want constant visual stimulation. Even if the slide you are talking about is complicated, reveal pieces at a time, which will be fresh visual stimulation. Give the audience enough time to absorb complicated graphs. If you are presenting a graph that may not be familiar to your audience, take the time to explain it clearly to them. Budget your time well enough so that you are not forced to rush through important explanations too rapidly.

If you have too many similar-looking slides in your talk (e.g., ten scatterplots in a row), the audience may start to lose interest. Can you compress the information into fewer slides without being too messy? Or, are all these similar slides even needed?

If a previous speaker had presented similar material to yours, do not feel compelled to plow through your material in exactly the same way you had planned. Skipping or condensing the material shows that you respect the audience, you are flexible, and that you are not in automaton mode.

I like to memorize roughly the first 10–30 sec of an oral presentation to ensure that my opening comes off cleanly. This builds momentum and self-confidence, helps me relax a bit, and makes a good first impression for the audience. Some football teams follow essentially the same approach in that they often script the first few plays of the game.
—Paul Markowski, Pennsylvania State University

26.5 FINISH STRONG

After delivering the body of the presentation, the time arrives to wrap up the talk. Having plunged to the depths for specialists (Fig. 24.1), surface and regroup with the summary of what has been learned. I have attended fascinating presentations where the speaker opened with a great motivation, plunged the depths expertly, but failed to surface when the talk ended, leaving the audience in the depths. I exit such talks not appreciating what I was supposed to have learned. Leave the conclusion slide up long enough for the audience to absorb the message, including, if possible, during the question-and-answer session.

Do not end with "That's all I have" or "I think I will stop now." End with a polite "Thank you." It is a signal to the audience to applaud.

26.6 HAVE A COMPELLING DELIVERY

Delivery of your talk involves the way you present yourself to the audience, and includes your style, personality, body language, voice, and use of props. Many people have their presentation personality, which is more outgoing than their regular personality. Others give their presentations as extensions of their regular personality. Whatever approach you take, be natural. Are you one for flair, or are you the down-home folksiness type? Your delivery also depends on the occasion, which would be more formal and serious for a memorial session than a university seminar on a Friday afternoon.

Be sincere, and show professional enthusiasm. Such enthusiasm is usually contagious with your audience. Above all, smile! You ought to be having fun giving your presentation. Exude confidence. Remember that you are the expert in what you are presenting. Avoid self-deprecating comments.

Speak clearly and slowly, especially to audiences where ESL speakers may be present. Vary your vocal intonations because monotone is sure to put the audience to sleep. Speak when standing up, opening up the diaphragm and airways, enhancing pronunciation and projection. If you have a naturally quiet voice and want to project more, go to the roof of a building or the middle of a forest, and work on speaking more assertively by reading some text aloud by shouting. Exaggerate the intensity of your voice. When you return, you will feel more comfortable speaking up at a normal level.

Avoid filler in your speech (e.g., "ummmm," "you know") or words you use repeatedly (e.g., "like," "basically," "my point is"). If you have ever recognized this problem in someone else, you know how distracting it can be. Filler arises to occupy awkward pauses where you feel compelled to say something while you think of the next thing to say. Breaking such habits is not easy, but slow down, think about what you say, and replace the filler with a small pause.

Your stance should be comfortable and upright, not shifting from foot to foot. If you have a podium, do not lean on it. Be mobile, but do not move as if you were pacing about a cage. Avoid nervous habits, especially with your hands, such as pulling your hair away from your face, scratching yourself, picking at your nails, or playing with your pen. Placing a hand in your pocket is a natural look, as long as you keep it still. Remove any keys or coins that might jingle when you walk or be an attraction for your hand to play with.

There are four ways to point at material on the screen: stick or telescoping pointer, hand, laser, or mouse-controlled arrow on the screen. If you use a pointer, choose a stick or telescoping pointer, if possible. Pointing the stick or using your hand to point on the screen is more active than waving a laser pointer. If you fidget with the stick, however, it will become a distraction. A stick is also a better choice if you shake when nervous because a laser pointer will exaggerate your shaking. If you use a laser pointer, make controlled motions with it. Underline or circle clearly what you want to emphasize. Turn it off when done pointing, and do not blind the audience with it. Using the mouse to move the on-screen arrow as a pointer allows you to continue to look at the audience without turning your head to point at the screen.

The advent of inexpensive digital voice recorders, camcorders, and Web cameras means you can record yourself speaking in rehearsal and play it back. Watch for annoying gestures that you make, nervous ticks, vocal filler, and other aspects of your presentation style that you would not have seen or heard otherwise. Work to improve your delivery by reviewing such recordings or getting feedback from others.

26.7 MAINTAIN EYE CONTACT

Face the audience, not the screen! Maintain eye contact with the audience, minimizing the time you look at and talk to the screen. Doing so shows that you care about the audience and allows you to gauge their responses. Choose only one person at a time, fix your gaze on that person for a second or two, and then shift eye contact to someone else. Do not broadly glance over the audience, and do not stare at someone for more than a few seconds—this will make the person uncomfortable. If you are talking to a large audience, shift your focus from one section of the room to another, picking out individuals within each section. An average speaker establishes eye contact about 40% of the time, so aim for better than 50% eye contact with the audience, and 90% during the introduction, conclusion, and other important parts of your talk.

26.8 WATCH THE TIME!

We give presentations for the *audience*, not for ourselves. When speakers purposefully exploit their position and continue speaking after their allotted time is up, they are violating this axiom. Never assume that going over the allotted time is acceptable. Doing so takes advantage of the other speakers' time, runs the conference behind, gives you a bad reputation, and is inconsiderate to the audience.

Ten or fifteen minutes may seem like a long time, but time flies when you are talking. Conference presentations are really time-management problems, so be prepared by focusing the content of your slides and rehearsing. In general, for audiences composed of native English speakers, the rule of thumb is one slide per minute. If the slides are complicated, or you talk slower than the average speaker, then you will need more time per slide.

On the podium, keep close track of time. Do not wait until near the end to determine if you are likely to run long. If a clock is not visible to you in the seminar room, take off your wristwatch, and place it on the podium. Or use the presenter's tools in PowerPoint, which counts your time. Turning your wrist to look at your watch during your talk reminds the audience to look at their watches, too. Keep them focused on your presentation.

THINGS NOT TO SAY TO YOUR AUDIENCE

- **I don't have that many slides so I will be finishing early.** Speakers who say this rarely ever do. By not letting your audience know you are going to run short, you can pleasantly surprise them. Plus, if you do go longer than you anticipate, you can still finish on time with the audience being none the wiser.

- **I didn't get as much work done on this project as I had hoped to.** The audience does not need to know your personal sagas. In fact, your results may be sufficient for them.

- **I am sorry for presenting preliminary work.** Conferences are for presenting work in progress. You do not need to apologize for it.

- **I know you are all looking forward to lunch.** If the audience wasn't thinking about how hungry they were, they most certainly are now. Keep them focused on your presentation, not on distractions.

- **Because time is limited, I won't be able to present everything.** You have the same time constraints as everyone else. Complaining is ungrateful.

- **You will not be able to see this, but . . .** You should have prepared your graphics better, otherwise the audience assumes that you do not care whether they can even see your results.

- **I will have to speed up, then.** When alerted that you are running short on time, the natural reaction is to speed up. Do not appear shaken. Confidently decide what you are going to do and pretend you had it planned all along.

- **I know I'm over time, but let me show this one more graph.** Ever notice how most of the time it is not just "one more graph"?

Leaving audiences wanting more is better than leaving them wanting less. Ending your presentation early, before your allotted time is up, is definitely acceptable. I learned that lesson at a concert festival in Albany. Five bands were to play, all for about equal time, or so I thought. Most bands stuck to about an hour in length, except for two. The second band played for an hour and 20 minutes, mostly slower numbers. When leaving the stage, the band received polite applause. The third band came up, ran through a cranking set of 40 minutes, then left with the fans screaming for more, even after the lights came up and the stage was being set for the next band. That's how you want to leave an audience.

26.9 HANDOUTS SHOULD NOT DUPLICATE YOUR SLIDES

For some meetings, you may want to provide handouts of your presentation to the audience. Typically, people just print out the slides and hand them out before the talk. If your slides are understandable without you, are *you* really needed? Instead, provide one or two pages of material with the essentials of your talk and a few key graphics. That way, the attendees can focus on your presentation, noting the links between your slides and the handout. Furthermore, surprises and other revelations within your talk are not compromised.

Alternatively, create two sets of slides (one with more writing on it intended as a handout, and the other the actual presentation), or write more content in the area for speaker notes and print them as part of the handout, too.

26.10 QUESTIONS AND ANSWERS

As in dealing with nervousness (Section 28.3), being prepared is the best way to address questions from your audience. Predict what the weaknesses of your presentation are and what might be reasonable (or unreasonable) questions that could be asked. When rehearsing your presentation, friends and colleagues can also help think of critical questions you could be asked.

Design your presentation to address these weaknesses and potential questions. Defend your research: do not allow weaknesses in your talk to be exposed by questioners. Some questions may require other evidence you were not prepared to show. To accommodate any predicted questions, place extra slides in your presentation after the conclusion slide. Chances are that you will not need them, but, if you do, such extra slides can add clarity and substance to your response should one of your predicted questions be asked.

After you are finished speaking, have said a polite "Thank you," and have listened to the glorious applause that erupts after your awesome presentation, ask "What questions do you have?," which is more engaging to the audience than "Any questions?" If you have been standing behind a podium, approach the audience to remove the barrier between you and the audience to appear more approachable. The audience will take some time to formulate their questions and get the nerve to ask, so the wait for the first question may seem interminably long. Be patient. If a minute or so of silence goes by (especially at an hour-long seminar) and you are feeling brave, you may ask yourself a particularly provocative question to get the ball rolling.

If asked a question, listen fully before formulating a response to ensure that you heard the question correctly. Often communication between two individuals fails because one party is already mentally plotting the rebuttal before completely hearing out the other. Whatever you do, address the question that is asked of you, not the one that you wished you were asked. Stay focused on addressing the question. At the end, you may ask the questioner whether you addressed the question.

If you get asked a difficult or unanswerable question, a pleasant response indicating that you do not have an answer at this time and that you will look into it can save you. Look away from any hostile questioners, addressing your answers to the audience as a whole. Breaking eye contact with the hostile individuals shifts the attention away from them. You may also use the opportunity to reiterate your conclusions, especially if you detect that the question is slightly off track. Do not go out of your way to do this for every question,

DEALING WITH QUESTIONS

Charles Doswell, Research Meteorologist, University of Oklahoma, and Consulting Meteorologist

Remember that you're likely to be the expert about the material you're presenting. Almost certainly, no one knows more about your work than you, so there's no real reason to be intimidated by the audience if you've done something worthwhile. If you are uncertain about its worth, then reconsider making the presentation!

Some questions might well stump you. Don't feel obligated to guess an answer unless you admit in advance that you're only guessing. Ignorance usually is forgiveable in technical presentations and if you simply don't know an answer, a simple "I don't know" can be an appropriate response. If you're uncertain, then admit your uncertainty. If you made an important mistake or omission, then admit it and be glad someone found your mistake so you can fix it. Honesty is definitely the best policy, as trying to handwave your way around a tough question only reduces your credibility.

Some audience members are on an ego trip and just want to show off how much they know. Remember that so long as you're the speaker, you're in control, and you should not willingly relinquish that control to someone in the audience. You might simply interrupt their interruption by asking "Excuse me, do you have a question or are you just making a statement?"

At other times, a questioner may not accept a simple answer and wants to engage you in a long argument. This might be acceptable in some circumstances, but it often monopolizes the question-and-answer time. In such a case, it is quite acceptable to terminate the discussion and suggest that you'll continue the discussion with him or her at the end of your allotted time, to allow others the chance to ask their questions.

You may get a question such as "Did you consider the Gezockstihagen effect?" or "Did you take into account the Hyperphantic Theorem?" No matter if you did or did not, a simple yes or no is probably not going to satisfy the questioner. Be prepared to justify why you did not account for what's likely to be his or her pet topic. If it should turn out that you never even heard of such a thing, say so, and be prepared to defer a lecture on the subject by the questioner to after your talk.

At times, a question can be confusing. It may be valuable to get the questioner to clarify the question. You might ask "Are you asking me about this-and-that or such-and-so?" Or you might respond with "If I understand your question, you're asking me to resolve the thermobaric flanxit issue. Is that correct?" Restating the question not only ensures that you're indeed answering the question as asked, but buys you some time to gather your thoughts on the question.

otherwise you may appear to be a slippery politician relying on your talking points.

Have a friend or colleague in the audience write down the questions that are asked. You may forget them, and they may be useful for improving your presentation in the future or for writing the formal publication.

Whatever you do, treat your questioner (and hence, by extension, your audience) with respect: "That was a really good question. Thank you for asking it."

What if I get no questions? Did people hate my talk? Probably not. Keep the following in mind:

- If it is late in the week, people may be tired.
- If it is before lunch or the end of a session, then people want to get lunch or coffee.
- If you gave such an astoundingly straightforward and well-presented talk, you answered all their questions in your talk.
- Your topic was not of interest to any in the audience (especially true at some conferences).
- People may not want to stir up trouble.
- People may not be asking questions of anyone. Sadly, this response is typical of many conferences (Errico 2000).
- In some cultures, asking questions is believed to be impolite.

After the session is over, try to stick around for a few minutes before running off to get coffee or go to lunch. Often, people may want to talk with you in person, so give them time to get to the front of the room to see you.

One final point to close this chapter. You are a unique individual. Make your presentations an extension of your personality. Think of new and fresh ways to present your science. Analogies, stories, props, and humor can go a long way to elevate you above the mediocrity of most scientific presentations. As your skills progress, vary the structure of your talks from the standard format (introduction, data, methods, results, conclusion) and express your creativity.

POTENT POSTER PRESENTATIONS

27

The other side of many scientific meetings are the poster sessions. In contrast to the relatively sedate monologues occurring at the podium, the poster sessions are boisterous and frenetic. How do you lure an audience amid the din and distractions? How do you communicate your results persuasively? This chapter addresses how to organize, assemble, present, and market your poster in a way to entice viewers, stimulate dialog, and enrich an active poster-session environment.

Poster sessions are a vital part of conferences and can be stimulating places to exchange scientific ideas. Sometimes viewed as the consolation prize for not getting an oral presentation, posters, if well constructed and well presented, can often be much more engaging and rewarding than oral presentations. I have heard the disappointment of colleagues who have spoken on what they thought was an exciting topic, only to receive middling interest or poor attendance from the audience, whereas a poster on the same topic would have congested the hallway with enthusiastic visitors.

With the pressure to make their meeting a success, conference committees sometimes select only the best or most well-known speakers to give oral presentations. Students and speakers not known to the committees may be less likely to receive talks. Thus, giving a good poster presentation by way of introducing yourself through your research to the conference committee members may be your way to more visibility at future meetings.

Corporate advertising that is simple, legible, and attention commanding sells. In the same way that consumers know Frosted Flakes because of Tony the Tiger, not because of the ingredients list or the nutritional information, the details of how your model was configured do not attract an audience—

the unexpected conclusion does. Imagine yourself walking through a poster session. Which posters will grab your attention and want to make you look? Certainly not the majority of ones that you see at a typical session, those that suffer from too much detail, too many words, and not enough *wow!*.

As discussed earlier, an oral presentation is necessarily condensed from a manuscript to meet time constraints. The difference between an oral presentation and a poster is an even greater commitment to focus. The best posters are characterized by a focused topic and a reduced amount of text and graphics relative to an oral presentation. This theme of focus and minimalism emerges repeatedly in this chapter.

27.1 TWO WAYS TO DESIGN A POSTER

There are two extremes of poster design. The first type of poster is the *self-discovery* poster (or less flatteringly, manuscript-on-the-wall poster). As the name implies, this poster contains the elements of a scientific manuscript: introduction, data, methods, . . . , conclusion. This poster is designed to be displayed without the presenter necessarily nearby. Usually such posters work best for conferences where the posters may hang for the few hours of the poster session, remain on display for most of the week, return home with the presenter, and then are put in the hallway at work. Without the presenter around, the poster needs to be largely self-explanatory with the viewers discovering the poster by themselves. Because large blocks of text do not attract anybody, results need to be explained in readable short-sentence or bullet form.

Figure 27.1 is an example of a well-constructed self-discovery poster. Below the title and the author list, a large box spans the poster and is boldly labeled "Introduction." The text briefly explains what the McICA radiation scheme is, its advantages, and a possible disadvantage. The introduction also says that McICA was installed in the ECHAM5 climate model and, then, boldly and in larger font, asks the central question of the poster, "What have we learned?" Except for a box with two references at the bottom, the remaining five boxes each highlight one principal result of McICA from its three years of tests. Each of the boxes is headlined by the principal result, contains one relevant graphic for support, and is summarized by one or two bullet points.

Although this poster was designed for a specialist audience because of the content and some terms and acronyms being undefined (ECHAM5, ISCCP, beta distribution), even nonspecialists can appreciate the layout and the principal results because the authors used nonspecialist language for the most part. I might recommend the font size be a little larger, especially on the figures, to enhance the readability of the poster from more than a meter away. I would also stress to the authors to reduce further the number of words on

Tests of Monte Carlo Independent Column Approximation in ECHAM5

Petri Räisänen & Heikki Järvinen

Finnish Meteorological Institute

Introduction

The Monte Carlo Independent Column Approximation (McICA) for computing domain-average radiative fluxes in GCMs separates the description of unresolved cloud structure from the radiative transfer solver, by dividing the cloud field into a set of subcolumns. One or more randomly selected subcolumns are then used for each point in the spectral integration.

This allows a very flexible treatment of subgrid-scale cloud structure in radiation calculations. The results are **unbiased** with respect the full Independent Column Approximation, but they contain conditional **random errors**. This "McICA noise" is the only potential disadvantage of McICA.

ECHAM5 provides an especially interesting testbed for McICA because it carries prognostic variables for the subgrid-scale probability distribution of total water content (Tompkins, JAS 2002). This allows us to derive subgrid-scale cloud variability directly from the resolved-scale model variables (Fig. 1).

After 3 years of tests of McICA in ECHAM5, what have we learned?

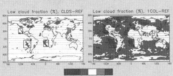

Figure 1. Cloud fraction C and PDF of condensate q_c are derived from the PDF of total water content q_t, subject to the assumption that the saturation specific humidity q_{sat} is constant within the grid box. Those points with $q_t > q_{sat}$ are cloudy and have condensate amount $q_c = q_t - q_{sat}$. In this example, q_t varies between 7 and 8 g/kg, $q_{sat} = 7.3$ g/kg, and α and β are shape parameters of the beta distribution.

---

1. The primary impact of McICA noise in ECHAM5 is a (very) slight reduction in low cloud fraction

Figure 2. Differences in low cloud fraction (percentage) between three versions of McICA in ECHAM5 simulations with prescribed sea-surface temperatures:

CLDS = a typical implementation of McICA
REF = a very low-noise reference version
1COL = a very high-noise implementation

---

2. The reduction in low cloudiness originates from a non-linear response of precipitation formation to random errors in radiative heating rates

- A very fast process: the biases related to McICA noise stabilize in a couple of days!

Figure 3. Impact of McICA noise on low cloud fraction, liquid water path, and large-scale precipitation rate for a **very high-noise version** of McICA (1COL), for a large ensemble of short (5-day) simulations. The upper row shows global mean values, and the lower row values for those regions dominated by marine stratocumulus clouds (marked with rectangles in Fig. 2). Mean values for the reference simulation are given in the upper-right corner of each panel.

---

3. When the sea-surface temperatures are allowed to adjust, McICA noise leads to a slightly warmer climate

- This is a straightforward response to the radiative perturbation related to reduced low cloudiness

- **For a typical implementation of McICA, the impact is small** (comparable to a 0.1 ?m increase in cloud-droplet effective radius!)

=> In practice, McICA noise is a very minor issue for ECHAM5

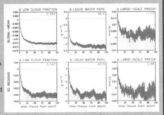

Figure 4. Impact on zonal-mean temperature and cloud fraction (a,b) due to McICA noise (for a **typical implementation of McICA**) and (c,d) due to a 0.1 ?m increase in cloud-droplet effective radius, in ECHAM5 experiments with a mixed-layer ocean model.

---

4. Clouds derived from the beta distribution of total water content feature too little subgrid-scale variability?

- At least, this is true for the subgrid-scale variations of total column optical thickness

Figure 5. Cloud variability parameter ? for (a) ECHAM5 (resolution T42L31) and (b) ISCCP data. ? is a measure of variations in **vertically integrated** optical thickness within the cloudy part of the domain (e.g., a GCM column).

---

5. For ECHAM5, cloud fraction is too often 0 or 1

- This is not unique to the beta distribution scheme for total water content. The problem is worse when an alternative (relative humidity -based) cloud fraction scheme is used.

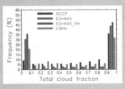

Figure 6. Frequency distribution of total cloud fraction for ISCCP data, ECHAM5, ECHAM5 with an RH-based cloud fraction scheme (ECHAM5_RH) and for global cloud-system resolving model data (CSRM; Khairoutdinov and Randall, Geophys. Res. Lett. 2001)

Interested in more details? Have a look at

Räisänen, P., S. Järvenoja, H. Järvinen, M. Giorgetta, E. Roeckner, K. Jylhä and K. Ruosteenoja, 2007: Tests of Monte Carlo Independent Column Approximation in the ECHAM5 atmospheric GCM. J. Climate, **20**, 4995-5011.

Räisänen, P., S. Järvenoja and H. Järvinen, 2008: Noise due to the Monte Carlo independent-column approximation: Short-term and long-term impacts in ECHAM5. Quart. J. Roy. Meteor. Soc, **134**, 481-495.

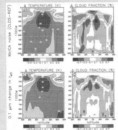

Fig. 27.1 A self-discovery poster. (Courtesy of Petri Räisänen.)

POSTER DESIGN BY SYNOPSIS

Valliappa Lakshmanan, Research Scientist, University of Oklahoma and NOAA/National Severe Storms Laboratory

Making posters comes down to a philosophy of what posters are and who the target audience is. I use posters to provide only a broad overview of techniques and results. Most of the viewers of my posters are interested in the topic, but are not performing research on that topic. So, putting too much detail in posters is pointless. This approach to making posters is very different from the approach to writing a journal article, which is written for someone who may discover the article by searching the literature for keywords. Such a person is knowledgeable in the field, is more likely to be performing research on the same topic, and is interested in the minute technical details of what we did.

The first step in creating a poster is to develop a one-minute synopsis of the research to be presented. Then, structure the poster as a whole to reflect this synopsis. Next, consider what you would say in a five-minute explanation of the research and make sure that the poster addresses those points without detracting from the one-minute explanation. Any detail beyond what you would explain in five minutes does not belong on the poster.

An example shows a poster describing our research on storm properties (Fig. 27.2). The one-minute explanation of this poster is that it describes the set of steps to extract storm-cell properties from radar imagery. Therefore, I designed the poster so that the main thing you see from a meter or two away are the arrows connecting the steps. The captions on the steps tell you what the steps are. The images themselves add the five-minute detail to the listing of the steps.

the poster. Other than those minor comments, this poster has a well-designed layout (the introduction box spans the whole top of the poster, boxes make it easy for the viewer to find the principal results) and accessible content (the introduction explains the purpose of the poster, the five principal results are in nonspecialist language).

The other type of poster is the *interactive poster* (Fig. 27.2). For such a poster to be successful, a presenter actively engages the visitors, interacts with them, and tailors the discussion to their interests. In the absence of the presenter, the viewer would get much less information from the poster. The organization of an interactive poster is much more flexible, using a minimum of words and lots of graphics. Interactive posters lure passersby to the poster with eye-catching graphics, a controversial statement, or stimulating conversation. Such posters can be the real highlight of a poster session, as large numbers of people are drawn to the debate.

Of course, posters of all kinds lie in between self-discovery and interactive posters. At the one extreme, a self-discovery poster allows the viewers to walk up, read it, and walk away, all the while the presenter standing there watching. At the other extreme, an interactive poster forces the presenter and the viewer to engage each other in conversation. Which is the better approach?

Automated Classification of Storms Based on Radar-Derived Storm Properties

Valliappa Lakshmanan, Travis Smith, Robert Rabin
University of Oklahoma & National Severe Storms Laboratory, Norman OK, USA

Goal To identify the storm type (supercell, linear, pulse storm or non-organized) in real-time.

Why? 1. Automated classification can be used to create climatology of storms across CONUS
2. The climatology can be used to create guidance for probabilistic warnings.

Technique

1. Some of the spatial grids input into the storm type algorithm: these are derived from multi-radar 3D grids created in real-time for all WSR-88D in CONUS

Reflectivity @ 11km   Reflectivity near ground   Reflectivity @ -20C

Prob. Of Significant Hail

VIL   Az Shear 0-3km

2. Pixels in the reflectivity composite field are clustered to find storms at different scales (20km², 160km², 480 km²). Properties are extracted from grids on left at these scales.

5. Storm type algorithm running in real-time. The results are shown visualized using Google Earth

4. Train decision tree on data

Part of automated storm type decision tree learned from data

MeanRef < 46 dBZ

Size < 477 km²2   AspectRatio < 3.6

Speed < 13 m/s   Line   Super cell   Line

Low-level-shear < 0.002/s

Line   Super cell

3. Human-training of storm-type algorithm, classifying storms into 4 types: supercell, line, pulse, unorganized

Future Plans 1. More categories of storms
2. A broader, more diverse training set
3. Build climatology in collaboration with NCDC

Can I try this on my data? Yes, you can! Download the software from
http://www.wdssii.org/ and run w2segmotionll

Please do stop me if you see me in the hallway! I'd love to address any questions or comments.

Partial funding for this research was provided under NOAA-OU Cooperative Agreement #NA17RJ1227

Fig. 27.2 An interactive poster. (Courtesy of Valliappa Lakshmanan.)

Given the variety of posters that can be created, coming up with detailed rules for their construction is more difficult than for journal articles or oral presentations. Nevertheless, given the guidelines presented here, I encourage you to explore your creativity.

27.2 CONTENT AND LAYOUT

Whether you are designing a self-discovery or interactive poster, the basic principles are the same. As with writing and oral presentations, the purpose of the poster needs to be considered before construction even begins. The one-minute synopsis discussed by Valliappa Lakshmanan in his sidebar (page 308) is the hook that you need to focus the poster and the bait to get people to visit. Make the content provocative. Because of the large number of posters at a conference all being presented at the same time, your title should be short and attention commanding to stand out from the others. Highlight the stimulating material on the poster—use a background color that makes it stand out or place a giant question mark or exclamation mark next to it.

Many of you feel an almost physical pain in deleting information from your poster. I know, I have felt it too. But what I have discovered, as have many others, is that it is possible to communicate almost as much in far fewer words and figures. —Warren Wiscombe, Atmospheric Radiation Measurement Chief Scientist, National Aeronautics and Space Administration

Visual allure is also an essential part of the bait. Because too much text turns away viewers, about 50% or more of the poster should be figures or photos. Perhaps a central image to the poster features a mysterious result demanding explanation, surrounded by other supporting observations. Or the poster focuses on the design of a new instrument, with data from its first field tests. Similarly, never crowd too much material on a poster—leave lots of empty space, otherwise the poster becomes too visually taxing for the audience.

Use a two- or three-column format, if necessary. The different sections should be made clear so that the reader knows how to naviagate the poster and in what order. Should the audience read across or down? Use vertical or horizontal lines or boxes to partition your poster into sections, which are helpful to viewers trying to navigate your poster. For example, Petri Räisänen used numbers 1–5 in Fig. 27.1, and Valliappa Lakshmanan used arrows in Fig. 27.2.

The conclusions should be readily apparent, whether spread out over the poster (e.g., Fig. 27.1) or in a single box entitled "Conclusions." When using a single box, most people pick the bottom right, although near the top is much more prominent because viewers do not have to bend over to see what is arguably the most important part of your poster. Use a differently colored box to highlight the conclusions, emphasize its location, and draw the audience directly to it.

Remember that minimal is better, especially for interactive posters (Table 27.1). First drafts of posters always tend to be too wordy. For that matter, so do second and third drafts. Draft out a poster, then cut and cut and cut. Use short,

Table 27.1   Things to avoid on a poster

Photos or other detailed patterns as background
Long blocks of text
Small unreadable graphs
Complicated tables
Equations (unless sparingly and elegantly done)
Pixelated graphics (insufficient resolution)
Chartjunk (decoration, logos, unnecessary graphics)
References (include on a separate page or in the extended abstract)

---

uncomplicated phrases, bullet points, and graphics. As with slides, minimize chartjunk such as excessive logos, allowing the most important material to stand out. The next poster session you attend, look at the quality of the posters and ask what it was that attracted you or scared you away from the poster. Adopt similar principles in your own design.

27.3 PUTTING IT ON PAPER

As with the electronic presentations, readability of a poster is key. Use dark type on a light-colored background to prevent the ink from smearing and to make the poster more readable; the background does not have to be white, but light blue, yellow, or light green are all good choices. To create a strong contrast for a conclusions panel, for example, try a bold color scheme of white text on a dark blue background. Avoid having the poster all be one boring solid color. Background photos are too garish and often clash—or worse, merge—with the color of the fonts. When you use color, do so with purpose and for boldness. A bright color or different font can also be used effectively to highlight important points on the slide.

Choose large font sizes and large graphics. For the title and authors, use 72-point font and larger for readability from about 5 m, and the rest of the poster should use fonts 36-point and larger. Figures and text should be readable from about 1–2 m. For section titles and headings, use 54-point font or larger and different colors or fonts for emphasis. Larger font sizes also have the additional benefit of forcing poster authors to use fewer words. As with slides, sans serif fonts are more readable at large scales. Warren Wiscombe of NASA recommends using Optima, Comic Sans, or **Arial Rounded** fonts on posters, which look better than the Arial and Helvetica fonts at such large sizes.

Placing publication-quality graphics on a poster is rarely successful because the graphics were designed for close-up reading in a journal, not for long-range viewing on a poster. As such, figures will almost always need to be redone, enhanced, and annotated to make them more readable at the larger scale. Text will have to be enlarged to proper font sizes, and lines may have to be thickened to survive the enlargement.

Keep figures simple and self-explanatory. Avoid long captions, if possible. Use arrows and annotations to illustrate key points. A relevant photo can be an icebreaker to attract an audience to your poster. If you are presenting a modeling study of flash floods, a large photo or radar image of one of your cases is appropriate. Some people put their own photograph at the top of the poster so that others can identify the presenter during the conference. In addition, put your e-mail address and Web page on the poster.

Minimize equations on the poster, but, if they are absolutely essential, put them on a line by themselves and make sure to use a font size larger than the text font size, so that the equation stands out. Explain or define every symbol. You may also consider drawing circles around terms with a line leading to a text box explaining the physical significance of the term.

As the Spiros G. Geotis Prize winner for the best student presentation at the 1997 AMS Radar Conference, Sabine Göke of the University of Helsinki advocates giving the audience something to interact with: something to touch, windows to open revealing important results, an experiment to run, or games to play. Have props or an instrument on hand to show off. Run an animation on your laptop. Hand out prizes or small candies. Be creative!

Whatever graphical flourishes and gimmicks you use to attract people to the poster, remember that the ultimate goal is to communicate the research. Substance should trump style.

For my first poster presentation in 1997, I attached transparencies to the top of graphs to compare model results and measured data. This interactive element encouraged members of the audience to touch my poster and increased their interest in discussing my findings.
—Sabine Göke, University of Helsinki

27.4 ASSEMBLING THE POSTER

Know the poster size before designing the poster. European posters tend to be vertical, whereas posters in the United States tend to be horizontal and a little larger. A bad outcome would be to design a large U.S.-style poster and get to the conference in Spain to find out that you have no space to hang it. If you are new to creating posters, find a design that you like as a starting point and ask permission from the author to borrow the style. Many posters are assembled in PowerPoint or Adobe software.

Before printing, look at the colors on a test poster (smaller size) because the printed colors may not look the same as on the screen. Thoroughly proofread the poster on the small paper, as well. Misspellings and typos are easier to spot on paper than on the screen.

There are primarily two ways to print the poster. The first way is to print on a large printer that uses large rolled paper. If your laboratory or university does not have such a printer, commercial printing and photocopying companies have that capability.

The other way, albeit used less often, is to print the poster on individual sheets of letter- or ledger-sized paper and tack the sheets on the poster board at the conference. The advantage of this way is the ease of carrying the poster on the plane, rearranging it on the poster panel, and making revisions. The disadvantages are that multiple pages need to be hung and the optional handout is more complicated to create.

27.5 AT THE POSTER SESSION

Visit the poster area before the session to determine where your poster is going to be. Bring a few extra copies of the extended abstract, business cards, copies of any relevant manuscripts, and copies of your poster on letter- or ledger-sized paper. You may also want a bottle of water for yourself.

What is a poster session like? It is crowded, but lonely; invigorating, yet tiring; it is a great time to catch up with friends, but it is also a bad time to catch up with friends. Some poster sessions have hors d'oeuvres available and alcohol flowing to encourage participation and loosen inhibitions.

In this environment, I think more natural interactions occur with people interested in your research than during oral presentations. Unfortunately, because poster sessions require direct human contact, poster sessions can also be intimidating for both the person presenting and the poster viewer. As the presenter, your job is to market your poster. Be enthusiastic about your research. Do not be shy.

If I am alone and someone walks near my poster expressing a slight interest, I politely ask whether I could walk them through the poster. Most people appreciate this invitation. Alternatively, if you are a poster session attendee and walk up to someone else's poster, but do not get an invitation to be led through the poster, ask for one: "Would you explain your poster to me?"

Use Valliappa Lakshmanan's one-minute synopsis to lure people to your poster, particularly for people you do not know. If they appear hooked by the research, work in some material from the five-minute plan. Save the full five-minute plan for people you have fully engaged and for your colleagues who you know will want to hear it all. By gauging the knowledge and experience of your visitors, you may be able to skip some of the introduction or motivation and get right into the results. Remember that people do not always find what they are looking for in your poster, so do not be disappointed if you give your one-minute summary, then they nod politely and walk away.

When presenting, engage your audience and walk them through the poster, using the visuals as a map. Point to the poster frequently, but look at them in

When you present materials you must do it in a way that gives the impression that you think is it some of the most important information in the world. If you aren't convinced of that, why should others pay attention? Make believe you are an actor. The larger the group, the more you must ham it up and seem larger than life. —Cliff Mass, University of Washington

the eyes. Giving the same spiel about your poster repeatedly will be redundant and tiring for you, but remember that each new person is seeing your work for the first time. Try not to ignore someone because you do not know them—they might be someone who could offer you a job later. Try to engage everyone standing at your poster. Even if others come up and are listening in on your conversation, make them feel included. They may have some insight into the discussion and want to contribute. Speaking directly at them from time to time as the discussion progresses is your invitation to them that it is okay for them to contribute. Or, at a convenient time, stop briefly and introduce yourself, and find out more about their interests.

When they are done listening, thank the viewers for staying at your poster. Make sure that you catch their names, and exchange business cards if the exchange was fruitful.

If you find it difficult to attract people to your poster, you might do what Warren Wiscombe recommends: cruise the poster session, grab people, perhaps even by the arm, and lead them to your poster, giving them the one-minute synopsis along the way to entice them. Your (presumably willing) target will likely feel a bit special, knowing that someone left their poster to find him or her to show important science to. If you have brought back one the leaders in the field or your National Science Foundation program manager, others may be curious to see what the fuss surrounding your poster is about. As Warren says, "when you go fishing in the aisles, bring back a BIG fish."

One downside of posters is that if you have a poster in a session, then likely there will be other posters on your topic also being presented at the same time. Because you want to be in front of your poster to receive questions, finding the time to look at others' posters may be difficult. You can either take time off from your own poster or visit the poster room at some other time during the day, hoping that if you have any questions, you can find the poster owner later. Focus your efforts on those that have piqued your interest from the conference program.

27.6 A VISION

I am convinced that we can change poster sessions to make them the highlight of the conference. Conference organizers would then treat the presentations in the poster session with respect rather than as second-class citizens. Poster presenters would create their posters with a minimum of text and a maximum of creativity, and they would entertain and challenge the audience with interesting research. And, the audience would engage and debate the poster presenters, dragging their colleagues to an interesting poster, exclaiming "This is one you just *have* to see!"

CHALLENGES TO DELIVERING YOUR PRESENTATION

Experienced speakers know that giving an effective presentation is more than just coming armed with a snappy, well-rehearsed presentation. External factors, such as an illness, audio-visual equipment mishaps, and audience antics, affect your ability to command the attention of the audience and to give a flawless presentation. This chapter gives tips on how to travel, prepare yourself to avoid nervousness, handle unforeseen problems, and deliver your best presentations.

Business travelers have stories about canceled flights, lost luggage, and driving into snow or ice storms. Professional speakers have stories about being late for their own presentation, rushing to find the auditorium, or having their electronic presentation not work exactly right. Unfortunately, a successful presentation hinges upon things that are out of your control. This chapter provides some guidance about how to minimize the external forces threatening your presentation.

28.1 MANAGING THE INCONVENIENCES OF TRAVEL

Whether it is the uncertainty about what toiletries I can carry onboard, the humiliation of being searched at airport security, not being fed well, being cramped on a plane, or not arriving at my destination on time because flights were late or canceled, air travel is not as much fun as it used to be. Given an important scientific meeting or a job interview, how do we minimize its impact on our lives, our moods, and our presentations?

Fly early to your destination in case flights are late or canceled. Arrive a day earlier, if possible, and do not schedule your talk the same day as your flight. Even a delay of an hour on one crucial leg of your flight itinerary may mean the difference between arriving in time for your presentation and missing it entirely.

Where possible, try to travel to the meeting with carry-on luggage only, especially if your flight has connections. If you must check luggage, consider putting a change of clothes in your carry-on bag in case your luggage gets lost. Wearing the same clothes for two or three days, especially at a meeting where you want to look (and smell) your best, can drop the spirits of even the most cheerful person.

Having your laptop computer with you at the meeting can allow you to work remotely. Furthermore, you can alter your talk based on conversations with colleagues, audience knowledge, other presentations, better ideas you have had, digital photos taken during week (remember your camera-to-laptop cable!), if you are given extra time to talk, or if you are reclassified from a poster to an oral presentation. Laptops also allow animations at your poster. In addition, I have folders on my laptop and memory stick with all my scientific articles, presentations, and my research, so if questions arise, I can freely answer questions or trade files with colleagues at the meeting.

Despite the obvious benefits of carrying your laptop, you risk theft, loss, and even confiscation. *The Washington Post* reported on directives from the U.S. Customs and Border Protection that allow agents to detain your laptop, memory stick, MP3 player, papers, or books, even those from U.S. citizens, "absent individualized suspicion." Therefore, back-up your laptop on an external storage system before you leave. Never put your presentation or laptop in checked luggage. Always have a copy of your talk on your computer and on a memory stick or CD accompanying you. You may also put a copy on a Web site so that you can download it if problems arise. If you are carrying a poster, use a carrying tube to protect the poster. Avoid losing your principal reason for going to the meeting.

The importance of a good night's sleep before your presentation cannot be overemphasized. When choosing a hotel, balance not only the costs, the amenities (e.g., free internet, free breakfast, workout room), and the convenience to the meeting venue (e.g., walking distance, easy access by public transportation), but also how quiet the hotel is likely to be and the other types of guests that may be found there. Sometimes you may not know that a group of high-school students on a field trip may be staying on the same floor as you, but the hotel management should make amends for your inconvenience.

28.2 PRESENTING IN A FOREIGN COUNTRY

When I moved to Finland and started giving scientific presentations, I found that my talks would run long. As someone who prides himself on generally ending on time, I repeatedly found it difficult to meet that standard in Finland. Eventually, I figured out that I was slowing down the rate at which I talked so that I was more easily understood. Therefore, accommodate your audience by considering the following changes to your presentation:

- If you had been using the one-slide-per-minute rule (Section 26.8), you may have to slow your rate of delivery down to more than one minute per slide with foreign audiences.
- . When presenting in a foreign country, the audience benefits from having more complete phrases and sentences on each slide, and for you to repeat a larger fraction of those words as you discuss the slide. If people do not hear you correctly (and hearing someone correctly if they talk fast or with a heavy accent is more difficult), they will at least be able to read the most important points. Adding more words does not mean to put paragraphs of text on the slide, however—you want the audience to focus on both you and the content of the slide.
- . Speak more simply. I avoid the colorful words and colloquial expressions I normally would use in favor of more common words.
- I try to look at the audience even more, making them feel more comfortable by being interested in their well-being. I can also see the vacant expressions on their faces if I start speaking too fast.
- In Finland, I often interact with more than just meteorologists. I speak to air-quality specialists, atmospheric chemists, aerosol physicists, engineers, and business professionals. Even the meteorologists have a different knowledge base than the meteorologists I know in the United States. So, I often need to present more background material or different material on some topics rather than just jumping right into my usual specialized material.

Finally, you may experience an extra twinge of nervousness when speaking to a foreign audience. Remember that at international conferences native English speakers are often the minority, so your audience probably understands how you are feeling quite well.

28.3 COMBATTING NERVOUSNESS

Imagine how Luke Howard felt on a December night in 1802. Howard was a modest, self-doubting pharmacist, speaking in a cold basement room to an audience, with some members eager to go to a dinner meeting later that night at the Royal Society. An amateur cloud-watcher, but an infrequent scientific speaker, Howard overcame his anxiety to deliver his presentation, "On the modifications of clouds." The talk was well received and the eventual publication

Table 28.1   Combatting nervousness

Be excited to show off your work.
Present material you are knowledgeable about.
Rehearse the talk until you are comfortable with it.
Focus on the point you want to communicate, not the presentation itself.
Get angry if fear limits your ability to deliver for the audience.
Visualize success.
Script the first few slides to ease you into the talk.
Use notes, if needed.
Exercise or take a walk to burn off excess adrenaline.
Arrive early, but not so early that the waiting increases the tension.
Take 30–60 minutes before the talk to mentally prepare.
Distract yourself by socializing with friends or talking with the audience.
Relax. Do some quiet meditation or small exercises to relax the muscles.
Avoid coffee, cola, or other caffeinated drinks.
Do not fidget. Place your hands at your side.
Take a few deep breaths before starting to speak.

There is one researcher, whose name I don't remember, who helps me to avoid being nervous. The reason is that he gave the absolute worst presentation I've ever seen. He had two bad transparencies, read all the text from paper, and didn't understand any questions. I always get confident before a presentation because I can't possibly suck that bad. —Vesa Hasu, Helsinki University of Technology

of that work in *Philosophical Magazine* led to the classification system of clouds that we use today (cumulus, stratus, cirrus, etc.).

Fear of public speaking is something nearly all of us have experienced. Total comfort, however, is not ideal because without some adrenaline, the talk may be perceived as flat and uninspired. What can be done to eliminate excess nervous energy and channel the beneficial part?

One cure underlies nearly all forms of anxiety: better preparation. Most of the ways in Table 28.1 to combat nervousness point to preparation to prevent you from failing miserably. For example, if you are having trouble remembering the right phrasing, write note cards or sticky notes, place them in front of you during the talk, and refer to them discreetly. The more nervous you feel, the more preparation you may need. Be careful though of excess nervous energy, which may cause you to look for things to do, such as drinking three cups of coffee. Too much caffeine may amplify your anxiety, even if you do not normally get nervous.

Wear comfortable clothes that give you confidence. Fidgeting with your clothes during your presentation or itching in that wool jacket is something you do not want. Wear shoes that you can be comfortable in, especially if your hotel is far from the conference center or you will be standing while presenting your poster. Do not wear that shirt with the button that has a habit of coming undone. Wardrobe malfunctions, or even the threat of one, can destabilize even the most confident speaker. If you have an important presentation, do something special for yourself. Buy a new shirt. Even if no one else notices, you will feel a little extra special when giving the presentation. I choose Jerry Garcia ties. The bright colors and creative designs—along with the irony of wearing the artwork of one of the counterculture's greatest icons as my corporate noose—boost my confidence a little bit more.

As with writing, maintain a positive mental attitude. Be proud to show off your newest research results to your friends and colleagues. Mentally visualize an excellent presentation and people congratulating you afterward. Visualizing failure and obsessing on it is not productive nor calming.

Finally, I emphasize these five points about anxiety and public speaking:

1. Scientists generally do not expect much from most speakers because we have become mostly jaded by the mediocre.
2. In general, the audience wants you to succeed. They want to come to a presentation and see something extraordinary.
3. Fear of speaking directs your focus internally instead of externally. The audience is the reason you are there; return the emphasis back to them.
4. The anticipation of the talk and the beginning of the talk are the worst times for anxiety. Most of your nervousness will subside once you are several slides into your presentation.
5. The audience is not likely to see most of your minor nervous habits (e.g., rapid heartbeat, fast talking, butterflies in the stomach). Therefore, if they do not see you are nervous, why let them think that you are?

28.4 AVOIDING AND MANAGING ILLNESS

Conferences and air travel are well known for disease transmission. Bringing people from all over the world and putting them in small confined spaces such as airplanes and meeting rooms means that anyone coughing or sneezing is likely to transmit his or her cold to you. If you get sick, do your best to take care of yourself. Get enough sleep. Carry a handkerchief, throat lozenges, pain-relief tablets, or whatever else makes you feel better. The day of your talk, drink an herbal tea (e.g., slippery elm, chamomile, ginger, peppermint) with some honey or lemon, gargle with salt water, or try some other home remedy. Minimize speaking during coffee breaks. Do whatever you can to save your voice before your presentation.

When I first started giving presentations, the very thought of it was extraordinary and intimidating, and I wanted a script. I practiced my presentations in front of my wife, and I wrote a lot of text on the graphics (the script). Now, I find I no longer need or want a script. Instead, I practice my talk silently once or twice to get the timing and the overall message (a gift to both my eventual listeners and to my wife), and then ad-lib when delivering to add a degree of freshness. —Paul Roebber, University of Wisconsin–Milwaukee

28.5 WHEN THINGS GO WRONG

Despite the best preparation, the smooth presentation you looked forward to giving may not happen. The projector bulb may burn out, the electricity may be cut, or a tornado siren may sound. Part of a swift recovery from such incidents is to remember the following:

- Take a breath. Do not react immediately to the problem.
- Do not panic, get angry, or cry. The professional response is to be cool and calm. Have a chuckle about it.
- Do not call excessive attention to or be fixated on the problem.
- Enlist the moderator or sponsor to fix the problem. Do not blame others for any problems, even through humor. You need them on your side to get the situation fixed, plus you may lose the support of the audience.
- Enact a reasonable back-up plan.

Problems with audio-visual equipment will disrupt your audience's concentration. Without the visual cues on the screen, the audience will watch the people working on the repair, not you. During this downtime, you have several options. One would be to do nothing. If the event is catastrophic enough, you may not have a choice but to wait for the problem to be fixed. Another would be change tack entirely. Do a survey of the audience about their thoughts on your topic. Recount a story that is relevant to your topic. A third option would be to pull out your back-up slides or notes, and plow ahead with the lecture.

Two real-life examples illustrate the last option. At the Ph.D. defense of one of my students, the projector bulb burned out with just a few slides left. She continued talking while I walked around the seminar room showing her slides on a laptop to the audience. Although a corny solution, it allowed her to finish and not be held hostage to the expired bulb. In a different situation, at the seminar for a faculty interview, the overhead transparency projector bulb burned out in the middle of the candidate's talk. While four faculty members tried to fix it, the candidate calmly turned to the chalkboard, picked up a piece of chalk, and continued lecturing off the top of his head. I was deeply impressed by the speaker's ability to regain control and demonstrate his confidence in the material despite this disruptive situation.

Even after the problem is fixed, regaining the audience's attention may take several minutes. You might need a snappy comeback or interesting graphic or photo to recapture them quickly. If you have something up your sleeve, pull it out. In those downtimes sitting on the plane or riding on the bus, think about how you might get yourself out of these situations. You may never need it, but if you do, your preparation will have paid off.

COMMUNICATING THROUGHOUT YOUR CAREER

IV

COMMUNICATION IN THE WORKPLACE

29

From this book, you might have the impression that most scientific communication is formal, through peer-reviewed manuscripts and conference presentations. We have an opportunity to practice for these formal communication vehicles through our informal hallway conversations, memoranda, e-mails, and group meetings. Unfortunately, these everyday forms of communication can end up being ineffective too, resulting in misunderstandings, lost opportunities, and wasted time. This chapter presents an introduction to improving the following types of professional communications: memoranda, résumés and curricula vita, cover letters, e-mail, and meetings.

At first glance, scientists and businesspeople might not have much in common. Ultimately, however, they both want to succeed in the marketplace of ideas and make some money along the way. Of course, businesses are not the only places where memoranda, e-mail, and meetings take place, but professional communication requires just as much forethought, concision, and precision as scientific communication. In this chapter, we briefly address a few common forms of communication in the workplace (which may be expanded to include the forecast office, laboratory, or university department, as well as the business office).

29.1 WRITING MEMORANDA

Your organization may ask you to write a summary of a recent site visit you participated in, a position paper on how the organization should prepare for a new funding initiative, an argument for why your customers should upgrade their software application, or an internal proposal for a large instrumentation purchase. Such documents (called *memoranda* or *memos*) may be required to inform, motivate, or persuade just like a scientific paper, but they will likely not have the same structure as a scientific paper. What are the characteristics of an effective memo?

If you were to ask 100 managers to list the characteristics of an effective memo, I suspect all 100 lists would include the same item: effective memos are short. Managers and other decision-makers—many of whom have mountains of reports, documents, and other memos waiting to be read on their desk—probably do not read memos longer than one or two pages in their entirety, if at all. Indeed, there might even be a relationship between how tall their pile is and how high up in the management structure they are. Therefore, the higher up your memo needs to travel, the more valuable that person's time is, and the shorter the memo should be to give it a better chance of rising to the top of the pile.

Here is how you can prepare your memo for just such an upward journey:

- Stay focused on material that your target audience needs to know. Avoid background information that they likely know already and details that they do not need to know.
- Use bullet points and numbered lists to list, summarize, and emphasize content.
- Make your memo shorter through précis of a longer document or of a first draft (Section 13.3).
- If action is required, give the recipient a concrete plan or a decision point with clear options.
- If the recommended action is not taken, what happens then and what is the incurred cost?
- Include the current date, and clearly identify the dates of any deadlines on the first page. However, avoid arbitrarily assigning a deadline that may run into higher-level priorities that you may be unaware of.

Two additional benefits arise from this approach. The first is that the short document forces you to present only the most salient points. The second is that you do not have to waste time writing a ten-page report that no one will read. Should your two-page memo be interesting to others, then they may ask for a longer report.

29.2 RÉSUMÉS VERSUS CURRICULA VITA

Some people use the terms *résumé* and *curriculum vitae* (popularly known as a *CV* or *vita*) interchangeably. They are not interchangeable. A résumé, most frequently used in the business community, is a one- or two-page document

Table 29.1   Information to place in a CV

Your name and contact information (mailing address, phone, e-mail, Web page)
Education (degrees, years, departments, universities, locations)
Experience (dates, locations, job titles, brief job descriptions, skills)
Offices and other volunteer leadership positions
Professional society memberships
Awards and honors
Teaching experience
Mentoring experience
Language skills and level of fluency (spoken and written)
Academic visits and other educational experiences
Research grants (pending and received)
Consulting experience
Journals and organizations for which you have served as a reviewer
Field-program experience
Books authored
Peer-reviewed articles (published, in press, submitted, in preparation)
Book chapters and encyclopedia entries
Other publications (non–peer reviewed)
Invited talks
Conference presentations and extended abstracts

that summarizes your career history, qualifications, and skills, and is tailored for a specific job to demonstrate that you are the best candidate for the position. In contrast, a CV is the complete documentation of your career and is more frequently used in the academic and research communities. I regularly update my CV and keep it on my computer and Web site. When I have needed a résumé, I can construct one by knowing the specific opportunity that I am applying for and adapting information from my CV.

Numerous Web sites and career books offer advice about creating effective résumés and CVs, but Table 29.1 provides a list of information that you may consider including in your CV.

29.3   PLANNING AND RUNNING MEETINGS

Bad meetings can sap the vital time and energy you need to have a productive day. Although you may not be in charge of many of the meetings you attend, make the most of it when it is your responsibility. Be known in your organization as the type of person who values your colleagues' time by hosting fewer, yet more productive, meetings. Before any meeting occurs, ask yourself these essential questions:

EFFECTIVE COMMUNICATION IN BUSINESS

Chris Samsury, Senior Director, The Weather Channel, Inc.

The fundamentals of effective communication extend to all aspects of business. The clarity, concision, and accuracy of our messages are critical. Just as with journal articles, successful business communication requires that you take advantage of the opportunity to tell the story you want to convey (with supporting data) and to get a little help from your friends (i.e., proofreading). Additionally, knowledge of your audience (technical or executive, customer or provider, supervisor or direct report) should be expertly applied in all communications to gain your desired results.

Mistakes in business communication are frequently made in two areas—the first area is the job search (e.g., cover letters and résumés), the second is communication on the job (e.g., e-mails).

Cover letters and résumés

Desirable jobs attract numerous qualified candidates. When hiring managers or recruiters receive tens, if not hundreds, of résumés, your cover letter and résumé must quickly capture their attention in a positive way. According to some reports, the average résumé is viewed for only 30 seconds before a decision is made by its reviewer. Confusing or complex sentences, misspellings, and misuse of grammar or vocabulary succeed in capturing attention, but very negatively.

Fair or not, such mistakes may cause an evaluator to be skeptical of your accomplishments or, worse yet, to discard your application altogether. Harsh? Yes. Short-sighted? Sometimes. Reasonable? Absolutely. Given the importance of attentiveness to detail in many meteorological jobs, sloppy writing can be interpreted as weakness in a key future job aspect. Though cliché, you do only get one chance to make a first impression. Make sure it is a good one.

Key suggestions:

▶ Don't blindly trust automated spellchecking. Although it can correct numerous mistakes, it does not flag misused words or homonyms. *Advise* versus *advice* is one of the more common mistakes I see. Incorrect punctuation is frequent as well (*it's* versus *its*). Always have someone you trust read your cover letter and résumé for organization and mistakes. In self-proofreading, we will often read what we hoped to write, not what we really did.

▶ Is a meeting appropriate for this situation? Could the issue be discussed informally over lunch, by dropping into someone's office for a chat, by phone, or by e-mail?

▶ What is the purpose of the meeting? Is it focused on just a few points that can be resolved during the meeting? Or, will the meeting collapse into gossip or whining?

▶ Who is required to participate in order for the meeting to occur? Who is expected to attend, but not essential for the meeting to take place? Who is invited, but not required to attend? Who is restricted from knowing about the meeting?

▶ How will the meeting be documented?

- Be confident but not arrogant. Few hiring managers will embrace a cover letter that directly says "You're crazy if you don't hire me." Let the readers draw that conclusion themselves from your clear and concise description of your accomplishments.
- Beware of overusing prose in paragraph form. Use, but don't overuse, bullets where you can, allowing readers to quickly and easily grasp the essence of your message. If readers have to work to figure out why they should hire you, often they won't.

E-mails

Just because it's easy and fast to type up and fire off an e-mail doesn't mean you should. Almost everyone in business has too many e-mails to respond to and too little time to do so. In busy environments littered with cluttered inboxes, you must capture your receiver's attention quickly and concisely if you want your question answered promptly or a problem solved.

It is also absolutely critical to remember that e-mails are easily forwarded and can be easily misunderstood. Though extremely efficient, electronic communication often eliminates the opportunity to provide immediate clarification or to alter your tone for the receiver. Nonverbal cues that you might pick up face-to-face are unavailable. Though often impractical, consider whether a phone call or getting together would be the more effective communication method for your message.

Key suggestions:

- Just as with journal articles, your title (i.e., subject) needs to be accurate, concise, and compelling.
- Make your e-mail organized and clear. Use bullets or numbered lists to quickly draw the reader to your important points.
- Get to your point quickly. Include necessary details or background but be concise.
- Include a call to action. Clearly and concisely define what you need from the recipient and when you need it. Say when you will follow up. Even if your note is simply informational and does not require action, let your reader know that.
- For your most important e-mails, if you have time, craft your note and save it, but don't send it immediately. Come back after a period of time and reread it aloud. Is it still what you want to say? Did you have to reread any of your sentences because they were confusing? If so, change them.
- If the information in your e-mail is important and not confidential, seek out a trusted peer or mentor to review it. You may get only one chance to make your point or request. Make the most of your opportunity.

To maximize the utility and effectiveness of the meeting further:

- Send out an agenda prior to the meeting to maintain focus.
- Be flexible in canceling a meeting lacking in attendance.
- Start and end the meeting on time to show that you value the time of your attendees.
- Be proactive in controlling the meeting to keep on topic and on schedule.
- Avoid spending too much time telling the attendees material they already know or could have read in a concise e-mail.
- Determine how and when attendees can ask questions or provide input.

- If appropriate, consider doing something out of the ordinary (e.g., food, icebreaker, off-site meeting location, games) to stimulate an otherwise lethargic group.
- At the end, summarize the items requiring follow-up (action items), who is responsible, and deadlines for action.

Most people dread going to meetings, seeing them as a waste of time. This attitude is unfortunate because effective meetings can lead to a much stronger organization. You know what bad meetings are like. Use your creativity to help make such meetings more productive and interesting.

29.4   WORKING EFFICIENTLY, WORKING SMARTER

The pace of life has accelerated with overnight delivery, e-mails, and instant messaging, leading some to feel like Newman from *Seinfeld*: "The mail never stops. It just keeps coming and coming and coming, there's never a let-up. It's relentless. Every day it piles up more and more and more! And you gotta get it out, but the more you get it out the more it keeps coming in." When you include writing proposals and memos, organizing and attending meetings, and participating in conferences, we may seem to spend more time talking about research than actually doing it.

To reverse this trend, make a concerted effort to ensure daily communications are shorter, more informative, and more directed to the intended audience. Shorter memos and e-mails take less time to write and less time to read, a win–win for everyone involved. Busy people do not need to be burdened by unnecessary meetings. If a meeting is absolutely required, have a concrete agenda to maintain focus and leave with an action plan. If the meeting can be conducted without electronic presentations, doing so will save everyone preparation time. Following the guidelines in this chapter will make the workplace not only more efficient, but leave more time for the fun stuff.

COMMUNICATION WITH THE PUBLIC AND MEDIA

30

Throughout this book, we have generally considered the audience of our work as being other scientists, a relatively small segment of society. In this chapter, however, we focus on communicating with the rest of the public. When the public is surveyed, they say that scientists are one of the most trusted occupations, yet a majority do not believe in the scientific consensus on global warming. How can we use our position of trust to inform a population that looks to us for answers on important questions facing society? What are the best approaches to communicating with the public, especially through the lens that most of them will meet us: the media?

As scientists, we are taught how to communicate with each other, but we are rarely taught how to communicate with the public. Such an attitude may be that our reward system is based on formal communication through the peer-reviewed literature rather than through *The New York Times*. Yet the public funds most of our research, and we should not view regular updates on our progress, reported through the media, as a burdensome task. Clearly, your latest improvement to an atmospheric radiation parameterization may not be what *Newsweek* is interested in, but you should be able to speak intelligently to others about the radiation balance in the earth's atmosphere and its relationship to the greenhouse effect.

More importantly, however, *we are* the public. Despite being scientists, albeit atmospheric scientists, we may rely on a newspaper article, not a journal article, to describe how colony collapse disorder, an illness that causes honey bees to abandon their hives, may be related to genetically modified foods. Whether we like it or not, even we receive much of our science through the media.

If you have seen media reports on a topic you know about and bristle at the language used (e.g., the "clash of air masses" that is responsible for the location of Tornado Alley), the unstated assumptions, and the overhyped predictions, then consider what a biologist might think when reading a newspaper article about colony collapse disorder. Consider what you read carefully, not only in the journals, but in the media as well.

If scientists are wary of journalists, then journalists are similarly distrustful of scientists. To develop their story, journalists rely on scientists explaining why their results are important to the public in a way that is clear, understandable, and free of jargon. Furthermore, if the story will air on TV or radio, the scientist must deliver these explanations in sound bites, clear concise explanations only seconds long. If scientists cannot communicate in their own language to other scientists, how could they even hope to communicate under these conditions?

Despite these differences, scientists and journalists share a lot of common:

- Both scientists and journalists tend to be skeptics.
- Both have strong egos and do not want to be wrong.
- Both are occasionally guilty of selectively interpreting the data.
- Both are dependent on gatekeepers (editors in both science and the media) who decide what gets published.
- Each group faces a language barrier keeping each from communicating with the other.

Given this common ground, perhaps there is hope to bridge the gap between scientists and journalists through dialogue and improved understanding. Whereas journalists need to be more knowledgeable about science and how it works, scientists should incorporate training in communication to the public as part of their education. As part of this training, scientists need to describe their research in words and concepts that the public can understand and be interested in.

Before a scientist communicates to the public (e.g., letters to the editor, interviews, articles in magazines), some ethical issues need to be considered. The American Geophysical Union (2006) developed a list of obligations for scientists publishing outside the scientific literature. These obligations require the scientist to be "as accurate in reporting observations and unbiased in interpreting them as when publishing in a scientific journal." If a new discovery is to be publicly announced (e.g., press conference, news release), the evidence should be strong enough to warrant publication in peer-reviewed scientific literature. The manuscript should be submitted, preferably, before the announcement or followed "as quickly as possible" afterward.

The primary way that scientists interact with the media is through interviews. Interviews may take the form of anything from a brief phone call, an e-mail exchange, a 30-minute visit, or a live shot on the evening news. Just like the question-and-answer session after a conference presentation, interviews can be quite intimidating, especially because the reporter is in control. On occasion, a reporter will approach an interview with an agenda (e.g., global warming is not occurring, the National Weather Service failed to warn for the most recent floods, this field research project is a waste of taxpayers' money). Taking control before the start of the interview is one way to put the advantage in your favor.

30.1 PREPARING FOR AN INTERVIEW

Whenever journalists call, be responsive. Find out their deadline. Let them know whether you have the time to talk with them by their deadline, and, if you do not, recommend someone else.

Before you accept the interview, interview the reporter. This preinterview can identify the format of the interview, identify potential problems that may arise, and help you decide whether you wish to participate. Find out the reporter's name, their employer (i.e., the media outlet), and who the audience of the media outlet is. A local television station will want a different angle than a national television network. *USA Today* caters to a different audience than *Time*.

Find out the topic of the interview and whether the reporter is knowledgeable already about the topic. Often, the reporter may not have a science background, so you may have to provide some introductory information first. Try to work with them on their level. For example, you might ask the reporter, "Are we talking about hurricane intensity forecasting only, or will there be questions about climate change, too? Who else is being interviewed for the story? When and where will the interview take place?" You can also use the preinterview to set time limits on the interview. Finally, ask when the story will air or the news article will be published.

Although interacting with the media can be a scary thought, be honored that you received some recognition of the importance of your work to the public. Take this opportunity to communicate your knowledge of and your passion for your work. Prepare your talking points before the interview. Talking points are the one, two, or no more than three most important points for the audience to receive from you. Craft these talking points with the interests, needs, and concerns of the audience in mind. Imagine your audience and what they need and want to know about this topic, and then write down your main messages that speak to the audience, perhaps on index cards for a quick review when needed. Talking points should consist of memorable sound bites

PRINT, RADIO, AND TELEVISION INTERVIEW TIPS

Stephanie Kenitzer, Public Relations Officer, American Meteorological Society

Lights. Camera. Action! You are ready for the media spotlight. Read on for specific tips about radio, television, and print interviews.

Radio

- Is live or is it taped? Always ask. If you stumble during an answer on a taped interview, you can always redo it. Just ask the reporter, "Can we do that over? I'd like to clarify my point." You only have one shot on live shows so make it your best by being prepared.
- Don't use your cell phone, a headset, or the speaker phone for radio interviews. The sound quality is not good for radio, and cell phones tend to run out of battery power at the most inopportune time.
- Always turn off your call waiting. Those beeps can be heard on the other end of the line.
- Standing up while talking on the phone makes you animated and energetic, which really comes through in a radio interview.

- Paint a picture for the listeners. Remember they can't see you or your work.
- Because you can't see the reporter's face (no non-verbal feedback), be sure to ask if the reporter understood your answer.
- If you are in a radio studio with a host, make eye contact with them instead of all the other bells and whistles that can distract you. If there is no host, be sure to find one focal point to avoid getting distracted.

Television

- Find out if you are doing a remote interview (where the interviewer is elsewhere in a studio) or if the interviewer will be there with you.
- If no interviewer is present, be sure to look straight at the camera. Looking off to the side or the ceiling is extremely distracting. If there is a person conducting the interview, then look at them directly. If you're uncertain, be sure to ask.
- Be animated. Go ahead, move those hands and gesture. Show the passion you have for your chosen field of research. Just don't forget to look at the camera or the interviewer, and don't stray from the microphone.
- Occasionally it will seem like the camera crew and the interviewer are crowding you as they lean in

lasting no more than 30 seconds that can be easily reprinted or broadcast. Be sure to mention your funding sponsors. For example, the National Science Foundation strongly encourages oral acknowledgement during all news media interviews about research they have funded.

Before the interview, rehearse! Have a colleague or public relations officer ask sample questions to practice responding and to refine your talking points. Better to stumble in front of friendly colleague than in front of a TV audience.

30.2 INTERACTING WITH THE PUBLIC

This chapter may feel horrifying to you, given all the guidelines about interacting with the media, but we interact with the public in many ways beyond

to get a closer shot. Rather than backing away, you can lean forward a bit into their space to help you feel more in control.

- Check your attire to make sure your tie is straight, your necklace clasp is in the back, your name tag is removed, your pens and pencils are not sticking out of your coat pocket, and no other such visual distractions are apparent.
- Speak in 30-second quotes or less if possible; it may take some rehearsing, but such short quotes are less likely to end up on the cutting-room floor.
- If you've been invited to a television talk show, ask for a chair that doesn't make you sink in too far. Sit toward the front of the chair to engage the host, and concentrate on the interviewer and the other guests.

Print

- Although face-to-face print interviews are easier to do than other types of interviews, beware of becoming too comfortable and relaxed. Assume everything you say is on the record unless you have a very specific agreement otherwise.
- Print interviews allow you to show visuals such as computer animations and graphics. If possible, let the reporter know how they can obtain a copy to accompany the story, but don't overwhelm them with too many statistics and details. Pick the best two or three that complement your message.
- As with radio and television reporters, print reporters need good sound bites for their story, so make your messages memorable.
- If you hear a reporter typing while doing a phone interview, slow down to let the reporter catch up.
- Don't be concerned if the reporter tapes the interview. The recording will enable him or her to review your comments if they need clarification and help the reporter write a more accurate article.

A few final reminders for all interviews: Never say "No comment"—it sounds like you have something to hide. It's better to tell a reporter that you don't know the answer or that you're unclear about the answer based on your current understanding of the information. If you are not comfortable with a line of questioning, simply tell the reporter that you are not the most appropriate person to be discussing that particular issue. And finally, always, always, always, answer the questions. "Do you have anything else to add?" is a bonus question that gives you a chance to repeat your key messages, and it may end up being your best quote.

media appearances. When you complete a research project that has societal impacts, have you tried to communicate it to the public? If you have a Web site for your research, have you designed it so that a layperson could understand the basic aspects of what you do and what it means for them?

You may be part of a club or a church that brings you in contact with other people who are not atmospheric scientists. What a wonderful opportunity to practice speaking to the public! Volunteer to give a talk at a community organization about weather forecasting, hurricane preparedness, climate change, or air pollution. Your local chapters of the Rotary, Lions, Sierra Club, Chamber of Commerce, etc., are always looking for speakers for their meetings. Visiting classrooms to talk about science, hosting school tours at work, and participating in science festivals can be excellent opportunities to pitch your science at

TOP TEN INTERVIEW PITFALLS

Keli Pirtle Tarp, Public Affairs Specialist, National Oceanic and Atmospheric Administration

10. Assuming previous knowledge

Do not assume the reporter knows or understands your work, your subject matter, your geographic area, your organization, or anything related to the interview.

9. Treating the interview like a conversation

Reporters try to make an interview feel like a casual conversation. It's not. You will have better quotes if you repeat the question in your answer. Emphasize your messages. Repeat yourself—it's okay to say the same thing several different ways.

8. Speaking off the cuff

If needed, ask the reporter if you can call him or her back in a few minutes. Then take that time to prepare.

7. Filling the uncomfortable silence

This is a common trick reporters use to get you to say something you shouldn't. Once you've said what you want to say, be quiet. Wait for the reporter—even if it's uncomfortable! Avoid adding something just to fill the silence.

6. Not considering the medium AND the audience

TV interviews are different from newspapers, which are different from radio. Your answers should reflect that.

5. Talking about a subject that is outside your area of expertise

Yes, you are well educated and know a little (or even a lot) about many topics. Stick to what you know best. If you're not the best source, feel free to refer the reporter to another person or organization. "I don't know" or "I don't know, but I will find out for you" are always acceptable answers. Don't speculate. Avoid "what if" questions.

4. Rambling and speaking in generalities

Reporters expect short, simple answers. Get to the point quickly and be precise. Practice before the interview for the best results.

3. Getting too comfortable with the reporter

Avoid jokes and sarcasm. Don't get lulled into saying something you shouldn't. Never assume the microphone is off. Don't let misstatements go unchallenged. Stay calm and relaxed, but not too relaxed. Be cooperative—but control the interview.

2. Being too scientific or technical

Avoid jargon and especially acronyms. Remember words that are extremely familiar to you and those you work with might be unknown to the reporter or the public.

1. Assuming something is off the record

Bottom line: If you don't want to see it in print or watch it on TV, don't say it!

---

an entirely different level, and even possibly encourage the next generation of young scientists. Even cocktail-party talk can go a long way toward honing your ability to communicate what you do with other people. Opportunities abound in your daily life to practice your skills at communicating science with the public.

FURTHERING YOUR JOURNEY

31

One of the attributes of successful people is their abiding quest for self-improvement. What steps can we take to be successful? What are the techniques to improve our writing and speaking skills? This chapter suggests ways to continue our own growth as effective communicators and as scientists.

Forecasters—whether for the National Weather Service, for the private sector, or in broadcasting—require continuing education to stay abreast of the latest developments, techniques, and numerical weather prediction models. Researchers thrive on lifelong learning and contributing to the growing body of knowledge in our world. Professors and instructors must learn the latest science to keep their students best prepared for their postgraduation world. Similarly, writers and speakers, or simply those who communicate their work through writing and speaking, must continue to expand and improve their skills. Even experienced scientists who are the strongest writers and presenters still have room for improvement. One of the characteristics of the best people in all these groups is, in fact, their passion to be better.

Reading this book, and the other recommended resources in the For Further Reading section, is certainly one step to be better. In this chapter, I present some ideas (certainly not an exhaustive list) for more ways to improve your ability to communicate.

31.1 WRITE MORE

All those books of the sort *So You Want to be a Writer* say the same thing: you must learn to write daily. Even if just a few hours a day and even if you throw away most of what you produce, you must get into a routine of sitting down and writing. Waiting for inspiration leads to just that: waiting. Discipline liberates creativity, whether for fiction or nonfiction. Writing should be part of one's day, just as eating dinner is. Doing so makes it easier to develop writer's muscles, so to speak. Regular workouts at the gym build up muscles; regular workouts on your PC develops the ability to find the right words, feel the rhythm of sentence structure, and gain greater confidence that the apparent total textual mess in front of you can be transformed with a bit of patience into prose that sings. The more you write, the easier it is to write.

The above was written by Robert Marc Friedman, professor of history of science at the University of Oslo. He has written books about the Bergen School of Meteorology, oceanographer Harald Sverdrup, and the Nobel prizes. He has written a screenplay about Vilhelm Bjerknes and plays about Albert Einstein and Lise Meitner. He is as busy as any of us, yet he understands that progress on big projects takes time, and all his projects took time—not only time to conceive, but time to write, and time to write *regularly*.

Chapter 5 already discussed quite a bit about motivating yourself for writing, ways to combat so-called writer's block, and how to prepare your writing environment, but now at the end of the book I highlight six things you can do *now* to improve your writing:

1. Take writing in your daily life seriously. If you are writing an e-mail to your friend, a letter to your father, or a grocery list, practice good writing habits. The practice on these texts will improve your manuscripts. Proofread your e-mails and other writing before sending them out, even if the consequences of a typo are negligible. Let your colleagues know that when they see a misspelling in your writing that it is the exception rather than the rule. Show professionalism in all contexts.
2. Maintain your list of weaknesses in your writing (Section 13.5). Make a habit of checking for your weaknesses and correcting them.
3. Use the information in this book as you write and revise. Table 3.1 presents the five characteristics of a good title, and "Final Checks of Your Manuscript" on page 169 provides a checklist for your manuscript before submission, to name just two. Need a transitional word? Try "Common Transitional Devices for Scientific Writing" on page 71.
4. First efforts do not need to be perfect. Write without inhibition.
5. Revise, revise, revise! Do not be afraid to delete the most perfect sentence if it does not fit the theme of the paragraph.
6. Use others to help you improve, whether it be the comments from a colleague, your university's writing center, or a professional manuscript editing service.

31.2 READ MORE

Only 40% of the U.S. students who responded to an AMS member survey said that they "read printed research literature on a daily or weekly basis." Reading the scientific literature can make you a better writer and scientist. Doing so can inspire new creative ideas, teach you new methods to incorporate into your research, provide examples of how others perform and write science, and make connections in your brain between disparate concepts.

Reading is active and forces you to think. Even if the topic is not your primary interest, seeing other types of science can expand your mind. When you read a fascinating paper, ask yourself what about it intrigued you? Could you replicate this in your own work? Don't limit yourself to scientific literature— reading quality nonscientific literature can also help you develop into a better writer.

31.3 GIVE MORE TALKS

Just as writing more frequently will make you a more fluid, comfortable, and prolific author, speaking more often will benefit your oral presentations. Give presentations about your research at your university or laboratory. Travel to other cities to visit with colleagues, and give a seminar in their department's weekly seminar series. Form an informal lunchtime discussion group where members talk about their latest research, the newest journal article, or their personal interests. Offer to give a talk at your local AMS or National Weather Association chapter meeting. Present a lightning safety lecture to your local outdoor club. Give a slide show of your trip to Venice to your family at Thanksgiving.

31.4 ATTEND MORE TALKS

Regularly attend your department or laboratory's seminar series. See the techniques that others use to give good presentations, and adapt them for your own presentations. At seminars and conferences, be an attentive and interactive audience member. Take notes, and ask questions. Talk with the speaker afterward. At conferences, invite new people to join you at lunch. Your lunch friend today may be a future research collaborator.

31.5 DEVELOP A PEER GROUP

At your institution or within your discipline, find out who your peers are who have the same scientific or personal interests as you do. Form an e-mail discussion list to chat about the current weather, the latest journal article on

TEACHING WRITING SKILLS IN A MEASUREMENTS CLASS

Petra Klein, Associate Professor, School of Meteorology, University of Oklahoma

At the University of Oklahoma, the School of Meteorology requires all students to take a course in meteorological measurements during the fall semester of their junior year. I have served as the instructor over the last five years, and one of the biggest challenges has been the lack of communication skills of the students upon entering the course.

Two major components of the course are (i) biweekly lab experiments and (ii) a semester-long project for which students install meteorological instruments, take measurements over a certain time, and then analyze their results. These hands-on activities are essential for teaching students how to work with state-of-the-art instrumentation and data-analysis techniques. Students also get a better understanding of typical measurement and exposure errors, which is important for training students to critically evaluate data quality. The lab experiments and project studies are generally well received by students, and, overall, we have received positive feedback from students.

However, both components require that students summarize their results in formal lab or project reports, and, after I started teaching the course for the first time in 2003, I soon noticed that the majority of students had poor writing skills and needed more instruction on how to write scientific texts. It appeared that most students had never written or even read scientific texts, did not know how to integrate and reference figures and tables in a text, and had no idea about how to conduct a literature review and correctly cite other people's works. It was also clear that the reports were mostly written in a rush and submitted without ever being proofread by their authors or their peers. I have thus worked on improving the descriptions of their lab and project studies and have created a detailed guide about the required format of their reports. I have also integrated help sessions focusing on communication skills into the course, for which I mostly use online resources.

Additionally, I require that students submit the literature review portion of their project reports as midterm reports early in the semester. This assignment serves three purposes: students get started early in the semester with their project studies and the related writing, students learn about the typical writing style of journal articles by reading published papers, and I can provide feedback on their writing early in the semester.

Last year, I have also collaborated with the university's writing center by having instructors from the writing center come to one of the help sessions and conduct a peer review with the student's midterm reports. The quality of the project reports has certainly improved over the last years, and the efforts in teaching better communication skills have started to pay off. However, to bring the students' communication skills to a level that they can successfully communicate in their future careers, teaching modules focusing on writing and communication should become an integral part of several courses of their curriculum.

drop-size distributions in drizzle, or the best location for backcountry skiing in the Wasatch Mountains.

Peer groups are natural places to discuss the scientific literature, perhaps at a brown-bag lunch meeting. A reading group composed of peers, but without supervisors or professors, will develop your skills in critical and independent thinking, reading, and debating. If necessary, follow-up questions can be written down and later asked of these senior people or other experts, or even e-mailed to the author.

Peer groups are also useful for identifying colleagues who would be willing and able to read and review your scientific work, whether it be for a conference abstract, thesis chapter, or scientific article. The immediate feedback from a constructive peer can go a long way toward helping raise the quality of science. And, such interactions do not even have to be chummy. Charles Doswell says, "A severe critic is your best friend in learning how to write well."

31.6 INCORPORATE COMMUNICATION SKILLS IN THE CLASSROOM

Good instructors know that the best education comes from *doing* rather than being lectured at. Thus, incorporating communication skills directly into the classroom through assignments that are stimulating and challenging can give students the practice they need to be better communicators. The skills do not need to be taught in a separate class. Practice for students in written and oral communication in most atmospheric science courses can be incorporated into the regular curriculum. Exercises may involve debates, class presentations, literature syntheses, group research projects, discussions of the reading assignment, or peer review of student writing and speaking assignments. Prof. Petra Klein discusses one example of how she incorporated communication skills in a measurements class (see Ask the Experts on page 338).

The turning point for me was in my undergraduate technical writing course at Penn State when I asked the instructor what would make me take writing seriously. (I was trying to do so, but wasn't succeeding.) His response was that the light bulb would go on when I reached a point in life when I would be held personally accountable for the writing of colleagues, employees, and students. —Dan Keyser, The University at Albany/State University of New York

31.7 INTERACT WITH MENTORS AND COLLEAGUES

Besides building support from your peer network, finding mentors can accelerate your growth as an effective scientific communicator and advance your career. Those of us in positions of authority got here because others helped us up to this level by providing career advice, critiquing our presentations, collaborating with us on research projects, supporting our proposals for field programs, and writing letters of recommendation. We need to return those favors. Find out who you enjoy interacting with and have lunch with them once a week. Interview your professors, department chair, and laboratory director to find out how they got to those positions. Spend time meeting with the seminar speaker, job interviewees, or visitors from other universities or laboratories.

Finding effective mentors does not even require personal interaction. Just listen to colleagues you admire in the classroom, at conferences, and in the hallways. Read their articles. Emulate those styles you like, avoid those you don't. Learn from positive *and negative* role models.

31.8 VOLUNTEER

Organizations are always looking for volunteers, especially hard-working and enthusiastic ones. Professional societies are natural places to direct your volunteering energy. Contact the editor of your favorite journal to volunteer as a reviewer, serve on a conference committee, or run for office in your local chapter of the AMS or National Weather Association. Nominate your professor for the department, university, or AMS teaching award. If you do these activities well, others will gain trust in you and your abilities. You never know what may become of these connections, whether it be speaking opportunities, collaborations, or even a job. Make sure, however, only to commit to what you will be able to follow through on.

31.9 DEVELOP YOUR OWN STYLE

Most professional musicians received some basic musical training in the fundamentals, learning scales and playing others' compositions. As they grew older and became more interested in writing their own music, they learned by watching other performers or listening to their music. Eventually, the best developed their own style—for example, the unique style that is immediately apparent as soon as you hear the first notes of a trumpet played by Miles Davis, the iconic tone of B.B. King's Lucille, or the intimate vocals of Billie Holiday.

Similarly, as a child, you learned the fundamentals of communication in your native language, refining it as you grew older. As a beginning scientist, your emphasis shifted to developing a set of skills to perform research, critique scientific hypotheses, and communicate results through peer-reviewed journal articles. Your earliest papers and presentations may have followed the approach recommended in Chapter 4: introduction, data, methods, results, discussion, and conclusion.

Just as musicians develop their own style, so must you. Start by modeling your work after those you admire. As your skill in writing and speaking improves, push the envelope. Be creative, and explore different approaches not covered in this book. Retain the basic principles, but flout these principles from time to time. "Our best stylists turn out to be our most skillful violators," Gopen and Swan (1990) said. "But in order to carry this off, they must fulfill expectations most of the time, causing the violations to be perceived as exceptional moments, worthy of note."

31.10 LEARN FROM YOUR MISTAKES

Continuing the musical analogy, musicians occasionally fail. Following the disappointing sales of his first two albums, Bruce Springsteen was likely to be cut by his record label if he failed to deliver his best performance with his third album. The result was the classic *Born to Run*. When one of my presentations fails to live up to my expectations, my next one is that much stronger because I try not to make those mistakes again. Failure can be a very strong motivator.

When I feel like I failed, sometimes other people may not even recognize it. I suspect that they are just being nice, but the reality is that, by being one of my harshest critics, I see mistakes in myself that others do not. I would rather exert this pressure for excellence on myself than to disappoint others.

31.11 CONCLUSION

To close this book, there are perhaps no better words about effective communication than the pithy advice on writing from the brother of atmospheric scientist Bernard Vonnegut:

1. Find a subject you care about.
2. Do not ramble, though.
3. Keep it simple.
4. Have the guts to cut.
5. Sound like yourself.
6. Say what you mean to say.
7. Pity the readers.

—Kurt Vonnegut

APPENDICES

V

COMMAS, HYPHENS, AND DASHES

Four of the most widely used, but most misunderstood, punctuation marks in scientific manuscripts are commas, hyphens, en dashes, and em dashes. This appendix is not intended to be a complete style guide for punctuation, but it will provide an overview of the do's and don't's of these four punctuation marks.

A.1 COMMAS

1. Commas are used to separate transitional or introductory words, phrases, or clauses from the rest of the sentence. If you can remove the word, phrase, or clause from the sentence and the sentence is still complete and readable, use a comma. Or, indicative of spoken language, commas should be used when there is a pause. A sentence with a subject composed of a gerund phrase (phrase starting with a verb + ing) or infinitive phrase (phrase starting with "to" + a verb) should not be separated from its verb by a comma.

   CORRECT: Therefore, the maximum possible wind speed increases at higher convective Reynolds numbers.
   CORRECT: On the other hand, the static stability increases with time.
   CORRECT: Flying through the stratocumulus cloud deck, the research aircraft was able to collect four hours of measurements of drop-size distributions.
   CORRECT: Choosing a structure for the wavelet is crucial to the success of the method.
   CORRECT: To obtain the best performance from the disdrometer requires careful calibration.

The world is populated by three types of people: comma-happy people, comma-reticient people, and people who know how to use commas properly.
—Unknown

2. Commas (and an appropriate conjunction) generally join two independent clauses, although short clauses may omit the comma.

CORRECT: The coarse-resolution model output captured the convective storm structure, and the high-resolution model output captured the gust front.
CORRECT: Convection initiation was expected by the forecasters on 6 May 1995, but only cumulus formation was observed.

3. Commas set parenthetical or nonrestrictive modifying phrases off from the rest of the sentence (nonrestrictive modifying phrases often start with "which"). If the phrase can be removed and the remaining words do not change the meaning of the sentence, then separate the phrase by commas. As a special case, use commas to separate items in a location, such as the name of the state when included with its city.

If you think commas are unimportant, consider this phrase, which was heard on a 24-h cable news channel on 30 March 2003: "Critics of the war plan, now being executed by the U.S. military. . . ."

CORRECT: The applicability of the idealized model to real atmospheric vortices, in which buoyancy gradients are important, is questionable.
CORRECT: If the analysis has large errors, or if it has moderate errors in regions where forecast errors grow quickly, then the resulting numerical forecast may be poor.
CORRECT: A preliminary calibration estimate brings the radar within 1–2 dB of its appropriate measurements, which is sufficient for the analyses in the present paper.
CORRECT: A preliminary calibration estimate that brings the radar within 1–2 dB of its appropriate measurements is sufficient for the analyses in the present paper.
CORRECT: Green Bay, Wisconsin, receives more snow in January than in any other month.

4. Use commas to set off items in a list. Many journals add a series comma after the last item in the list to eliminate any misinterpretation.

CORRECT: Section 2 summarizes the numerical model, simulation methods, and analysis methods.

5. Place a comma within a list of adjectives or adjective phrases. Use a comma if the adjectives or adjective phrases can be correctly replaced with "and."

CORRECT: Storms east of New Zealand are embedded in a stronger, more zonal flow than those to the west.

A.2 HYPHENS

Hyphens are mainly used in three situations:

- in between numbers not representing a range, such as zip codes, phone numbers, technical report numbers, aircraft, or instrumentation: Boston, MA 02108-3693, (617) 555-3223, P-3 aircraft, WSR-88D;
- splitting up words across a line break (most word processing software does this automatically and reasonably well; when in doubt, refer to a dictionary); and
- compound words.

Compound words provide the greatest opportunity for the misuse of hyphens. Simple rules apply in some situations:

- Hyphenate spelled-out numbers: "twenty-nine," "two-hundred and thirty-four."
- Hyphenate phrases connected together as modifiers: "cloud-to-ground lightning," "cause-and-effect relationship."
- Hyphenate single capital letters joined to nouns: "X-ray," "T-bone," "H-factor."
- Hyphenate the values and units if they modify a noun: "500-hPa wind," but "the wind at 500 hPa."
- If you have two or more words that modify the same hyphenated term, hyphenate all the modifers: "lower- and upper-level potential vorticity anomalies," "700-, 500-, and 300-hPa temperatures."
- Adverbs ending in -ly do not need to be hyphenated with their modifying adjective: "slowly moving thunderstorm," "widely used parameterization," but "well-known equation." Furthermore, "the equation is well known" is not hyphenated because "well known" does not directly precede its noun.
- Hyphenate prefixes to clarify meaning: "recount" (to describe) versus "re-count" (to count again).
- In general, do not hyphenate a prefix, unless the word is a proper noun, three of the same consonants would appear together, or the prefix is "ex-" (meaning "former"): "reexamine," "reinvestigation," "trans-Atlantic," "shell-like," "ex-hurricane."

Beyond these rules, proper hyphen use depends upon context, current fashion, press style, and even the decisions of the copy editor. For example, as language evolves and new words are created, nouns that were initially two words may become hyphenated then become one word: blackbody, meltwater, landmass, snowmelt, and streamflow. In contrast, nouns that are currently hyphenated include lift-off, degree-day, and clear-cut. When used as adjectives, some words may be combined as one (e.g., leeside winds, brightband melting) or hyphenated (e.g., real-time model, along-shore flow, lake-effect snow). Even more confusing, some prefixes and suffixes require hyphens (e.g., half-barb, upper-level flow), whereas others do not (e.g., postfrontal, nonlinear, semigeostrophic). Because of the subtleties and vagaries of proper hyphenation, your best approach is to check the style guide of the journal, a dictionary, or *The Chicago Manual of Style* (The University of Chicago Press 2003).

A.3   EN DASHES

The en dash is used to combine two items of equal standing, to compare opposite items, and to hyphenate a compound adjective in which one part consists of two words or a hyphenated word. The en dash possesses the same width as a lowercase *n*, hence the name. In general, do not place spaces around the en dash.

- the National Centers for Environmental Prediction–National Center for Atmospheric Research (NCEP–NCAR) reanalyses
- 0000–1200 UTC 15 December 1987
- skew$T$–log$p$ chart
- Kain–Fritsch convective parameterization scheme
- relative humidities of 30%–70%
- a climatology during 1970–1999
- pp. 112–119
- the 800–600-hPa layer
- air–sea interaction
- Nobel Prize–winning research

The en dash also serves as a minus sign: −14°C.

The en dash is entered as two dashes (--) in TeX and LaTeX, except in math mode where a single dash will produce a minus sign. Generally, in Microsoft Word, you can produce an en dash with option-hyphen on a Mac and by inserting the en dash from the Symbol menu in Windows.

A.4   EM DASHES

An em dash is used to set off contrasting information, examples, or interruptions from the rest of the sentence. Historically, the em dash is the same width as an uppercase *M* and is twice as wide as an en dash. Many publishers do not place spaces around the em dash. Functioning similar to a comma, the em dash indicates a much more severe break in the flow of the sentence. It packs a punch and is meant for emphasis. Because the em dash is so powerful, use it infrequently to maintain its vitality.

CORRECT: The maritime air—flowing westward from the Atlantic toward Madison County—had dewpoint depressions of only a few degrees.

CORRECT: In the dry case, neutral stability is defined based on only one thermodynamic variable—potential temperature.

CORRECT: Although there are many ways to adapt observations, in this study only two simplified sample adaptive strategies—one idealized and the other a more realizable approximation to the idealized strategy—are tested.

CORRECT: Such tests must be carried out over a period at least as long as the radiative-subsidence timescale—about 30 days—governing the water vapor adjustment time.

The em dash is entered as three dashes (---) in TeX and LaTeX. Generally, in Microsoft Word, you can produce an em dash with shift-option-hyphen on a Mac and by inserting the em dash from the Symbol menu in Windows.

COMMONLY MISUSED SCIENTIFIC WORDS AND EXPRESSIONS

Just as Strunk and White have their list of commonly misused words and expressions, a list for the atmospheric sciences has long been needed. Some of the entries were written by others; many are my own irritations. Some people may agree with nearly all of these entries; others may agree with few, if any. Whatever your opinion, I invite you to at least think about how you are using these words and expressions.

Accuracy versus skill. When describing the quality of forecasts, the notions of accuracy and skill often are treated as synonymous, but they are not. *Accuracy* refers to the correspondence between forecasts and observations, with increasing accuracy associated with increasing correspondence. *Skill*, on the other hand, is associated with the relative performance of the forecasting system in question, when compared to some baseline forecasting system. Baseline systems often used for measuring skill include climatology, persistence, and model output statistics (MOS) forecasts; the idea is to measure the improvement (or lack thereof) of the system in question compared to the baseline system. An accurate forecast is not necessarily skillful, and vice versa. —*Charles Doswell*

Activity (convective, electrical, hurricane, lightning, thunderstorm). "Activity" is an imprecise word in these contexts. Be specific about the measure: number of cloud-to-ground lightning flashes, total flash rate, number of supercells, frequency of hurricane passage, etc.

Analysis of a vector quantity. When creating a gridded analysis from or interpolating a vector quantity, perform the analysis on each vector component (e.g., u and v for a horizontal wind field) not on magnitude and direction (e.g., wind speed and direction) (Doswell and Caracena 1988).

Causing. Be careful using this phrase in some contexts. Usually we do not know the chain of cause and effect in the atmosphere, although we often infer it. "Associated with" is a better option. Similarly, read *Statistical association does not imply cause and effect* on page 363.

Chaos/random. These two terms have very specific scientific meanings, so casual use of these terms should be avoided (e.g., "chaotic or random cloud patterns"). Use "poorly organized" or "disorganized" instead.

Cold-type occlusions, existence of. The cross-frontal difference in static stability not near-surface temperature is what creates the three-dimensional structure of a cold- or warm-type occlusion (Stoelinga et al. 2002). Because warm fronts tend to be much more stable than cold fronts, the three-dimensional structure of a warm-type occlusion will be favored to develop, irrespective of the near-surface temperature difference across the occluded front. Thus, cold-type occlusions should be quite rare, if they exist at all (Schultz and Mass 1993).

Collaboration versus coordination. These two terms can be misused in either an operational-forecasting environment or a research-and-development environment. *Collaboration* refers to the intellectual process having a collective goal of producing the best possible forecast or forecast product by the interaction of two or more weather information sources. In contrast, *coordination* is the obligatory communication to ensure the forecasts and products from two or more sources meet a minimum standard of acceptance from users. —*Neil Stuart*

Condensation occurs because cooler air cannot hold as much water as warmer air. Condensation and evaporation are always occurring regardless of temperature—what matters is whether the rate of condensation exceeds the rate of evaporation. The Clausius–Clapeyron relation states that the saturation vapor pressure of the atmosphere increases with temperature. Thus, when everything else is held constant, as the temperature increases, the rate at which increasingly energetic water molecules evaporate is more likely to exceed the rate of condensation. When the air temperature drops below the dewpoint temperature, the rate of condensation exceeds the rate of evaporation, and water droplets form. These processes occur regardless of the volume or pressure of the air. Thus, water vapor is not "held" by the air.

Convective initiation. Use *convection initiation* instead.

Convective temperature. The convective temperature is the surface temperature that corresponds to the elimination of any convective inhibition associated with ascending low-level parcels, usually by insolation. Presumably, use of this term implies that deep convection initiation is delayed until the convective temperature is reached, after which deep convection begins. If this were a valid concept, then deep convection should begin by clouds flashing into existence over big chunks of real estate, all at the same time. Instead, deep convection usually commences as isolated convective clouds, perhaps at a few places along a line, usually well before the attainment of the convective temperature. Sometimes, the convective temperature is reached and nothing happens. The use of "convective temperature" seems to imply that deep convection is initiated solely by elimination of the inhibition through solar heating. Because the reality is quite different, the concept of the convective temperature is not valuable in forecasting, and perpetuates an improper understanding of deep convection initiation. Thus, this term should not be used. —*Charles Doswell*

Correlate/correlation. Often authors will refer to "correlation" when they really mean a relation, an association, or a correspondence between two phenomena. Reserve "correlate" when you mean it in a mathematical sense, as when you calculate a linear correlation coefficient. In general, use "relate," "relation," or "correspond" instead.

Correlation, linear. See the sidebar "Misuses of Linear Correlation" on page 121.

Data. "Data" is always plural. "Datum" is the singular form, but I think saying "data point" sounds better.

Data, model output as. Some scientists are uncomfortable with model output being called "data." Reserve the use of "data" for observations, not the output from models.

Date/day. Do not use the word "day" as a substitute for "date."

INCORRECT: The day of the tornado in Lone Grove, Oklahoma, was 9 February 2009.

CORRECT: The date of the tornado in Lone Grove, Oklahoma, was 9 February 2009.

Dates and times. Use the standard format for dates and times, wherever possible: 1200 UTC 10 December 1994. Avoid the 12/10/94 or 12.10.94 formats because of the ambiguity of whether the date is December 10 (U.S. format) or October 12 (European format). Do not use the syntax "1200 UTC on December 10th," which contains more characters than is needed.

Difluence does not equal divergence. In meteorology, α represents the angle of the wind direction with the convention that wind from the north is 0° and the angle increases in a clockwise direction. In a natural coordinate system (s, n) where s is the direction along the flow and n is the direction normal to the flow (and to the right of the wind), divergence $\nabla_h \cdot \mathbf{V}_h$ is given by:

$$\nabla_h \cdot \mathbf{V}_h = V\,\frac{\partial \alpha}{\partial n} + \frac{\partial V}{\partial s}$$

where V is the wind speed. Difluence, the spread of streamlines downstream, is only the first term $V \, \partial\alpha/\partial n$ in the expression for divergence. Therefore, difluence cannot be equivalent to divergence, although they clearly are related. —*Charles Doswell*

Divergence/convergence does not cause vertical velocity. See also the entry for *causing*. Divergence of the horizontal wind $\nabla_h \cdot \mathbf{V}_h$ and vertical velocity ω are connected by the Law of Mass Continuity. In pressure (p) coordinates, this takes the form:

$$\frac{\partial \omega}{\partial p} = -\nabla_h \cdot \mathbf{V}_h$$

The simultaneous existence of ascent, with convergence at its base and divergence at its top, is a necessary consequence of mass continuity. Mass continuity is a diagnostic equation and contains no time derivative of vertical velocity. Hence, it cannot identify causes for vertical wind. —*Charles Doswell*

INCORRECT: Low-level convergence along the front caused strong ascent to occur.

INCORRECT: Deep moist convection resulted when a region of upper-level divergence became superimposed over a region of low-level convergence.

CORRECT: Ascent is associated with upper-level divergence and low-level convergence.

Dynamics. This term is often used to describe physical processes vaguely without actually stating what those processes are. Replace such expressions with a more physical description.

DRAFT: The strong dynamics of the rapidly developing extratropical cyclone . . .

IMPROVED: A strong short-wave trough in the jet stream was responsible for the rapid development of the extratropical cyclone.

Equations, formulas, and theories (generality of). Theories, equations, or empirical formulas are often developed for specific circumstances with a given set of assumptions, or based on limited datasets. As such, caution should be exercised if extending these theories, equations, or formulas to situations outside of their original intent.

False alarm rate versus false alarm ratio. Often people not being careful will refer to the false alarm ratio as the false alarm rate. Do not confuse the two! The *false alarm ratio* is the number of false alarms divided by the number of forecasted events, whereas the *false alarm rate* (also known as the *probability of false detection*) is the number of false alarms divided by the number of times the event did not happen (e.g., Wilks 2006, Section 7.2.1; Barnes et al. 2009).

Fog burning off: Popular as a colloquialism, this phrase misrepresents the physical processes involved in the elimination of fog and should not be used in a scientific context.

Forcing: Although an imprecise term at best, "forcing" is most troubling when used in connection with diagnostic equations, such as the omega equation, where the terms on the right-hand side are referred to as "forcing terms." Forcing is clearest in the context of an applied force resulting in an acceleration, where some process derives a time-dependent response. Thus, terms on the right-hand side of the horizontal momentum equation, such as the pressure gradient force, would be appropriately described as forcing. In the quasigeostrophic system, vertical velocity is not forced—it is merely required for consistency with the changes that are occurring to the geostrophic flow. *—Chris Davis*

Frequency. When using this word, ensure that the units are in "per time," such as the number of events per unit time. Otherwise, the expression is just a "number of events," not a frequency.

Front, definition of. A front is characterized by a horizontal gradient in density (temperature). Therefore, analyzing fronts should be performed using temperature or potential temperature only. (Virtual temperature, which accounts for the effect of moisture on the density of air can also be employed.) Including moisture in the definition of fronts, even indirectly through variables such as equivalent potential temperature or wet-bulb potential temperature, runs the risk of weakening the definition of a front—analyzing features that may not be temperature gradients, but merely moisture gradients, and implying a frontogenetical circulation when none may exist. For more on proper frontal analysis, read Sanders and Doswell (1995) and Sanders (1999).

Frontogenesis, as a measure of the intensity changes of a front. Frontogenesis is the Lagrangian rate of change of the horizontal temperature gradient (Petterssen 1936; Keyser et al. 1988). Thus, air parcels approaching a front from the warm sector experience an increasing temperature gradient, or positive frontogenesis. Petterssen frontogenesis says little about what the temperature gradient along the front is doing in time because even fronts where the temperature gradient is weakening experience positive Petterssen frontogenesis. A proper analysis of frontogenesis to explain the strengthening or weakening of a front would require a new formulation, a quasi-Lagrangian, or front-following, form of the frontogenesis function (cf. Schultz 2007 vs Markowski and Stonitsch 2007). Thus, the value of Petterssen frontogenesis is to objectively determine where active frontogenesis is occurring, not whether a front is strengthening or weakening.

Ignoring the shades of gray that exist in the natural world is one hallmark of bad science; employing multiple definitions for the same term is another. —Corfidi et al. (2008. p. 1301)

Frontogenesis, use of the tilting term. The Miller (1948) expression for frontogenesis to assess the physical processes acting to change the magnitude of the potential temperature gradient includes a tilting term. Some people have calculated the tilting term, then said that the complete frontogenesis expression can be used to assess the regions of vertical velocity. This approach is incorrect. Petterssen (1936) frontogenesis is the correct expression used to estimate the areas favorable for ascent (Keyser et al. 1988); the tilting term is not included.

Frontogenesis, warm or cold. Consider the term *warm frontogenesis* that some have tried to coin as an abbreviation for "frontogenesis along a warm front." This term does not make scientific sense because frontogenesis does not have a sign of cold or warm, only positive or negative (frontogenesis or frontolysis). To be precise, write out the phrase completely: "frontogenesis along a warm front."

Froude number. The Froude number Fr is classically defined as the ratio of the flow speed U to the phase speed of linear shallow-water waves, $\sqrt{gH}$, where g is gravity and H is the fluid depth. By comparison, in stratified flow over an obstacle, the quantity Nh/U is often referred to as either the Froude number or its inverse, where N is the Brunt–Väisälä frequency and h is the obstacle height. (Because of the ambiguity about Fr or its inverse, always define Fr for the readers.) Scaling analysis shows that Nh/U is the sole nondimensional parameter controlling two-dimensional hydrostatic flow forced by a steady wind U in an atmosphere with constant N. In fact, Nh/U is best referred to as a measure of nonlinearity because the perturbation wind u' is proportional to Nh in the linear limit. In contrast to the classically defined, shallow-water Froude number, associating Nh with the phase speed of a significant internal gravity wave mode is difficult.

A third context in which the Froude number arises is when a strong inversion is present and a reduced-gravity shallow-water Froude number is computed as $U/\sqrt{g'H}$, where H is the height of the inversion, $g' = g\Delta\theta/\theta_0$, $\Delta\theta$ is the potential temperature jump across the inversion, and θ_0 is a potential temperature representative of that in the inversion layer. Empirical observational and modeling evidence suggests that when the inversion is sufficiently strong, and the static stability below and above the inversion is sufficiently weak, $U/\sqrt{g'H}$ governs nonlinear flows in a manner at least qualitatively similar to that played by the classically defined, shallow-water Froude number. Nevertheless, the precise numerical value of the reduced-gravity shallow-water Froude number should not be overemphasized, because vertical wind shear and the finite thickness of the inversion layer introduce considerable uncertainty in its evaluation. In addition, at least one example involving coastally trapped waves exists in which the phase speed of linear disturbances in the presence of a strong inversion does not agree with the reduced-gravity shallow-water phase speed (Durran 2000a, Fig. 9). —*Dale Durran*

Gravity currents, cold fronts as. Despite the widespread use in the literature of equations to calculate the theoretical speed of a gravity current, Smith and Reeder (1988) argue that any similarity between the theoretical speed and observed speed of cold fronts is superficial. Thus, a close correspondence between the two is not evidence for a front being a gravity current. See also *Morphological similarity does not equal dynamical similarity.*

Greenhouse effect. The name, greenhouse effect, is unfortunate, for a real greenhouse does not behave as the atmosphere does. The primary mechanism keeping the air warm in a real greenhouse is the suppression of convection (the exchange of air between the inside and outside). Thus, a real greenhouse does act like a blanket to prevent bubbles of warm air from being carried away from the surface. This is not how the atmosphere keeps the Earth's surface warm. Indeed, the atmosphere facilitates rather than suppresses convection. —*Alistair Fraser*

Greenhouse gases behave as a blanket and trap radiation. At best, the reference to a blanket is a bad metaphor. Blankets act primarily to suppress convection; the atmosphere acts to enable convection.

As rapidly as the atmosphere absorbs energy it loses it. Nothing is trapped. If energy were being trapped (i.e., retained), then the temperature would of necessity be steadily rising (a temperature increase unrelated to global warming). Rather, on average, the mean temperature is constant and the energy courses through the system without being trapped within it.

The correct explanation is remarkably simple and easy to understand; namely, the surface of the Earth is warmer than it would be in the absence of an atmosphere because it receives energy from two sources: the sun and the atmosphere. —*Alistair Fraser*

Instability, conditional, convective, and potential. *Conditional instability* occurs when the environmental lapse rate lies between the dry- and the moist-adiabatic lapse rates, or the saturated equivalent potential temperature (θ_e^* or θ_{es}) decreases with height. *Potential* or *convective instability* occurs when the equivalent potential temperature (θ_e) decreases with height. Conditional instability is one of the three ingredients of deep, moist convection, and is therefore the proper instability to be considered for such situations (e.g., Johns and Doswell 1992). Potential instability is usually considered as instability over a layer that is released when lifted in slab ascent (e.g., Bryan and Fritsch 2000). Schultz and Schumacher (1999), Sherwood (2000), and Schultz et al. (2000) discuss the differences between and the origins of these terms.

Instability, presence of versus release of. The presence of an instability does not imply that it will be released (e.g., Sherwood 2000). Therefore, conditional symmetric instability bands are not a proper term. It is more accurate to say, "bands associated with the release of conditional symmetric instability in the presence of frontogenesis," acknowledging the presence of the instability and moisture, as well as the lifting mechanism.

Jet streaks, locations of severe weather. The four-quadrant model of a straight jet streak (e.g., midtropospheric ascent in the right-entrance and left-exit regions, descent in the left-entrance and right-exit regions) is often invoked as evidence of a preference for severe weather occurrence in the ascent regions, but Rose et al. (2004) and Clark et al. (2009) show that severe weather can occur in any quadrant of a straight jet streak, particularly in both quadrants of the jet-exit region and the right-entrance region. Curvature further accentuates the differences between expected locations of severe weather from the model alone and observed locations. That the four-quadrant model does not solely explain the formation and locations of convective storms is not surprising given that such storms are also influenced by low-level convergence (e.g., along surface fronts) and favorable environments of convective available potential energy and vertical shear of the horizontal wind, not just the synoptic-scale vertical velocities associated with the jet streak. This result is a reminder that convective storms result from the superposition of several ingredients, not just synoptic-scale ascent alone.

Julian day. The Julian day is the number of days since 1 January 4713 B.C. Thus, the Julian day corresponding to 22 August 2008 is 2,454,700, not 235. Use *day of the year* instead (also called *ordinal date*).

Lightning (bolt, flash, strike, and stroke). There is a hierarchy of terms from general to specific when referring to lightning or lightning processes. *Lightning* is the most general, and the entire phenomenon of lightning includes the processes involved in the formation of the channel itself, the associated light, and the acoustic properties of thunder, to the end of the time of the travel of the last thunder from the lighting channel. The more specific term *flash* refers to a single interconnected discharge. A *cloud-to-ground flash* is often defined by the point where the flash strikes the surface of the earth as located by a lightning mapping system. The colloquial terms *bolt* and *strike* have no specific scientific meanings.

There are two categories of lightning type: *intracloud lightning* (preferably called *cloud flashes*, because intracloud flashes technically mean a flash completely within a cloud, but are often used to mean any flash that fails to strike ground) and *cloud-to-ground lightning* (which are often called *ground flashes* for variety and brevity).

Lightning stroke is more specific still, but with two different usages. First, it can refer to the return stroke, which is the bright surge back up the channel after the downward propagating leader connects with the ground.

A cloud-to-ground flash contains one or more return strokes; the average is three or four return strokes per flash. When you see lightning, the light may seem to flicker. Those are return strokes running up and down the channel of the first stroke. Second, the word *stroke* by itself should mean the combination of downward leader and return stroke, several of these being possible in a given flash. —*Don MacGorman and Ron Holle*

Low-level jet. The definition of the term low-level jet is precise, but its usage is sloppy in the literature. *Low-level jet* simply means that a low-level maximum exists in the vertical profile of wind speed. Usually various criteria for the maximum value are given, along with criteria about the decrease in wind speed above the level of maximum wind. From this definition, it is not surprising that low-level jets are relatively common. However, some have used the term loosely, leading to a myriad of problems and confusion. In essence, more information is needed than "low-level jet" to know what kind of meteorological phenomena is being discussed because jets at low levels may be due to a variety of reasons. As argued by Reiter (1963), Stensrud (1996), and Doswell and Bosart (2001), a distinction should be made between *low-level jet streams* and *nocturnal low-level wind maxima*, where possible. Low-level jet streams have mesoscale or synoptic-scale horizontal extent with strong horizontal shears along their edges, are associated with synoptic-scale processes, and have little diurnal variability. They may be barrier jets related to orography. In contrast, the nocturnal low-level wind maxima possess a strong diurnal cycle that low-level jet streams do not possess. —*David Stensrud*

Moisture flux convergence. Although the term appears in the conservation of water vapor equation, the divergence of water vapor flux is not a useful expression for determining convection initiation (Banacos and Schultz 2005). Near-surface mass convergence is a more appropriate quantity to examine.

Morphological similarity does not equal dynamical similarity. Observations of the midtropospheric flow around convective storms often appear as though the storm is an obstacle, and there have been studies making extensive use of these observations to make statements about the vorticity source for the counterrotating vortices seen on the flanks of the updraft. Although an interesting analogy, the morphology of the flow does not necessarily mean that the flow dynamics are identical to those associated with solid obstacles embedded in a fluid flow (Davies-Jones et al. 1994, commenting on Brown 1992).

When there really is a solid obstacle in the flow, vorticity is generated in the viscous boundary layer associated with the solid obstacle. This vorticity is shed into the wake of the flow and is the source of the vorticity in the counterrotating vortices. Thus, even if the ambient flow is completely uniform with no ambient vorticity, obstacle flow will generate these vortices.

Severe thunderstorms are associated with environmental flows having considerable vertical shear and, therefore, considerable vorticity about a horizontal axis. The counterrotating vortices associated with severe thunderstorms arise from tilting of this substantial ambient vorticity. Thus, the similarity in appearance to flow around an obstacle is only coincidental. —*Charles Doswell*

Normals, calculation of. Every 30 years the international meteorological community produces a document of the "normal" climate for all of the nations of the world. The effort originated from the International Meteorological Committee in 1872 to assure comparability between data collected at various stations. Thirty years is used to calculate the average climate, most often on a monthly or annual basis, for the official normals as specified by the World Meteorological Organization, and these values are updated every ten years. Although averaging over 30 years will help to filter out short-term fluctuations, this number of years appears to be defined arbitrarily, perhaps because of the rule of thumb in sampling theory suggesting 30 independent samples can be used to arrive at a well-behaved sampling distribution through the Central Limit Theorem. Such an interpretation is incorrect, as the closeness of the parent distribution to normal is related to the required sample size. Moreover, independence of the adjacent years as well as stationarity and homoscedasticity in climate on the 30-year scale is assumed (e.g., stations' data records are assumed to be homogeneous). There is nothing special about 30 years in computing average weather conditions. In fact, averaging for periods less than 30 years can offer advantages (e.g., Huang et al. 1996; Scherrer et al. 2005). —*Michael Richman*

Northward/southward. To foster writing free of geographical bias, replace "northward" and "southward" with their hemispheric-neutral siblings, "poleward" and "equatorward" (page 102).

Numerical prediction. What people really mean when they use this term is "dynamical prediction," but statistical prediction methods are numerical, also. —*Dan Wilks*

Objective versus subjective methods. Because so-called "objective" methods involve subjective decisions, do not use the terms "subjective" and "objective" (page 210). Instead, use the terms "manual" and "automated."

Observed/seen. Unless you have direct measurements of the quantity, reword.

DRAFT: Cyclonic vorticity advection at 500-hPa was observed throughout Montana and Wyoming.

IMPROVED: Cyclonic vorticity advection at 500-hPa occurred throughout Montana and Wyoming.

DRAFT: Precipitation was not seen in the simulation.

IMPROVED: The simulation did not produce precipitation.

Obstacle flow around a convective storm. See *Morphological similarity does not equal dynamical similarity*.

Overrunning. This term is generally applied to the physical process responsible for precipitation falling on the cold side of a surface front. This term lacks any insight into the physical process responsible for the ascent, and so should be eliminated from scientific discussion.

Percent/percentage. *Percent* is the unit for a particular measure (%), whereas *percentage* is synonymous with "fraction" or "portion." Do not use "percent cloud cover," instead use "cloud cover in percent" or "percentage of cloud cover."

Positive vorticity. As with *northward/southward*, replace "positive vorticity" and "negative vorticity" with their hemispheric-neutral siblings, "cyclonic vorticity" and "anticyclonic vorticity."

Propagate. *Propagate* is often used in the meteorological literature as a technical-sounding word for *move*. Almost always use the word *move* instead. Movement is advection plus propagation. Consider a boat in a river. If the boat has no motor or sail, then the boat moves downstream at the speed of the river—the boat is advected by the river. If the boat has a motor or sail and moves against the river's flow, then the boat is propagating relative to the river. Similarly, consider a feature that is not an object, such as a squall line. The propagation of the feature may involve subsequent development of the convective cells in the warm air ahead of the squall line, or the propagation component. But the movement of the feature is the addition of the propagation component and the advection component. Therefore, writers should be precise about whether they mean the total motion of the squall line or the propagation component alone.

Radar reflectivity factor. Strictly speaking, *radar reflectivity* and *radar reflectivity factor* are two different parameters (e.g., Rinehart 2004, pp. 90–91). The parameter that nearly all meteorologists use (radar reflectivity factor, with units of dBZ) is independent of wavelength of the radar beam. Thus, 50 dBZ as measured by two different, yet identically calibrated, radars should characterize precipitation in the same way. In contrast, radar reflectivity depends on the radar wavelength and has different units (cm^{-1}). Furthermore, the radar equation assumes a spherical water drop and Rayleigh scattering. Should these conditions not be met (as in clear air where the scatterers may be birds, insects, or gradients in the index of refraction), the qualifier "equivalent" should be used.

Random. See *chaos/random*.

Reradiation/reemission. One often hears the claim that the atmosphere absorbs radiation emitted by the Earth (correct) and then reradiates or reemits it back to Earth (false). The atmosphere radiates because it has a finite temperature, not because it received radiation. When the atmosphere emits radiation, it is not the same radiation (which ceased to exist upon being absorbed) as it received. The radiation absorbed and that emitted do not even have the same spectrum and certainly are not made up of the same photons. The terms reradiate and reemit are nonsense. —*Alistair Fraser*

Resolution. When describing the *resolution* of a model, the grid intervals (in space and time) typically are cited. Strictly speaking, features on the scale of the grid intervals are not resolved by the model. The smallest features that can be said to be resolved in any meaningful sense of the term are those at twice the model's grid interval, and even at that scale, the amount of information about such small features is pretty limited (Doswell and Caracena 1988). Thus, this terminology should be discouraged. See also the published comments by Pielke (1991, 2001), Laprise (1992), and Grasso (2000a). —*Charles Doswell*

Yet another term is the *effective resolution*, defined by Walters (2000, p. 2475) as "the minimum wavelength the model can describe with some required level of accuracy (not defined)" (Laprise 1992; Walters 2000; Skamarock 2004). Therefore, because no precise definition of resolution exists (e.g., Durran 2000b; Grasso 2000b), choose "grid spacing," "grid increment," "grid separation," or "grid interval," instead of "resolution."

Severe storms. To be precise, refer to "severe convective storms." See also *thunderstorm*.

Severe weather, definition of. In the United States, "severe" weather has a specific definition as applied by the NOAA/Storm Prediction Center (Galway 1989): any tornado, hailstones with diameter greater than ¾ in. (1.9 cm), or convective wind gusts with speeds greater than 50 kt (25.7 m s^{-1}). A generic term to discuss weather that has a high impact on society is "hazardous weather" or "high-impact weather." The term "violent weather" is too colloquial.

Short-wave. "Short-wave" (waves in the jet stream) should always be followed by "trough" or "ridge."

Significance/significant. Only use "significant" in the context of statistical significance or *significant severe weather* (see entry). To do so otherwise may confuse the reader.

Significant severe weather, definition of. Significant severe weather is defined as hail 2 in. (5.1 cm) or larger in diameter, wind gusts of at least 65 kt (33.4 m s^{-1}), or tornadoes with F2 intensity or larger (Hales 1988).

State. "State" means "to declare definitively," which is a much stronger definition than the way that most people use "state," as a synonym for "say." Use "state" specifically for where a strong declaration is needed as in "to state a hypothesis" (Section 10.2.1).

Statistical association does not imply cause and effect. If event A is strongly associated with event B, it is tempting to presume that A explains B or vice versa. As a somewhat contrived (but still useful) example, it is easy to show that nearly every criminal has, at one time or another, eaten at least one pickle. If we did a statistical analysis of the data, there might well be a near-perfect correlation between crime and having eaten at least one pickle. Does it make sense to infer that pickles cause crime? Perhaps we could do a study that showed that nearly all noncriminals had eaten at least one pickle, as well, demonstrating that pickles are unlikely to be the source of criminal behavior (or we have a large number of unrecognized criminals). If an association can be shown, then it might be a clue to causality, but there should be a plausible causal connection before pursuing the issue in detail. Is there a plausible reason that explains why eating a pickle would lead to a life of lawlessness?—*Charles Doswell*

t **test.** Formally known as *Student's* t *test*, not "the student *t* test." Student was the penname of author William Sealy Gosset, who published the test in 1908 (Student 1908).

Temperatures, cold and warm. Temperatures are not warm or cold—they are high or low. Air (the object) is warm or cold. See page 99. Other examples of inconsistencies between an adjective and its noun exist as well. For example, change broad/narrow spectral width to large/small spectral width, fast/slow velocity to large/small velocity, long/short wavelength to long/short waves or large/small wavelength, and deep/shallow boundary layer height to deep/shallow boundary layer or high/low boundary layer height. If the noun is a measurement or quantity, then adjectives such as "high," "low," "fewer," "more," etc. are preferred. Qualitative adjectives should be reserved for physical objects.

Theory. Reserve the word *theory* for a time-tested idea, framework, or conceptual model that has unified observations and theory, and makes testable predictions about the future (e.g., baroclinic instability theory, Milankovitch theory). Do not use the word to describe someone's results or speculation from a previous paper ("Smith's theory"); use "hypothesis" instead.

Thunderstorm. The term *thunderstorm* is not necessarily synonymous with *convective storm* or *severe convective storm*. Although thunder and lightning may be present in many convective storms, they are not requirements.

Trigger. Triggering is not a synonym for "lifting," especially when applied in the context of thunderstorms. Thunderstorm initiation requires moisture, instability, and lift. For a thunderstorm to form, somewhere within the atmosphere, a parcel exists that has buoyancy if lifted far enough to attain

Theories are good for the intellect, but are no more useful than a bit of practical experience. Perhaps the most significant thing about them is that they have been accorded quite unjustified status by engineers. —R. S. Scorer (2004)

its Level of Free Convection (LFC; beyond which it is buoyant and can accelerate upward with no further lift required). For this to take place, three things are required: moisture, conditional instability, and some process to lift a nonbuoyant parcel to its LFC. Presumably, the notion of lifting as a trigger assumes the presence of moisture and instability sufficient to allow some parcel to have an LFC, and it is only awaiting the lift.

In the absence of any one ingredient of the necessary triad, no thunderstorms will occur. So which is the trigger? If any two are present, in the absence of the third, the thunderstorms await the missing ingredient as a trigger. For example, moisture and lift often occur in the absence of conditional instability—its arrival could then logically be considered a trigger! To avoid an incorrect impression of how convection works, we should forgo the idea of a trigger completely. —*Charles Doswell*

TRMM rainfall. Because the Tropical Rainfall Measuring Mission (TRMM) does not directly measure rainfall or hydrometeors, the term "TRMM rainfall" is misleading. The TRMM Microwave Imager measures upwelling microwave radiation in several bands, and then those measurements are input to algorithms from which estimates of instantaneous rain rates, hydrometeor profiles, and other geophysical variables are calculated. These variables are then mapped and issued as products labeled by the algorithm(s) used to produce them. The best terminology to use is the proper product names (1B11, 3B42, 3B43, etc.) when referring to specific products or "TRMM-based products" in a more general context. —*Karen Mohr*

UTC. For the convenience of the reader, define any local time conventions (e.g., LST or local standard time) in UTC: UTC = LST + n hours.

Vertical motion. Use the term *vertical velocity* instead. Generally, we do not say "horizontal motion" when referring to the wind, so why would we say "vertical motion?"

Vorticity, definition versus equation. The vorticity vector $\vec{\omega}$ is defined by $\vec{\omega} \equiv \nabla \times \vec{v}$ where $\vec{v}$ is the three-dimensional velocity vector. This expression is merely a definition of a kinematic quantity of the flow. In contrast, a vorticity equation (there are many different versions of them) is derived from the equations of motion (i.e., Newton's second law as applied to fluids). A vorticity equation describes how the vorticity at a fixed point (or of a parcel, if a Lagrangian version of vorticity equation is being considered) changes with time in response to various dynamical processes (e.g., tilting, stretching, diffusion, baroclinic generation). Thus, to "analyze the vorticity equation" means to diagnose the time tendencies of vorticity through the various processes, not to perform the trivial calculation of $\nabla \times \vec{v}$. —*Alan Shapiro*

Vorticity generation by shear. Consider the boring scenario of an environment in which the v and w velocity components are zero, and the u

velocity component is positive (westerly wind) and increases with height, $\partial u/\partial z > 0$ such that the shear vector is westerly (points toward the east). In this case, the only nonzero component to the vorticity vector $\vec{\omega}$ is the y component $\partial u/\partial z$. Thus, there is shear in this flow and there is also vorticity, and neither the shear nor the vorticity "generated" the other—they are both present in the environment and are associated with the same u velocity field.

On the other hand, suppose that a thunderstorm begins to grow in the same environment considered above. In this case, the vertical velocity field associated with the developing updrafts can tilt the environmental vorticity (y component of vorticity) into the vertical, thus generating vertical vorticity. Since the environmental vorticity is associated with the environmental wind shear, one can say that the shear does play a role in the generation of the vertical vorticity. —*Alan Shapiro*

Why. "Philosophy and theology explore the *why* of nature; science deals with *how*." (Lipton 1998, p. 25).

> **DRAFT:** CDI does not explain why mammatus only appears locally on some regions of the anvil and not over the entire anvil.
> **IMPROVED:** CDI is an inadequate explanation for mammatus that only appear locally on some regions of the anvil and not over the entire anvil.

> **DRAFT:** Why the formation of the aerosol particles varies with solar radiation has not been determined.
> **IMPROVED:** How the formation of the aerosol particles varies with solar radiation has not been determined.

NOTES

PREFACE

xi   "What does not destroy me": Friedrich Nietzsche's (1844–1900) quote is from *The Twilight of the Idols* (1899). Yes, apparently the comma is there in the original.

xi   "In writing the journal article that resulted from my Ph.D. thesis": The result of all those phone calls with Dan Keyser was Schultz et al. (1998).

INTRODUCTION: AN INCOHERENT TRUTH

xxvi   Geerts (1999): The clarity of a paper was defined as a "measure of the readibility of the abstract and conclusions." The increasing number of words and figures was also found by Johnson and Schubert (1989) and Jorgensen et al. (2007).

xxvi   "47 atmospheric science journals": The data on rejection rates at atmospheric science journals were compiled from a survey I performed in 2006–2008 (Schultz 2010). Some journals in other disciplines, however, have much higher rejection rates. For example, the 27 journals of the American Psychological Association have rejection rates of 36%–92%, with most reporting around 75% (www.apa.org/journals/statistics). The principal reason for this variation in rejection rates across disciplines is *consensus*, defined as a measure of the shared "conceptions of appropriate research problems, theoretical approaches, or research techniques" (Hargens 1988), with the physical sciences having higher consensus than the social sciences.

Nevertheless, some journals restrict submissions even though they may have scientific merit. For example, *Nature*, with a rejection rate of 91.45% in 2006, admits having "to decline many papers of very high quality but of insufficient interest to their specific readership." In fact, 60% of the 10,000 submitted manuscripts each year at *Nature* are not even sent out for review (Nature 2006). Similarly, *Science* says that, "priority is given to papers that reveal novel concepts of broad interest," (www.sciencemag.org/about/authors/prep/gen_info. dtl) returning a large fraction of submitted manuscripts to authors without having been peer reviewed. Each journal decides on its own editorial policy, and rejection rates sometimes can reflect that policy. If you look hard enough, you can sometimes find sources online for the rejection rates in other journals.

xxvii "A study conducted by the College Board's National Commission on Writing": More information available online at www.writingcommission.org.

CHAPTER 1: THE PROCESS OF PUBLISHING

4 Fig. 1.1: The history of these early scientific journals is discussed at rstl. royalsocietypublishing.org and www.sil.si.edu/libraries/Dibner/newacq_2000. htm.

7 "'Accept as is' occurs in less than 1% of papers submitted to AMS journals.": In 2006, there were 2353 submissions to the eight scientific journals of the AMS, and only 21 of those manuscripts were accepted as is, many that were previous rejections that were corrected to the previous reviewers' satisfaction, according to David Jorgensen, Publications Commissioner.

8 "How this decision is determined varies by the editor and the paper": Schultz (2009) has a discussion of how editors make decisions, including a simple model describing editor behavior.

8 "Copy editors correct grammar and style of the text": An ode to copy editors can be found in Henige (2005).

9 "the decreasing number of comments": Those expressing concern with the decreasing number of comment–reply exchanges include Errico (2000) and Schultz (2008).

CHAPTER 2: SHOULD YOU PUBLISH YOUR PAPER? QUESTIONS TO ASK BEFORE YOU BEGIN WRITING

13 "publon": A publon is analogous to a photon, the minimum quantum of energy. A publon would be the minimum quantum of knowledge constituting publishable material (Feibelman 1993, p. 40). Batchelor (1981, p. 6) also discussed whether there was a minimum quantity of publishable scientific information.

18 "two to six times more citations": Numerous sources discuss the advantage of open access. The best are S. Lawrence (2001), Antelman (2004), Harnad and Brody (2004), Hajjem et al. (2005), Eysenbach (2006), and Swan (2007). An overview of open access can be found online at Peter Suber's page: www. earlham.edu/~peters/fos/overview.htm.

CHAPTER 3: WRITING AN EFFECTIVE TITLE

23 "When including the word 'using' in titles": The problems with the word "using" are discussed further by Day and Gastel (2006, p. 42).

24 "Assertive sentence titles annoy some scientists": Rosner (1990) and Day and Gastel (2006, p. 42).

26 Section 3.4, Examples: The first paper is Schultz and Schumacher (1999), the

second is Xu and Emanuel (1989), the third I made up, and the last one is Schultz et al. (2004).

CHAPTER 4: THE PARTS OF A SCIENTIFIC PAPER

29   "Some scientific writing books take a more conservative stance": Specifically, Day and Gastel (2006, 4–5).

31   "from a study of managers at the Westinghouse Corporation": More information is available from Souther (1985).

33   "Even if your target journal does not require an abstract": Not having an abstract means that abstracting services will either index your article without an abstract or write their own abstract for your article. Fulda (2006) argues that neither alternative is ideal.

34   "Consider the following introduction": Condensed slightly from Schultz and Steenburgh (1999).

35   "resounding banality": More on writing the opening sentence and paragraph is discussed in Dixon (2000): newsarchive.asm.org/sep00/animalcule.asp.

37   "a box set of music": The analogy between a box set and a literature review was first discussed in Schultz (2004).

41   "Incomplete, incorrect, or inappropriate methods, once in the literature, are hard to eliminate, with later researchers often citing past bad work": For the prevalance of this in the medical community, see Altman (2002).

42   "Some people argue that negative results should not be included": A partial list of papers presenting principally negative results include MacKeen et al. (1999), Doswell et al. (2002), Richter and Bosart (2002), and Schultz et al. (2007b).

44   "Other times they simply want to do more research using the same or slightly modified methods": Another way to view these statements is what Marc Abrahams, editor of the *Annals of Improbable Research*, calls the *genug–genug* question (*genug* is "enough" in German): When is enough enough? He asked whether any paper has ever ended with a statement that the question is now answered and no further research is necessary. In response, Prof. Graham de Vahl Davis of the University of New South Wales pointed out that the California Supreme Court ruled in 1984 that "appellate judges do not have the luxury of waiting until their colleagues in the sciences unanimously agree that on a particular issue no more research is necessary. Given the nature of the scientific endeavor, that day may never come." Because science apparently will never conclude that enough research has been done on a topic, in your paper, you do not need to state the obvious.

CHAPTER 5: THE MOTIVATION TO WRITE

49   "The scariest moment": King (2000, p. 274): *On Writing: A Memoir of the Craft*.

51   "In fact, most technical writers do not write linearly": Roundy and Mair (1982).

CHAPTER 6: BRAINSTORM, OUTLINE, AND FIRST DRAFT

57   "Turtle or rabbit?": The turtle/rabbit analogy comes from Alley (2003, pp. 241–242); a similar analogy (slow/quick) is discussed by Booth et al. (2003, p. 190).

58   "avoid the tendency to 'fall in love with your own text'": Said by Jaakko Kukkonen, Finnish Meteorological Institute.

CHAPTER 7: ACCESSIBLE SCIENTIFIC WRITING

61   Section 7.3, Structuring Logical Arguments: Other sources have more on logical arguments. The components of a logical arguments and what constitutes evidence derives from Shermer (2002, chap. 3), Booth et al. (2003), U.S. Air Force (2004, chap. 5), and Weston (2009).

CHAPTER 8: CONSTRUCTING EFFECTIVE PARAGRAPHS

65   "Effective paragraphs possess two primary characteristics: unity and coherence": The importance of picking one theme per paragraph and breaking up any paragraph with more than one theme is discussed particularly well by Williams (2004, p. 35). Coherence, on the other hand, is well handled by Wilkinson (1991, p. 438).

67   "In this way, writing is linking up information in a logical, flowing manner": The link between the old information and the new information is also called the *coherent-ordering principle*: "Context is built before new points are introduced" (McIntyre 1997, p. 207).

71   Section 8.3, Coherence Between Paragraphs: This text comes from Schultz and Schumacher (1999).

73   "Sections and Subsections": Some of this sidebar derives from Alley (1996, pp. 37–40).

CHAPTER 9: CONSTRUCTING EFFECTIVE SENTENCES

75   "sentences are the vehicle that delivers the message": Williams (2004, p. 34) refers to the sentences as the hammer that drives the message home.

75   "Woof woof woof.": Of thousands of jokes submitted to an online joke archive (www.laughlab.co.uk), this was the Web site creator's favorite (Wiseman 2008, p. 220).

76   Section 9.1, Active Voice Versus Passive Voice: Wilkinson (1991, p. 74) provides some good discussion of active and passive voice. More examples of how to change passive sentences into active ones are described in U.S. Air Force (2004, p. 74).

79   "Choose active verbs rather than their noun forms.": The case for eliminating superfluous words is argued particularly strongly in Perelman et al. (1998, pp. 244–245) and Ebel et al. (2004, p. 39).

81   "Disagreements begin when considering the following situation": The use of the present tense to describe actions in the past is referred to as the *historical present*.

81   "These generalizations, however, are not supported by everyone": Day and Gastel (2006, p. 192) argue that previously published literature should be discussed in the present, whereas the counter quote comes from Wilkinson (1991, p. 78).

83   "If the word 'than' is present": Incomplete comparisons using "than" are discussed by Cook (1986, pp. 69–72, 197–198) and Strunk and White (2000, p. 59).

85   "Negative information is more difficult for people to comprehend": Why negative phrases are more difficult to understand is discussed by many (e.g., Clark and Chase 1972; Tweney and Swart 1977; Podsakoff et al. 2003; Gorin 2005; Leenaars et al. 2006).

85   Section 9.7: Misplaced modifiers are discussed in more detail by Cook (1986, pp. 43–46), Perelman et al. (1998, p. 261), Strunk and White (2000, pp. 11 and 20), and Day and Gastel (2006, p. 186).

CHAPTER 10: USING EFFECTIVE WORDS AND PHRASES

89   "the little word 'about' deserves more respect": Day and Gastel (2006, p. 203).

90   "hides our lack of knowledge of the science": A thorough discussion on how language can obscure meaning in science is found in P. A. Lawrence (2001).

90   Section 10.2.1: Alley's (1996, p. 77) discussion of denotation and connotation inspired this section.

90   "authors commonly overuse 'state'": Strunk and White (2000, p. 58).

94   "Consider the word 'role'": To illustrate how empty and overused the word "role" is in scientific papers, P. A. Lawrence (2001) provided a list of adjectives used with the word "role" from a quick scan of the literature: "major, pivotal, key, global, potent, leading, important, principal, vital, critical, regulatory, endogenous, master, multiple, controlling, fundamental, special, dual, basic, specific, essential, novel, evolving, potential, new, changing, active, central, functional, counteractive, prominent, very specific, very important and essential, legitimate, biological, physiological, integral, more important role than previously suspected."

96   Section 10.2.5: P. A. Lawrence (2001) asks why we hype our results in scientific papers. His response is that, because everyone else does it, we must do so as well in order to not fall behind.

99   Section 10.3.2: Schall (2006, p. 80) discusses the rules for writing numbers in scientific manuscripts.

100   Section 10.4.1: Gender bias is covered in more detail in the following sources: Cook (1986, pp. 90–94), Perelman et al. (1998, p. 281), Strunk and White (2000, p. 60), Williams (2004, p. 83), and Schall (2006, p. 60).

CHAPTER 11: FIGURES, TABLES, AND EQUATIONS

103 "the natural sciences are among the most figure-intensive sciences": Cleveland (1984) surveyed 57 journals from various disciplines and found that the *Journal of Geophysical Research* contained the most area devoted to graphs (over 30% of the length of the articles).

103 "the Stüve diagram": The Stüve diagram was not the earliest thermodynamic diagram. In 1915, Sir Napier Shaw developed the tephigram, which is still used primarily in the United Kingdom and the British Commonwealth countries.

104 Fig. 11.1: Fig. 1 from Clayton (1911).

104 Fig. 11.2: adapted from Schultz and Knox (2007), their Fig. 7a and caption.

106 "worldwide acceptance of the Norwegian cyclone model": See, for example, Davies (1997).

106 "the crystal clear drops [of water] seem more refreshing": This quote was documented in Friedman (1989, p. 200).

109 Fig. 11.7: data from the Sea Ice Trends and Climatology from Scanning and Multichannel Microwave Radiometer (SMMR) and Special Sensor Microwave Imager (SSM/I) NASA dataset downloaded from the National Snow and Ice Data Center (available online at nsidc.org/data/smmr_ssmi_ancillary/area_extent.html).

114 Fig. 11.8: adapted from Fig. 11c in Schultz and Trapp (2003).

115 Section 11.6: More discussion of color can be found in Anderson (1999, p. 288), Tufte (1990, pp. 82–83), and Spekat and Kreienkamp (2007).

117 "Warmer colors (red, yellow, and orange) will appear to jump out": The discussion of visual perception of warmer and cooler colors derives from Kosslyn (2007, p. 106).

120 Fig. 11.10: Fig. 5 from Hanna et al. (2008).

120 Fig. 11.11: adapted caption and Fig. 7 from Kingsmill and Crook (2003).

122 Fig. 11.13: adapted from Fig. 2 in Roebber et al. (2003).

123 Fig. 11.14: figures based on data from Hayden et al. (2007).

124 Fig. 11.15: adapted from data presented in Fig. 5a in Schultz et al. (2007a).

126 Section 11.7.4: Box-and-whisker plots were developed by Tukey (1977, pp. 39–43).

126 Fig. 11.16: Fig. 16 from Heinselman and Schultz (2006).

127 Fig. 11.17: Fig. 3e and caption from Novak et al. (2008).

129 Fig. 11.18: Fig. 9e from Wakimoto and Martner (1992).

130 Fig. 11.19: adapted from Fig. 12 in Schultz and Knox (2007).

131 Fig. 11.20: Fig. 2b from Martner et al. (2007).

132 Section 11.7.10: Tufte (1990, chap. 4) discusses more about small multiples.

132 Fig. 11.22: Fig. 8 from Chattopadhyay et al. (2008).

133 Fig. 11.23: Fig. 9a from Wang et al. (1995).

135 Table 11.2: adapted from Table 1 in Shapiro (2005).

139 Section 11.13: Montgomery (2003, pp. 134–136).

141 Section 11.16: More on punctuation and equations can be found in Wilkinson (1991, Section 6.10).

CHAPTER 12: CITATIONS AND REFERENCES

144   "a shield to strengthen your arguments": Cronin (2005).

144   Section 12.2: This passage on blocking originates from Bals-Elsholz et al. (2001).

149   "Because cold advection can occur": This passage on bent-back warm fronts comes from Schultz et al. (1998).

151   "Dr. Richard Tyson of Newcastle University": An unpublished document entitled "A Workshop on Referencing and Construction of Bibliographies" by Dr. Richard Tyson provided perspectives on some of the information in this chapter.

154   Table 12.1: Adapted from the Monash University Language and Learning Online Web site, www.monash.edu.au/lls/llonline/quickrefs/22-referencing-internet.xml.

155   Simkin and Roychowdhury (2003): Simkin and Roychowdhury (2003) derive their estimate on the percentage of papers that are not read by the citing author from the frequency of misprints in reference lists. Such an approach is not a direct measure of this percentage, but because their approach yielded such a large number is indicative of how pervasive the problem may be.

    Using a different approach, the results of a self-reporting survey to the members of the Mini-AIR mailing list, sponsored by the *Annals of Improbable Research* at www.improb.com, found that only 66% of authors and coauthors read all of their cited sources, 32% read just a summary, and 2% read only the titles of the cited works. Following up on this survey, the *Annals* then asked whether the respondents had read every one of the research papers on which they were listed as a coauthor. Eighty-six percent of the authors replied yes, 6% replied no, and 8% were unsure.

    Having a copy of your cited articles in hand also can alleviate the problem of incorrect citations copied from others' sources. Blanchard (1974) provides a humorous example of why it is important to create the references yourself for your own manuscripts.

CHAPTER 13: EDITING AND FINISHING UP

158   "typed on a scroll": The book *On the Road: The Original Scroll* has a contribution by Howard Cunnell on the writing of the book. In 2008 and 2009, the scroll went on a world tour: www.ontheroad.org.

158   "To the contrary, this transition between the first complete draft and the first revision is critical to the development of the manuscript.": This discussion is derived from Ebel et al. (2004, p. 38).

160   Section 13.3: Fairbairn and Fairbairn (2005, pp. 40–47) provide an extensive discussion of précis, including examples.

160   "The National Weather Service (NWS) is now in the midst of a major paradigm shift": This example comes from Mass (2003).

161   "The spatial and temporal occurrence of large (at least 2 cm in diameter) hail in Finland": This abstract comes from Tuovinen et al. (2009).

168   Section 13.6: Lipton (1998, p. 50) has more information about why the author has the responsibility for submitting a proper manuscript.

169   "Final Checks of Your Manuscript," "Lines numbered in margin": You may wish to add line numbers for manuscripts that you submit whether or not the journal requires it, making the reviewers' task easier. In Microsoft Word for Windows, enabling line numbers is under the layout tab. For Microsoft Word for Mac, enabling line numbers is under Format, Document, Layout. For La-TeX, usepackage{lineno} and linenumbers*.

171   "An ever-increasing amount of literature is being published": Some journals have to reject an increasing number of manuscripts to keep the number of published pages each year relatively constant because of the increasing length of papers (Drummond and Reeves 2005) and their fixed budgets and staff. Journals are stressing shorter papers for practical and economic reasons. For example, the AMS implemented the first limit on the length of papers not requiring editor approval in 1991 (10,000 words or about 35 pages) and the second limit in 2001 (7500 words or about 26 pages).

172   "2006 Ig Nobel Prize in Literature": The *Annals of Improbable Research* [**12** (6), p. 17] reported on the 2006 Ig Nobel Awards ceremony.

173   "Use a spell checker or grammar checker": Schall (2006, pp. 44–45) has more about the strengths and weaknesses of grammar checkers.

CHAPTER 14: THE RESPONSIBILITIES OF AUTHORSHIP

175   "One active scientist can typically write one or two papers a year": This idea about the importance of collaboration for increasing your output comes from Prof. Markku Kulmala, University of Helsinki.

177   "If a coauthor is willing to take credit for the article": An author willing to take credit for an article should also accept responsibility for it as well (Sigma Xi 1986, pp. 24–25). The suggestion to ask a trusted colleague who is not a coauthor for comments comes from Sigma Xi (1986, p. 27).

178   "In one article published by the AMS": The paper with the authors contributing equally is Hobbs and Rangno (1985).

178   "Lead authorship could result": How to determine lead authorship is discussed in Houk and Thacker (1990) and Wilson (2002).

180   Section 14.3: "Guidelines to Publication of Geophysical Research" (American Geophysical Union 2006) was largely adopted from the American Chemical Society statements and defines ethical standards by which editors, authors, and reviewers should follow. More information is available online at www.agu.org/pubs/pubs_guidelines.html. Reproduced by permission of American Geophysical Union.

CHAPTER 15: SCIENTIFIC ETHICS AND MISCONDUCT

183   "Consider the following cases": The retraction of Case 1 was published by Taylor & Francis (2006). The correction of Case 2 was never published, and the

wronged author dropped his grievance. The retraction of Case 3 was published by Wu et al. (2004).

Unfortunately, retraction does not necessarily mean the end of the line for most of these articles. Budd et al. (1998) looked at 235 retracted articles from the biomedical literature, of which 86 were retracted for misconduct. They examined 299 of the 2034 citations of the retracted articles after the retractions were published. In only 19 of the 299 cases (6%) was the retraction noted by the authors. The rest either implicitly or explicitly considered it valid research.

184   "The National Science Foundation (NSF) has defined misconduct. . . .": More information available online at www.nsf.gov/oig/resmisreg.pdf.

186   "Studies document the positive association between sponsors' interests and the outcomes of research": Bekelman et al. (2003), Ridker and Torres (2006), and Lesser et al. (2007). This sidebar was based on the results of a workshop conducted by The Center for Science in the Public Interest at the Fourth National Integrity in Science Conference: Rejuvenating Public Sector Science. More information is available online at www.cspinet.org/integrity/conflictedscience_conf.html.

188   "one highly publicized case": Articles discussing this particular case can be found at Brumfiel (2007) and the following locations: www.math.columbia.edu/~woit/wordpress/?p=638 and arxiv.org/new/withdrawals.aug.07.html. The response by one of the authors of the retracted papers was published in *Nature* (Yilmaz 2007).

189   "Nevertheless, as many as 8% of biomedical articles are believed to be duplicate publications": Rosenthal et al. (2003), Matinson et al. (2005), and Errami et al. (2008).

CHAPTER 16: GUIDANCE FOR ENGLISH AS A SECOND LANGUAGE AUTHORS AND THEIR COAUTHORS

191   "Until thirty years ago or so, most scientists receiving Ph.D.s in the United States had to demonstrate proficiency in a foreign language before graduating": The importance of knowing a foreign language in atmospheric science is evidenced by the fact that several popular atmospheric science journals were originally published in languages other than English. In fact, some [e.g., *Meteorologische Zeitschrift* (German), *Journal of the Meteorological Society of Japan*] still publish the abstracts in their native language, despite the articles being published in English.

As an illustration of how simple English may be for native speakers, but challenging for ESL speakers, consider this sentence: "If the unstable layer is destabilized further, then the layer will possess more instability." Look at the different prefixes and suffixes that the word "stable" possesses. Or, consider the difference in the pronunciation of the word "separate" when used as a verb (sep-*uh*-reyt) and when used as a adjective (sep-er-it). Is it any wonder that English can be so challenging?

In fact, the problem is even greater. Here are three examples. The Chinese language does not have an explicit syntax for singular and plural nouns. Many

Asian and some eastern European languages do not have definite articles (*the*) or indefinite articles (*a* and *an*). Other languages use subject–object–verb order in the sentence, rather than subject–verb–object order as is common in English. These examples are just a sampling of the differences in the language that make communication challenging for ESL authors.

191 "ESL": Different terms describe multilingual authors. Within countries where English is the native or primary language (e.g., Australia, Canada, United Kingdom, United States), ESL (English as a Second Language) is slowly giving way to ESOL (English as a Second or Other Language) or EAL (English as an Additional Language). EFL (English as a Foreign Language) is used within countries where English is taught but is not the native language. For convenience and brevity, I use the term ESL for all such authors.

The number of articles published by ESL authors is increasing as well. For example, less than 4% of articles published in six AMS journals in the 1970s had authors with Chinese surnames compared to more than 13% by 2004 (Li et al. 2007). At American Geophysical Union journals, the percentage was even higher in 2004 (more than 25%).

192 Section 16.1: This discussion of high- and low-context cultures derives primarily from Brown et al. (1995).

194 Table 16.1: The information in Table 16.1 derives from many sources: Brown et al. (1995), Dong (1998), Pagel et al. (2002), Jaakko Kuukonen at the Finnish Meteorological Institute, and Mary Golden at *Monthly Weather Review*, as well as my own editorial experience.

198 "We are led to the conclusion that ESL scientists have been dealt a Faustian bargain": A Faustian bargain is to make a deal where the devil gets the bargainer's soul after a period in which the bargainer achieves great knowledge, wealth, and power.

CHAPTER 17: PROOFS, PUBLICATION, AND LIFE THEREAFTER

204 "Perhaps enlist a friend": Lipton (1998, p. 19) suggests using two people to check proofs.

207 "50% of the published articles never get cited": The percentage of articles that have never been cited has been increasing. De Solla Price (1965) estimated that 10% of articles published in 1961 were never cited, increasing to 50% more recently (Garfield 2005).

CHAPTER 18: METHODS AND APPROACHES TO WRITING FOR THE ATMOSPHERIC SCIENCES

210 Section 18.2: Kalkstein et al. (1987, p. 728) has more on the objective versus subjective debate.

213 "Since the advent of mesoscale models as tools for synoptic meteorologists": The use of mesoscale models as tools for synoptic meteorologists was presciently discussed by Keyser and Uccellini (1987).

216 "One caveat about any modeling study is that the model may get the right answer for the wrong reason": An example of getting the right answer for the wrong reason can be found in Pfiefer and Gallus (2007).

218 "How many events do you need to collect in order to claim some measure of representativeness for your climatology?": Doswell (2007) discusses the issue of sample size in the context of the U.S. tornado database.

CHAPTER 19: EDITORS AND PEER REVIEW

228 "Specifically, these obligations require editors to judge each manuscript on its scientific content": These obligations are detailed by the American Chemical Society (ACS) in "Ethical Guidelines to Publication of Chemical Research," pubs.acs.org/instruct/ethic.html. Although originally formulated by the ACS, these obligations have been adopted by other professional societies such as the American Geophysical Union. "Guidelines to Publication of Geophysical Research" (American Geophysical Union 2006) was largely adopted from the ACS statements and defines ethical standards by which editors, authors, and reviewers should follow. More information is available online at www.agu.org/pubs/pubs_guidelines.html. Reproduced by permission of American Geophysical Union.

CHAPTER 20: WRITING A REVIEW

231 Section 20.3: How to review a manuscript is covered in several articles: Smith (1990), Wilson (2002), Benos et al. (2003), and Provenzale and Stanley (2005).

239 "Much has been written about the strengths and weaknesses of peer review": *Nature* hosted a debate on peer review (available online at www.nature.com/nature/peerreview/debate/index.html). Furthermore, the literature abounds with authors who have criticized the peer-review system (e.g., Campanario 1998; Balaram 2002; Wilson 2002; Beck 2003; Frey 2003; Kundzewicz and Koutsoyiannis 2005). For example, Bhatia (2002) suggested that the names of the reviewers who support publication of the articles be published, therefore making the reviewers more accountable. Some journals (e.g., *Electronic Journal of Severe Storms Meteorology*) already do this. Should a bad paper be published, the reviewers' names would be linked to the paper. Reviewers who recommended rejection would not have their names published. The *Atmospheric Chemistry and Physics* open peer-review model is discussed by Pöschl (2004) and Koop and Pöschl (2006).

239 Section 20.6: Recently, the drug manufacturer Pfizer sued the *New England Journal of Medicine* for access to the normally confidential information about the names of the reviewers, their comments, correspondence, and other documents. Pfizer's motion to open up the journal's records includes the statement that "The public has no interest in protecting the editorial process of a scientific journal." Upon first glance, this quote sounds bold and unacceptable, but upon further reflection, begins to sound more true, unfortunately. More on Pfizer vs

the *New England Journal of Medicine* can be found in Kennedy (2008) and at these links:

- arstechnica.com/journals/science.ars/2008/02/17/aaas-ethics-in-scientific-publishing
- seekingalpha.com/article/68446-pfizer-vs-the-new-england-journal-of-medicine-a-significant-legal-showdown.

239   "Democracy is the worst form of government": The complete quote comes from Churchill's speech in the House of Commons (11 November 1947): "Many forms of Government have been tried and will be tried in this world of sin and woe. No one pretends that democracy is perfect or all-wise. Indeed, it has been said that democracy is the worst form of government except all those other forms that have been tried from time to time."

CHAPTER 21: RESPONDING TO REVIEWS

243   "Comments from reviewers usually fall into one of four categories": The four categories of comments from reviewers is based on an initial list of three by Valiela (2001, pp. 140–141).

249   "Persistence and Precedence": Further examples of Nobel Prize winners who have had papers rejected can be found in the book *Rejected: Leading Economists Ponder the Publication Process* (Shepherd 1994), digested in the journal article Gans and Shepherd (1994).

CHAPTER 22: HOW SCIENTIFIC MEETINGS WORK

256   Weather and Society * Integrated Studies: For more information on WAS*IS, visit their Web page www.sip.ucar.edu/wasis.

CHAPTER 24: ACCESSIBLE ORAL PRESENTATIONS

265   "Science in the early nineteenth century used to be the rock shows of today": Hamblyn (2001, chap. 1) discusses the public spectacle that was the scientific presentation in the early nineteenth century. Imagine getting that kind of welcome at your next scientific presentation!

268   "You may want to inform, persuade, confront, inspire, educate": The different types of presentations are discussed by Alley (2003, pp. 37–43).

269   "The human brain has a high capacity to process information": Aarabi (2007, pp. 30–31) discusses this limitation of the narrow information channel connecting the speaker and the audience.

269   "Unfortunately, to connect the speaker's brain to the brains of the audience members requires communication through a narrow channel that delivers information both verbally and visually": Atkinson (2008, pp. 40–47) discusses

more on managing the flow of information between the spoken and visual channels.

270 "Your goals as a speaker are to connect with the audience, hold their attention on your topic, and help them remember it": Kosslyn (2007, p. 3) says that these three goals "virtually define an effective presentation."

270 "How effectively you can do that is determined by the presentation quality, the presenter quality, and the audience quality": Aarabi (2007).

CHAPTER 25: CONSTRUCTING EFFECTIVE ORAL PRESENTATIONS

273 "Bashing PowerPoint seems to be all the rage": I am unaware of the origins of the phrase "Death by PowerPoint," but the phrase seems ubiquitous on the Web with over 82,000 hits on google.com. "PowerPoint Phluff" was a term coined by Edward Tufte for the flashy transitions and animations, "PowerPoint is Evil" was written by Edward Tufte for *Wired* magazine (11.09, their September 2003 issue). "Power(Point) Ballad" is a tongue-in-cheek homage on youtube.com (www.youtube.com/watch?v=hq-JaaUkcSw). *Why Most PowerPoint Presentations Suck* is by Rick Altman (2007).

274 "Imagine if Abraham Lincoln had given the Gettysburg address by Power-Point": See online at norvig.com/Gettysburg.

277 "Do not close with . . . the list of references from your talk": If you wish, keep such a list at the end of the talk, but save it if you need to refer someone to a specific paper. Alternatively, you could put citations to important papers within the body of the talk, with just enough information so that people know what journal to look in: "Hoskins et al. (1985, *QJRMS*)."

279 Section 25.4: Several sources make compelling recommendations that the titles be headlines (e.g., Alley 2003, pp. 125–129; Atkinson and Mayer 2004).

281 "visibility can vary": Rasmussen et al. (1999).

282 "The indiscriminate use of bullets": Alley (2003, p. 138) says bullets can be distracting and recommends limiting their use.

CHAPTER 26: DELIVERING COMPELLING ORAL PRESENTATIONS

292 "Four Ways of Delivering Presentations": Alley (2003, p. 47) classifies these four types of delivery as the *source of the speech.*

299 "We give presentations for the audience not for ourselves": Aarabi (2007, chap. 4) has excellent material on the realities of the audience. Atkinson and Mayer (2004) discuss the importance of the audience processing the talk through both verbal and visual means.

301 "Attempting to have slides serve both as projected visuals and as stand-alone handouts makes for bad visuals and bad documentation": Garr Reynolds's quote comes from his Web site: presentationzen.blogs.com/presentationzen/2005/11/the_sound_of_on.html.

CHAPTER 28: CHALLENGES TO DELIVERING YOUR PRESENTATION

316 "*The Washington Post* reported on directives from the U.S. Customs and Border Protection": *The Washington Post*, 1 August 2008, p. A01, See online at www.washingtonpost.com/wp-srv/content/article/2008/08/01/laptops.html.

317 "Imagine how Luke Howard felt": This story is derived from Hamblyn (2001). In those days, "modification" would have meant "classification."

CHAPTER 29: COMMUNICATION IN THE WORKPLACE

325 Section 29.3: More on planning and running meetings can be found in U.S. Air Force (2004).

CHAPTER 30: COMMUNICATION WITH THE PUBLIC AND MEDIA

329 "When the public is surveyed, they say that scientists are one of the most trusted occupations, yet a majority do not believe in the scientific consensus on global warming": A Harris Poll in 2006 indicates that scientists are the third most trustworthy occupation, behind doctors and teachers, with 77% of those surveyed saying that they would trust scientists to tell the truth. (See online at www.harrisinteractive.com/harris_poll/index.asp?PID=688.) Yet, only 48% say that most scientists think that global warming is happening. (More information available online at environment.yale.edu/news/Research/5317/americans-consider-global-warming-an-urgent-threat/.)

330 "The American Geophysical Union (2006) developed a list of obligations for scientists publishing outside the scientific literature": The AGU document emulates a similar document by the American Chemical Society: www.agu.org/pubs/pubs_guidelines.html.

331 Section 30.1: Much of this section was written with the help of Stephanie Kenitzer, Public Relations Officer of the AMS.

CHAPTER 31: FURTHERING YOUR JOURNEY

335 "Furthering Your Journey": Some may be wondering why the title of this chapter is called "furthering," rather than "farthering." As Strunk and White (2000, p. 46) discuss, "farther" is the better word for distance and "further" is better used for time or quantity. In this case, I am referring to the advancement of your career. Used in this way, it is not about making the journey longer, but making it richer.

336 "Robert Marc Friedman": Robert Marc Friedman's books are entitled *Appropriating the Weather: Vilhelm Bjerknes and the Construction of a Modern Meteorology*, *The Expeditions of Harald Ulrik Sverdrup: Contexts for Shaping an Ocean Science*, and *The Politics of Excellence: Behind the Nobel Prize in Science*. His penchant for writing drama comes from his undergraduate studies in drama and theater at New York University while working on his degree in geophysics.

In graduate school, he studied the history of science at The Johns Hopkins University, writing his doctoral dissertation on Vilhelm Bjerknes and the Bergen School of Meteorology.

337 "Only 40% of the U.S. students": The report of the AMS member survey on students was published by Stanitski and Charlevoix (2008).

340 Section 31.10: Failure as a motivating factor for improvement is further discussed by Aarabi (2007).

341 "Bernard Vonnegut": Bernard Vonnegut (1914–1997), emeritus professor at The University at Albany/State University of New York, discovered that silver iodide served as an effective nuclei to form ice in a cloud, leading to the birth of cloud seeding. His later research evolved to cloud electrification and lightning, challenging the existing theories of charge electrification (Vonnegut 1994). He wrote over 190 scientific articles and holds 28 patents. A biographical sketch can be found at www.deas.albany.edu/deas/bvonn/bvonnegut.html.

341 "Kurt Vonnegut": This guidance comes from an advertisement Vonnegut did for the International Paper Company in 1980 called "How to Write With Style" and was reprinted in his 1999 book *Palm Sunday: An Autobiographical Collage.* You can find a copy of the original ad, reprinted in *IEEE Transactions on Professional Communication* (1981, PC-24, No. 2, pp. 66–67) and at public.lanl.gov/kmh/pc-24-66-vonnegut.pdf.

APPENDIX B: COMMONLY MISUSED SCIENTIFIC WORDS AND EXPRESSIONS

351 Appendix B: Much of this chapter was either written by or inspired by Charles Doswell. His Web page listing his pet peeves should be required reading for all atmospheric science students (www.flame.org/~cdoswell/peeves/Pet_Peeves.html). Although I do not agree with all of his peeves, they inspire me to ensure that my writing is as precise as it can be.

352 "Condensation": For further explanation, see Alistair Fraser's Bad Meteorology Web page at www.ems.psu.edu/~fraser/Bad/BadClouds.html.

353 "Date/day": The date/day issue was raised by Montgomery (2003, p. 121).

354 "Equations, formulas, and theories": The issue of equations, formulas, and theories being used inappropriately is described nicely by Scorer (2004, p. 366): "The value of the insight gained from the study of theories leading to formulas is illustrated by the fact that generally the inventors of formulas are less ready to use them than people who did not invent them. Recognizing all the limitations, some researchers have made comparisons between observations and the predictions of various formulas, the idea being that the formulas probably represent the efforts of the best thinkers in the subject and therefore merit use when possible. I have been flattered that on one or two occasions some of my formulas have been included in such texts, but the experience of such honour is a bad stimulus because it clouds the judgement. In almost all cases my formulas have been used quite out of context yet instead of being glad when they have been deemed unsuitable as they undoubtedly were, I was glad when one was

once judged to give reasonable guidance (though not quite as good as another person's). One does not mind being erroneously recommended, but it is still galling when one's formula is stated to be unreliable when it has been tested out of context. One knows it would be 100 per cent correct if Nature ever contrived circumstances exactly to fit its assumptions!"

360 "Observed/seen": The precision with *observed/seen* is described by Lipton (1998, p. 41).

361 "Overrunning": See Charles Doswell's Web page for further discussion of this term: www.flame.org/~cdoswell/overrun/overrunning.html.

362 "State": The precision with *state* is discussed by Strunk and White (2000, p. 58).

363 "Statistical association": Kinsman (1957) also has a nice discussion and example of inferring cause and effect from a statistical relationship.

FOR FURTHER READING

THREE ITEMS OF ESSENTIAL READING

1. Everyone should own a copy of *The Elements of Style* by Strunk and White (2000). Ranked twenty-first on the list of the 100 Best Works of 20th Century English-Language Nonfiction by the Modern Library, this classic is powerful enough to teach anyone how to write better. The book is short enough to read in an evening (about 100 pages), comes in the size of a paperback novel, and is less than $10.

2. After Strunk and White, every scientific writer should read Gopen and Swan (1990), "The science of scientific writing." Gopen and Swan (1990) argue that by understanding the science of how readers read, authors can improve their own writing. Of greatest importance is the coherence between sentences. Examples, both before and after editing, show how to apply their techniques.

3. *Presentation Zen: Simple Ideas on Presentation Design and Delivery* by Reynolds (2008) and the accompanying Web site, www.presentationzen.com, provide a fresh way to approach the process of planning and constructing presentations. Although the Zen approach may not work for every slide in your scientific presentation, the simplicity of design and the distinctive approach will have the same impact on your presentations as *The Elements of Style* does for your writing.

HIGHLY RECOMMENDED READING ON WRITING

Cook (1986): *Line by Line: How to Edit Your Own Writing* delivers a thorough accounting of the editing process. The book deals mainly with sentence-level revisions and contains numerous examples.

Day and Gastel (2006): *How to Write and Publish a Scientific Paper*, 6th ed., is one of the most popular and thorough resources on this topic. Earlier editions were authored by Day alone and are just as good.

The Journal of Young Investigators: A Guide to Science Writing (2005) is a great one-stop resource for undergraduate and graduate students on preparing their first scientific journal article, especially the material on the parts of a scientific manuscript. See their Web site at www.jyi.org.

Montgomery (2003): *The Chicago Guide to Communicating Science* is a well-written, informative, and motivational book. Chapter 5 "Writing very well: Opportunities for creativity and elegance" is exceptional, providing the next step to scientific-writing brilliance after *Eloquent Science*.

Orwell (1945): "Politics and the English Language" fights dying metaphors, pretentious diction, and meaningless words, among other aspects of poor writing. Similar to *Animal Farm* and *1984*, which were written by the same author, the article reads as if it could have been published today instead of 1945.

Perelman et al. (1998): *The Mayfield Handbook of Technical & Scientific Writing* is another popular book that deals more with the mechanics of preparing and writing a scientific paper than the other sources listed here. This book presents excellent material on modes of paragraph development with examples and other rules of grammar. It is available online at www.mhhe.com/mayfieldpub/tsw/home.htm.

Schall (2006): *Style for Students* has clear explanations, lots of examples, tables on active verbs, and one of the best discussions of how to cite references that I have found. The book has been updated and is available online at www.e-education.psu.edu/styleforstudents.

U.S. Air Force (2004): *The Tongue and Quill*. Despite being authored by the military, this guide can be useful for anyone. The manual stresses communication philosophy and connecting with the audience. Several chapters address e-mail, punctuation, capitalization, and grammar.

Williams (2006): *Style: Ten Lessons in Clarity and Grace*, 9th ed., or any edition, forces further clarity and grace in your writing beyond prescriptive rules.

RECOMMENDED READING ON WRITING

Alley (1996): *The Craft of Scientific Writing*. Given the motivational challenges that we all face when writing, I found precious little information in most scientific writing books on this topic. Chapter 17 "Actually sitting down to write," however, provides a good deal of information.

Alley (2000): *The Craft of Editing: A Guide for Managers, Scientists, and Engineers* focuses on how to edit others' work, especially when in a supervisory role or a collaboration.

Anderson (1999): *Technical Communication: A Reader-Centered Approach* is a broad and detailed book for all types of technical communication, not just journal articles. The best material in this book is determining your audience, defining your objectives for a manuscript, planning persuasive strategies, brainstorming, free-writing (a form of brainstorming), writing the first draft, and defining the criteria for classification schemes.

Ebel et al. (2004): *The Art of Scientific Writing* is a thorough and academic, albeit not particularly practical, book. The best sections are those on decisions an author must make prior to publication, and on citations. One chapter discusses how to acquire, build, and manage your own literature collection.

Fairbairn and Fairbairn (2005): *Writing Your Abstract: A Guide for Would-Be-Conference Presenters*. Imagine a whole book about writing conference abstracts! Here is one book that can be read in one sitting, has numerous examples and writing exercises, and offers a five-minute daily writing workout to stimulate the reluctant writer.

Lipton (1998): *The Science Editor's Soapbox: An Aid for Writers of Scientific and Technical Reports*. This self-published guide from a former editor for the American Society for Horticultural Science's *HortScience* presents a collection of his essays on effective scientific writing. This book is applicable to other scientists, not only those in horticultural science.

Wilkinson (1991): *The Scientists' Handbook for Writing Papers and Dissertations* presents a thorough, academic analysis of the different sections of a scientific manuscript, drawing examples from numerous sciences.

Williams (2004): *Sin Boldly! Dr. Dave's Guide to Writing the College Paper* is aimed more at writing effective essays in college courses, not necessarily scientific papers. It is a non–politically correct, entertaining read. Although I would argue with some of his admonitions, his points are presented clearly.

RECOMMENDED READING ON ORAL PRESENTATIONS

Aarabi (2007): *The Art of Lecturing* is an exceptional resource for those who give university lectures, but the lessons also apply to giving presentations. One of the strengths of this book is the discussion of how the audience receives and processes information.

Alley (2003): *The Craft of Scientific Presentations* provides much insight into scientific presentations through ten critical errors that many speakers make. This book also includes some of the best published material on dealing with nervousness and questions. What I like most about his book are the examples of good and bad habits drawn from Nobel Prize winners and other less-celebrated scientists.

Altman (2007): *Why Most PowerPoint Presentations Suck* is a light-hearted and readable book that dives a bit deeper into the mechanics of using PowerPoint to enhance your presentations.

Benka (2008): "Who is listening? What do they hear?" The editor-in-chief of *Physics Today* describes his revelation on presentations: "It's the audience, stupid!"

Heath and Heath (2007): *Made to Stick: Why Some Ideas Survive and Others Die* discusses the six factors that make ideas "sticky." Applying these to your science and presentations can help make them more memorable: simplicity, unexpectedness, concreteness, credibility, emotional, and stories.

Kosslyn (2007): *Clear and to the Point: 8 Pyschological Principles for Compelling PowerPoint Presentations* provides a thorough documentation of how the style and structure of our slides determines whether and *how* the audience recognizes and remembers our presentation.

The Oceanography Society (2005): *Scientifically Speaking* is a good all-purpose resource for poster and oral presentations, as well as practical advice about answering questions.

RECOMMENDED READING ON DOING RESEARCH

Booth et al. (2003): *The Craft of Research* has chapters about defining research questions, making good arguments, providing evidence, and writing up your research. This is one of the most accessible books I am aware of that deals with the practical side of doing research.

Valiela (2001): *Doing Science: Design, Analysis, and Communication of Scientific Research* discusses, among other aspects of scientific research, the design of research studies, with particular emphasis on proper statistical analyses.

Weston (2009): *A Rulebook for Arguments* delves into 45 rules for constructing effective arguments. Many of these rules are discussed in various places throughout this book, but, here, they are collected in a short readable 88-page guide.

PUBLICATION AND ITS DISCONTENTS

American Geophysical Union (2006): "Guidelines to Publication of Geophysical Research" defines ethical standards for editors, authors, and reviewers. It is available online at www.agu.org/pubs/pubs_guidelines.html.

Batchelor (1981): "Preoccupations of a journal editor." Twenty-five years after founding and editing *Journal of Fluid Mechanics*, Batchelor describes his experiences being a journal editor in wonderfully eloquent prose that is to be envied by all. He discusses such wide-ranging issues as the scope of journals, communicating science through publications, the statistics of rejection, the review process and the role of reviewers, and the future of journals. In this last section, he presages the current way that manuscripts are stored and distributed electronically. Clearly, a forward-thinking scientist!

Benos et al. (2007): "The ups and downs of peer review" discusses the modern history of peer review (since the 1970s) and its merits and demerits.

Editorial Board of the *Electronic Journal of Severe Storms Meteorology* (2006): *Guide for Authors, Reviewers, and Editors*. This document provides further insight into how editors craft decisions and is available online at www.ejssm.org.

Errico (2000): "On the lack of accountability in meteorological research." Frustrated with what he perceived to be the failure of three means of scientific evaluation (competition of proposals for grants, questioning in public presentations, and peer review of publications), Errico provides his observations and recommendations, calling for community-wide action.

Geerts (1999): "Trends in atmospheric science journals: A reader's perspective." Geerts created simple measures based on abstracts and the conclusion sections of articles in several atmospheric science journals through time, showing that articles are getting longer and less reader-friendly.

Goudsmit (1969): "What happened to my paper?" Although some aspects are outdated, Goudsmit describes what life is like in the editorial office of *The Physical Review* in 1969. Back then, editors for *The Physical Review* handled 1200 articles a year, passing decisions on most without peer review. By comparison, editors at *Monthly Weather Review* today each oversee about 50–80 articles a year and nearly all go out for peer review.

Hames (2007): *Peer Review and Manuscript Management in Scientific Journals.* Although many books are aimed at improving scientific writing, few actually describe the details of the publication process. This book is aimed at editors, but provides an incredible behind-the-curtain look on the peer-review process.

Jorgensen et al. (2007): "The evolving publication process of the AMS" provides an explanation of how the current publication process works at the AMS.

Wilson (2002): "Responsible authorship and peer review." This article provides the most succinct discussion of authorship, the discontents with peer review, and guidelines for reviewers.

Eugene Garfield's Web site has extensive references and links to papers on citation indices and their use. Available online at garfield.library.upenn.edu.

Of course, always follow the specific advice and guidelines for peer review for your target journal in the Instructions to Authors.

ILLUSTRATIONS AND GRAPHING

If you liked the conceptual model in Fig. 11.4, more examples of effective illustrations can be found in any of the beautifully illustrated books by Edward Tufte: *Envisioning Information* (1990), *Visual Explanations* (1997), and *The Visual Display of Quantitative Information* (2001). His books draw from all fields of science and communication to illustrate concepts behind effective communication through figures. Chapter 6 of *The Visual Display of Quantitative Information* admonishes the reader to maximize the data-to-ink ratio, and in doing so, redesigns many types of common graphs. Some people may argue his purist approach goes too far, but the text is worth reading to appreciate minimalism and effectiveness in scientific graphics.

Another author whose books you may find useful is William Cleveland: *Visualizing Data* (1993) and *The Elements of Graphing Data* (1994). As one reviewer on amazon. com writes, "Tufte shows you why it's important to do graphs well. Cleveland shows you *how.*" *Elements* goes into much more detail about the construction of figures than this chapter does.

Color schemes appropriate for scientific research are discussed in Light and Bartlein (2004) and on this Web page by the Department of Geography, University of Oregon, geography.uoregon.edu/datagraphics/color_scales.htm.

ETHICS AND MISCONDUCT

Sigma Xi, The Scientific Research Society, publishes two books related to scientific ethics. *Honor in Science* (Sigma Xi 1986) covers such topics as data manipulation, working in a cooperative research environment, whistleblowing, and dealing with unethical situations. It also provides a compelling discussion of how authorship issues can vary among different disciplines and lays out strict terms for authorship in multiple-authored papers.

The Responsible Researcher: Paths and Pitfalls (Sigma Xi 1999) is largely organized by the different stages of a scientific career (e.g., graduate students, postdoctoral fellows, junior faculty, senior faculty), offering general views of ethical challenges at each stage.

Miguel Roig's "Avoiding plagiarism, self-plagiarism, and other questionable writing practices: A guide to ethical writing" contains excellent material on self-plagiarism, which the material in this book was based upon. An especially useful section discusses student–faculty interactions regarding authorship. See the Web site at facpub.stjohns.edu/~roigm/plagiarism/Index.html.

The U.S. Office of Research Integrity maintains a list of resources at ori.dhhs.gov.

Discussions of ongoing scientific plagiarism cases can be found at plagiarism-main.blogspot.com.

Many colleges and universities have Web pages that define plagiarism, provide examples, and describe the consequences for proven violations. Funding agencies also provide similar information to researchers.

ENGLISH AS A SECOND LANGUAGE

Brown et al. (1995): *Technical Writing Guide for Nonnative Speakers of English*, a companion guide to Anderson (1999), provides good information for ESL students and their instructors about cultural differences and how those can be leveraged to improve scientific communication.

Campbell (1995): *ESL Resource Book for Engineers and Scientists* is probably the best published resource on science writing for ESL authors.

Day (1995): *Scientific English: A Guide for Scientists and Other Professionals* provides excellent information on scientific English, useful to both native English–speaking and ESL authors.

COMMONLY MISUSED WORDS AND EXPRESSIONS

Elaborations on some of the entries in Appendix B and other meteorological misuses can be found on these Web pages:

- Pet Peeves by Charles Doswell: www.flame.org/~cdoswell/peeves/Pet_Peeves.html
- Bad Science by Alistair Fraser: www.ems.psu.edu/~fraser/BadScience.html

If you find these alphabetical lists of words and expressions useful to you, the following books have longer lists of other English-language words that are commonly misused. Although such lists originated with Strunk, the longest and most useful list is in Perelman et al. (1998).

- Alley (1996): *The Craft of Scientific Writing*, 3rd ed., Appendix B
- Day (1995): *Scientific English*, 2nd ed., Appendix 2
- Lipton (1998): *The Science Editor's Soapbox*
- Perelman et al. (1998): *The Mayfield Handbook of Technical & Scientific Writing*, Chapter 14
- Schall (2006): *Style for Students*, Chapter 4
- Strunk and White (2000): *The Elements of Style*, 4th ed., Parts III and IV

▶ U.S. Air Force (2004): *The Tongue and Quill*, pp. 78–79

▶ Williams (2004): *Sin Boldly!*, Chapter 13

CAREER GUIDANCE

Feibelman (1993): *A Ph.D. Is Not Enough* focuses on how to develop your scientific career and is the best quick-read available for students.

Fiske (1996): *To Boldly Go: A Practical Career Guide for Scientists*. This guide published by the American Geophysical Union discusses career opportunities for scientists, especially in nontraditional routes. The book also includes specific information about résumés, curriculum vitae, and cover letters.

Schall (2006): *Style for Students*. This manual, now online, provides a lot of information about preparing résumés, curriculum vitae, and cover letters. Available online at www.e-education.psu.edu/styleforstudents.

Schall has published two other books: *Writing Personal Statements and Scholarship Application Essays: A Student Handbook* (www.e-education.psu.edu/writingpersonal statementsonline) and *Writing Recommendation Letters: A Faculty Handbook*, both of which can be ordered online at www.ichapters.com.

COMMUNICATING WITH NONSCIENTISTS

Working with politicians is covered in *Working With Congress: A Practical Guide for Scientists and Engineers* by Wells (1992).

Stephanie Kenitzer, Public Relations Officer of the American Meteorological Society, recommends the following other useful books:

Ward (2008): *Communicating on Climate Change, An Essential Resource for Journalists, Scientists, and Educators* helps lower the barriers to communication between these groups and is published by the Metcalf Institute. More information available online at www.metcalfinstitute.org.

Communicating Uncertainty: Media Coverage of New and Controversial Science edited by Friedman et al. (1999), *Creating a Climate for Change: Communicating Climate Change and Facilitating Social Change* by Moser and Dilling (2007), and *Selling Science: How the Press Covers Science and Technology* by Nelkin (1995).

Books specifically about journalism include *The Elements of Journalism: What Newspeople Should Know and the Public Should Expect* by Kovach and Rosenstiel (2001) and *The Elements of Online Journalism* by Rosales (2006).

To see the best popular science writing, look for the annual series called *The Best American Science Writing*. The 2008 collection was edited by Nasar and Cohen (2008).

TICKLING YOUR FUNNY BONE

Globus and Raible (1994): "Fourteen ways to say nothing with scientific visualization" focuses mainly on producing computer-generated animations and graphics, yet provides a tongue-in-cheek look at "producing pretty pictures while avoiding unnecessary illumination of the data."

Kohn (2003): "How to make a scientific lecture unbearable." See www.improbable.com/news/2003/mar/unbearable_lecture.html.

Oxman et al. 2004: "A field guide to experts" lists the types of people that you might run into at a conference.

More on the types of people who might ask questions at your presentation and the types of people who might encounter at a poster session is beautifully described by the blogger Orac at Respectful Insolence: oracknows.blogspot.com/2005/03/field-guide-to-biomedical-meeting.html, and oracknows.blogspot.com/2005/04/field-guide-to-biomedical-meeting.html.

Plotkin (2004): "How to get your paper rejected." The author knows what he is talking about—the manuscript was rejected by six other journals before being published.

Sand-Jensen (2007): "How to write consistently boring scientific literature." Sand-Jensen lists ten recommendations for writing boring scientific publications.

STYLE AND REFERENCE GUIDES

Authors' Guide is the regularly updated manual for publications at the AMS. Part I deals specifically with the publications of the AMS and the publishing process. Part II addresses manuscript preparation and submission. Several appendices address accepted abbreviations and correct spellings of atmospheric science terms. A companion to the *Authors' Guide* is the *Comprehensive Reference Guide* for citation and reference formats for AMS journals. Both are available on the Authors' Resource Center section of the AMS Web site (www.ametsoc.org).

Unless specifically noted, the AMS follows the press style in *The Chicago Manual of Style*, 15th ed. (The University of Chicago Press 2003), the standard reference for press style in many U.S. publishing houses.

The AMS *Glossary of Meteorology* (Glickman 2000) is useful to check the spelling and definitions for atmospheric science terms. It is also available online at amsglossary.allenpress.com/glossary.

LET'S GO SURFIN' NOW

www.eloquentscience.com offers additional resources to supplement this book, a blog with questions and answers, and a long list of links to other useful Web sites.

owl.english.purdue.edu: The Online Writing Laboratory at Purdue University is one of the most comprehensive online resources for writing. My favorite pages are transitional devices, writer's block, and ESL resources.

www.languageisavirus.com has numerous techniques and tools for inspiring the creative process in writers. Although this site caters to literary writing and poetry, some of the techniques discussed here can be used to open the floodgates in your scientific writing, as well.

www.usingenglish.com/articles is a quality source of information for English grammar.

www.ucar.edu/commsci/esl.html: The University Corporation for Atmospheric Research has a Web site for the Communicating Science group with links to many resources for ESL authors.

en.wikipedia.org/wiki/Category:Free_plotting_software is a clearinghouse for free plotting packages for your computer that may help enhance the quality of your figures.

ams.allenpress.com: The AMS Journals Online page offers online access to 10 different AMS journals, including *Monthly Weather Review* back to Volume 1 in 1873.

www.doaj.org: The Directory of Open Access Journals lists over 20 open-access journals in atmospheric science and climatology.

publications.copernicus.org: Copernicus offers open-access journals for atmospheric sciences and geophysics.

OTHER VALUABLE READING

Atkinson and Mayer (2004): "Five ways to reduce PowerPoint overload" is a concise article on how to improve your electronic presentations.

Blanchard (1974): "References and unreferences" admonishes authors not to blindly copy references from another paper and provides a disheartening, if humorous, instance of why verifying the entries in the reference list is absolutely essential.

Boote and Beile (2005): "Scholars before researchers: On the centrality of the dissertation literature review in research preparation" provides a thorough and sophisticated discussion of the literature review, its necessity, and the criteria of a high-quality literature review.

Flower and Hayes (1977): "Problem-solving strategies and the writing process" provides a detailed discussion of brainstorming as a problem-solving strategy for writing.

Geerts (1999): "Trends in atmospheric science journals: A reader's perspective" focuses on the abstract and conclusion and the importance of clarity and content within these parts of the manuscript.

Hamill (2007): "Toward making the AMS carbon neutral: Offsetting the impacts of flying to conferences" motivates conference-goers to consider their impact on the environment.

King (2000): *On Writing: A Memoir of the Craft*. Novelist Stephen King provides an entirely different perspective on writing in the second half of this autobiography.

Schultz et al. (2007a): "Factors affecting the increasing costs of AMS conferences" explains why the financial costs of attending scientific meetings are so high and rising at several times the rate of inflation.

Shermer (2002): *Why People Believe Weird Things* provides a list of 25 problems in logical thinking that allow people to make false arguments, believe in pseudoscience, and maintain outdated beliefs in his Chapter 3. If you are looking to find possible flaws in your or others' arguments, this material might be useful to you.

Stohl (2008) explores the surprisingly large carbon footprint of scientists attending conferences.

Sun and Zhou (2002): "English versions of Chinese authors' names in biomedical journals: Observations and recommendations." Non-Chinese authors may be curious about how the names of Chinese authors should be cited. Such information and a proposal for Chinese authors to deal with the inconsistencies can be found this article.

REFERENCES

Aarabi, P., 2007: *The Art of Lecturing: A Practical Guide to Successful University Lectures and Business Presentations.* Cambridge University Press, 157 pp.

Ahmed, S. M., C. A. Maurana, J. A. Engle, D. E. Uddin, and K. D. Glaus, 1997: A method for assigning authorship in multiauthored publications. *Fam. Med.,* **29** (1), 42–44.

Alley, M., 1996: *The Craft of Scientific Writing.* 3rd ed. Springer, 282 pp.

Alley, M., 2000: *The Craft of Editing: A Guide for Managers, Scientists, and Engineers.* Springer, 159 pp.

Alley, M., 2003: *The Craft of Scientific Presentations.* Springer, 241 pp. [Resource section available online at: www.writing.engr.psu.edu/csp.html.]

Altman, D. G., 2002: Poor-quality medical research. What can journals do? *J. Amer. Med. Assoc.,* **287,** 2765–2767.

Altman, R., 2007: *Why Most PowerPoint Presentations Suck.* Harvest Books, 271 pp.

American Geophysical Union, 2006: Guidelines to publication of geophysical research. [Available online at www.agu.org/pubs/pubs_guidelines.shtml.]

American Meteorological Society, 2008: Authors' guide. 4th ed. Version 20081103. Amer. Meteor. Soc., 83 pp. [Available online at www.ametsoc.org/PUBS/ Authorsguide/pdf_vs/authguide.pdf.]

American National Standards Institute, Inc. 1979: *American national standard for writing abstracts.* ANSI Z39.14-1979. American National Standards Institute, Inc.

Anderson, P. V., 1999: *Technical Communication: A Reader-Centered Approach.* 4th ed. Harcourt Brace, 643 pp.

Anscombe, F. J., 1973: Graphs in statistical analysis. *Amer. Stat.,* **27** (1), 17–21.

Antelman, K., 2004: Do open-access articles have a greater research impact? *Coll. Res. Libr.,* **65,** 372–382.

Archambault, É., and V. Larivière, 2009: History of the journal impact factor: Contingencies and consequences. *Scientometrics,* **79,** 635–649.

Atkinson, C., 2008: *Beyond Bullet Points.* Microsoft Press, 349 pp.

Atkinson, C., and R. E. Mayer, 2004: Five ways to reduce PowerPoint overload. [Available online at www.sociablemedia.com/PDF/atkinson_mayer_powerpoint_4_23_04.pdf.]

Baibich, M. N., and Coauthors, 1988: Giant magnetoresistance of (001)Fe/(001)Cr magnetic superlattices. *Phys. Rev. Lett.,* **61,** 2472–2475.

Balaram, P., 2002: Editorial discretion and indiscretion. *Curr. Sci.,* **83,** 101–102.

Bals-Elsholz, T. M., E. H. Atallah, L. F. Bosart, T. A. Wasula, M. J. Cempa, and A. R. Lupo, 2001: The wintertime Southern Hemisphere split jet: Structure, variability, and evolution. *J. Climate,* **14,** 4191–4215.

Banacos, P. C., and D. M. Schultz, 2005: The use of moisture flux convergence in forecasting convective initiation: Historical and operational perspectives. *Wea. Forecasting,* **20,** 351–366.

Barnes, L. R., D. M. Schultz, E. C. Gruntfest, M. H. Hayden, and C. Benight, 2009: Corrigendum: False alarm rate vs false alarm ratio? *Wea. Forecasting,* **24,** 1140–1147.

Batchelor, G. K., 1981: Preoccupations of a journal editor. *J. Fluid Mech.,* **106,** 1–25.

Beck, M. E., Jr., 2003: Anonymous reviews: Self-serving, counterproductive, and unacceptable. *Eos, Trans. Amer. Geophys. Union,* **84,** 249.

Bekelman, J. E., Y. Li, and C. P. Gross, 2003: Scope and impact of financial conflicts of interest in biomedical research: A systematic review. *J. Amer. Med. Assoc.,* **289,** 454–465.

Benka, S. G., 2008: Who is listening? What do they hear? *Phys. Today,* **61,** 49–53.

Benos, D. J., K. L. Kirk, and J. E. Hall, 2003: How to review a paper. *Adv. Physiol. Educ.,* **27,** 47–52.

Benos, D. J., and Coauthors, 2007: The ups and downs of peer review. *Adv. Physiol. Educ.,* **31,** 145–152.

Bhatia, C. R., 2002: Transparency in editorial discretion. *Curr. Sci.,* **83,** 927.

Binasch, G., P. Grünberg, F. Saurenbach, and W. Zinn, 1989: Enhanced magnetoresistance in layered magnetic structures with antiferromagnetic interlayer exchange. *Phys. Rev. B,* **39,** 4828–4830.

Bjerknes, J., 1919: On the structure of moving cyclones. *Geofys. Publ.,* **1** (2), 1–8.

Blanchard, D. C., 1974: References and unreferences. *Science,* **185,** 1003.

Boote, D. N., and P. Beile, 2005: Scholars before researchers: On the centrality of the dissertation literature review in research preparation. *Educ. Researcher,* **34** (6), 3–15.

Booth, W. C., G. G. Colomb, and J. M. Williams, 2003: *The Craft of Research.* 2nd ed. University of Chicago Press, 329 pp.

Bosart, L. F., 1983: Analysis of a California Catalina eddy event. *Mon. Wea. Rev.,* **111,** 1619–1633.

Brown, R. A., 1992: Initiation and evolution of updraft rotation within an incipient supercell thunderstorm. *J. Atmos. Sci.,* **49,** 1997–2014.

Brown, R. M., H. Y. Kim, and R. L. Damron, 1995: *Technical Writing Guide for Non-Native Speakers of English.* Harcourt Brace, 98 pp.

Brownstein, K. R., 1999: Extremalyze über alles. *Ann. Improb. Res.,* **5** (2), 3.

Brumfiel, G., 2007: Turkish physicists face accusations of plagiarism. *Nature,* **449,** 8.

Bryan, G. H., and J. M. Fritsch, 2000: Moist absolute instability: The sixth static stability state. *Bull. Amer. Meteor. Soc.,* **81,** 1207–1230.

Budd, J. M., M. E. Sievert, and T. R. Schultz, 1998: Phenomena of retraction: Reasons for retraction and citations to the publications. *J. Amer. Med. Assoc.*, **280,** 296–297.

Campanario, J. M., 1998: Peer review for journals as it stands today—Part 2. *Sci. Comm.*, **19,** 277–306.

Campbell, E., 1995: *ESL Resource Book for Engineers and Scientists.* Wiley, 336 pp.

Campbell, P., 2008: Escape from the impact factor. *Ethics Sci. Environ. Polit.*, **8,** 5–7.

Chattopadhyay, R., A. K. Sahai, and B. N. Goswami, 2008: Objective identification of nonlinear convectively coupled phases of monsoon intraseasonal oscillation: Implications for prediction. *J. Atmos. Sci.*, **65,** 1549–1569.

Clark A. J., C. J. Schaffer, W. A. Gallus Jr., and K. Johnson-O'Mara, 2009: Climatology of storm reports relative to upper-level jet streaks. *Wea. Forecasting,* **24,** 1032–1051.

Clark, H. H., and W. G. Chase, 1972: On the process of comparing sentences against pictures. *Cogn. Psychol.*, **3,** 472–517.

Clayton, H. H., 1911: A study of clouds with data from kites. *Ann. Astron. Observatory Harvard Coll.*, **68,** 170–192.

Cleveland, W. S., 1984: Graphs in scientific publications. *Amer. Stat.*, **38,** 261–269.

Cleveland, W. S., 1993: *Visualizing Data.* Hobart Press, 360 pp.

Cleveland, W. S., 1994: *The Elements of Graphing Data.* Hobart Press, 297 pp.

Cook, C. K., 1986: *Line by Line: How to Edit Your Own Writing.* Houghton Mifflin, 219 pp.

Corfidi, S. F., S. J. Corfidi, and D. M. Schultz, 2008: Castellanus and elevated convection: Ambiguities, significance, and questions. *Wea. Forecasting,* **23,** 1280–1303.

Cronin, B., 2005: A hundred million acts of whimsy? *Curr. Sci.*, **89,** 1505–1509.

Curtis, R., 1996: *How to Be Your Own Literary Agent.* Houghton Mifflin, 285 pp.

Davies, H. C., 1997: Emergence of the mainstream cyclogenesis theories. *Meteor. Z.*, **6,** 261–274.

Davies-Jones, R., C. A. Doswell III, and H. E. Brooks, 1994: Comments on "Initiation and evolution of updraft rotation within an incipient supercell thunderstorm." *J. Atmos. Sci.*, **51,** 326–331.

Day, R. A., 1995: *Scientific English: A Guide for Scientists and Other Professionals.* 2nd ed. Oryx Press, 148 pp.

Day, R. A., and B. Gastel, 2006: *How to Write and Publish a Scientific Paper.* 6th ed. Cambridge University Press, 302 pp.

Demuth, J. L., E. Gruntfest, R. E. Morss, S. Drobot, and J. K. Lazo, 2007: WAS*IS: Building a community for integrating meteorology and social science. *Bull. Amer. Meteor. Soc.*, **88,** 1729–1737.

de Solla Price, D. J., 1965: Networks of scientific papers. *Science,* **149,** 510–515.

Devine, E. B., J. Beney, and L. A. Bero, 2005: Equity, accountability, transparency: Implementation of the contributorship concept in a multi-site study. *Amer. J. Pharm. Educ.*, **69,** 455–459.

Dixon, B., 2000: How not to build bridges. *Amer. Soc. Microbiol. News.* [Available online at newsarchive.asm.org/sep00/animalcule.asp.]

Dong, Y. R., 1998: Non-native graduate students' thesis/dissertation writing in science: Self-reports by students and their advisors from two U.S. institutions. *Engl. Specific Purp.,* **17,** 369–390.

Doswell, C. A. III, 1991: Comments on "Mesoscale convective patterns of the southern High Plains." *Bull. Amer. Meteor. Soc.,* **72,** 389–390.

Doswell, C. A. III, 2007: Small sample size and data quality issues illustrated using tornado occurrence data. *Electron. J. Severe Storms Meteor.,* **2** (5), 1–16.

Doswell, C. A. III, and F. Caracena, 1988: Derivative estimation from marginally sampled vector point functions. *J. Atmos. Sci.,* **45,** 242–253.

Doswell, C. A. III, and L. F. Bosart, 2001: Extratropical synoptic-scale processes and severe convection. *Severe Convective Storms, Meteor. Monogr.,* No. 50, Amer. Meteor. Soc., 27–69.

Doswell, C. A. III, and D. M. Schultz, 2006: On the use of indices and parameters in forecasting severe storms. *Electron. J. Severe Storms Meteor.,* **1** (3), 1–14.

Doswell, C. A. III, H. E. Brooks, and R. A. Maddox, 1996: Flash flood forecasting: An ingredients-based methodology. *Wea. Forecasting,* **11,** 560–581.

Doswell C. A. III, D. V. Baker, and C. A. Liles, 2002: Recognition of negative mesoscale factors for severe-weather potential: A case study. *Wea. Forecasting,* **17,** 937–954.

Drummond, C. W. E., and D. S. Reeves, 2005: Reduced time to publication and increased rejection rate. *J. Antimicrob. Chemother.,* **55,** 815–816.

Durran, D. R., 2000a: Small-amplitude coastally trapped disturbances and the reduced-gravity shallow-water approximation. *Quart. J. Roy. Meteor. Soc.,* **126,** 2671–2689.

Durran, D. R., 2000b: Comments on "The differentiation between grid spacing and resolution and their application to numerical modeling." *Bull. Amer. Meteor. Soc.,* **81,** 2478.

Ebel, H. F., C. Bliefert, and W. E. Russey, 2004: *The Art of Scientific Writing.* 2nd ed. Wiley-VCH, 595 pp.

Editorial Board of the *Electronic Journal of Severe Storms Meteorology,* 2006: Guide for authors, reviewers, and editors. *Electronic Journal of Severe Storms Meteorology,* 15 pp. [Available online at www.ejssm.org/EJSSM-Guide.pdf.]

Einstein, A., and L. Infeld, 1938: *The Evolution of Physics.* Cambridge University Press, 319 pp.

Efron, B., and R. J. Tibshirani, 1993: *An Introduction to the Bootstrap.* Chapman and Hall, 436 pp.

Errami, M., J. M. Hicks, W. Fisher, D. Trusty, J. D. Wren, T. C. Long, and H. R. Garner, 2008: Déjà vu—A study of duplicate citations in Medline. *Bioinformatics,* **24,** 243–249.

Errico, R. M., 2000: On the lack of accountability in meteorological research. *Bull. Amer. Meteor. Soc.,* **81,** 1333–1337.

Eysenbach, G., 2006: Citation advantage of open access articles. *PLoS Biol.,* **4,** e157. doi:10.1371/journal.pbio.0040157.

Fairbairn, G., and S. Fairbairn, 2005: *Writing Your Abstract: A Guide for Would-Be-Conference Presenters.* APS Publishing, 128 pp.

Feibelman, P. J., 1993: *A Ph.D. Is Not Enough! A Guide to Survival in Science.* Addison-Wesley, 109 pp.

Fiske, P. S., 1996: *To Boldly Go. A Practical Career Guide for Scientists.* American Geophysical Union, 188 pp.

Flower, L. S., and J. R. Hayes, 1977: Problem-solving strategies and the writing process. *Coll. Engl.,* **39,** 449–461.

Frey, B. S., 2003: Publishing as prostitution?—Choosing between one's own ideas and academic success. *Public Choice,* **116,** 205–223.

Friedman, R. M., 1989: *Appropriating the Weather: Vilhelm Bjerknes and the Construction of a Modern Meteorology.* Cornell University Press, 251 pp.

Friedman, S. M., S. Dunwoody, and C. L. Rogers, Eds., 1999: *Communicating Uncertainty: Media Coverage of New and Controversial Science.* Lawrence Erlbaum, 296 pp.

Fulda, J. S., 2006: What happens when the author does not provide an abstract. *J. Scholarly Publishing,* **37,** 136–144.

Galway, J. G., 1989: The evolution of severe thunderstorm criteria within the Weather Service. *Wea. Forecasting,* **4,** 585–592.

Gans, J. S., and G. B. Shepherd, 1994: How are the mighty fallen: Rejected classic articles by leading economists. *J. Econ. Perspect.,* **8,** 165–179.

Garfield, E., 1955: Citation indexes for science: A new dimension in documentation through association of ideas. *Science,* **122,** 108–111.

Garfield, E., 1972: Citation analysis as a tool in journal evaluation. *Science,* **178,** 471–479.

Garfield, E., 2005: The agony and the ecstasy—The history and meaning of the journal impact factor. *Int. Congress on Peer Review and Biomedical Publication,* Chicago, IL, American Medical Association, 22 pp. [Available online at www.garfield.library.upenn.edu/papers/jifchicago2005.pdf.]

Garfield, E, 2006: The history and meaning of the journal impact factor. *J. Amer. Med. Assoc.,* **295,** 90–93.

Geerts, B., 1999: Trends in atmospheric science journals: A reader's perspective. *Bull. Amer. Meteor. Soc.,* **80,** 639–651.

Glickman, T. S., Ed., 2000: *Glossary of Meteorology.* 2nd ed., Amer. Meteor. Soc., 855 pp.

Globus, A., and E. Raible, 1994: Fourteen ways to say nothing with scientific visualization. *IEEE Comput.,* **27** (7), 86–88.

Gopen, G. D., and J. A. Swan, 1990: The science of scientific writing. *Amer. Sci.,* **78,** 550–558. [Available online at www.americanscientist.org/issues/feature/the-science-of-scientific-writing/1.]

Gorin, J. S., 2005: Manipulating processing difficulty of reading comprehension questions: The feasibility of verbal item generation. *J. Educ. Meas.,* **42,** 351–373.

Goudsmit, S. A., 1969: What happened to my paper? *Phys. Today,* **22** (5), 23–25.

Grasso, L. D., 2000a: The differentiation between grid spacing and resolution and their application to numerical modeling. *Bull. Amer. Meteor. Soc.,* **81,** 579–580.

Grasso, L. D., 2000b: Reply. *Bull. Amer. Meteor. Soc.,* **81,** 2479.

Hajjem, C., S. Harnad, and Y. Gingras, 2005: Ten-year cross-disciplinary comparison of the growth of open access and how it increases research citation impact.

IEEE Data Eng. Bull., **28** (4), 39–47. [Available online eprints.ecs.soton. ac.uk/12906/.]

Hales, J. E., Jr., 1988: Improving the watch/warning program through use of significant event data. Preprints, *15th Conf. on Severe Local Storms,* Baltimore, MD, Amer. Meteor. Soc., 165–168.

Hamblyn, R., 2001: *The Invention of Clouds: How an Amateur Meteorologist Forged the Language of the Skies.* Farrar, Straus and Giroux, 403 pp.

Hames, I., 2007: *Peer Review and Manuscript Management in Scientific Journals.* Blackwell, 293 pp.

Hamill, T. M., 1999: Hypothesis tests for evaluating numerical precipitation forecasts. *Wea. Forecasting,* **14,** 155–167.

Hamill, T. M., 2007: Toward making the AMS carbon neutral: Offsetting the impacts of flying to conferences. *Bull. Amer. Meteor. Soc.,* **88,** 1816–1819.

Hanna, J. W., D. M. Schultz, and A. R. Irving, 2008: Cloud-top temperatures for precipitating winter clouds. *J. Appl. Meteor. Climatol.,* **47,** 351–359.

Hargens, L. L., 1988: Scholarly consensus and journal rejection rates. *Amer. Sociol. Rev.,* **53,** 139–151.

Harnad, S., and T. Brody, 2004: Comparing the impact of open access (OA) vs. non-OA articles in the same journals. *D-Lib Mag.,* **10** (6). [Available online at www.dlib. org/dlib/june04/harnad/06harnad.html.]

Hayden, M. H., S. Drobot, S. Radil, C. Benight, E. C. Gruntfest, and L. R. Barnes, 2007: Information sources for flash flood warnings in Denver, CO and Austin, TX. *Environ. Hazards,* **7,** 211–219.

Heath, C., and D. Heath, 2007: *Made to Stick: Why Some Ideas Survive and Others Die.* Random House, 291 pp.

Heinselman, P. L., and D. M. Schultz, 2006: Intraseasonal variability of summer storms over central Arizona during 1997 and 1999. *Wea. Forecasting,* **21,** 559–578.

Henige, D., 2005: Commas, Christians, and editors. *J. Scholarly Publishing,* **36** (2), 58–74.

Hinrichs, G., 1888: Tornadoes and derechos. *Amer. Meteor. J.,* **5,** 306–317, 341–349.

Hobbs, P. V., and A. L. Rangno, 1985: Ice particle concentrations in clouds. *J. Atmos. Sci.,* **42,** 2523–2549.

Hobbs, P. V., J. D. Locatelli, and J. E. Martin, 1990: Cold fronts aloft and the forecasting of precipitation and severe weather east of the Rocky Mountains. *Wea. Forecasting,* **5,** 613–626.

Hobbs, P. V., J. D. Locatelli, and J. E. Martin, 1996: A new conceptual model for cyclones generated in the lee of the Rocky Mountains. *Bull. Amer. Meteor. Soc.,* **77,** 1169–1178.

Holton, J. R., 1992: *An Introduction to Dynamic Meteorology.* 3rd ed. Academic Press, 511 pp.

Hoskins, B. J., and F. P. Bretherton, 1972: Atmospheric frontogenesis models: Mathematical formulation and solution. *J. Atmos. Sci.,* **29,** 11–37.

Houk, V. N., and S. B. Thacker, 1990: The responsibilities of authorship. *Ethics and Policy in Scientific Publication,* CBE Editorial Policy Committee, Eds., Council of Biology Editors, 181–184.

Hovmöller, E., 1949: The trough and ridge diagram. *Tellus,* **1,** 62–66.

Huang, J., H. M. van den Dool, and A. G. Barnston, 1996: Long-lead seasonal temperature prediction using optimal climate normals. *J. Climate,* **9,** 809–817.

International Committee of Medical Journal Editors, 2003: Uniform requirements for manuscripts submitted to biomedical journals. *J. Amer. Osteopath. Assoc.,* **103,** 137–149.

Johns, R. H., and W. D. Hirt, 1987: Derechos: Widespread convectively induced windstorms. *Wea. Forecasting,* **2,** 32–49.

Johns, R. H., and C. A. Doswell III, 1992: Severe local storms forecasting. *Wea. Forecasting,* **7,** 588–612.

Johnson, R. H., and W. H. Schubert, 1989: Publication trends in American Meteorological Society technical journals. *Bull. Amer. Meteor. Soc.,* **70,** 476–479.

Jolliffe, I. T., and D. B. Stephenson, 2003: *Forecast Verification, A Practitioner's Guide in Atmospheric Science.* Wiley, 240 pp.

Jorgensen, D. P., R. M. Rauber, K. F. Heideman, M. E. Fernau, M. A. Friedman, and A. L. Schein, 2007: The evolving publication process of the AMS. *Bull. Amer. Meteor. Soc.,* **88,** 1122–1134.

Journal of Young Investigators, 2005: *Writing Scientific Manuscripts. A Guide for Undergraduates. Journal of Young Investigators,* 44 pp. [Available online at www.jyi.org/resources/rs.php.]

Kalkstein, L. S., G. Tan, and J. A. Skindlov, 1987: An evaluation of three clustering procedures for use in synoptic climatological classification. *J. Climate Appl. Meteor.,* **26,** 717–730.

Kennedy, D., 2008: Confidential review—or not? *Science,* **319,** 1009.

Keyser, D., and L. W. Uccellini, 1987: Regional models: Emerging research tools for synoptic meteorologists. *Bull. Amer. Meteor. Soc.,* **68,** 306–320.

Keyser, D., M. J. Reeder, and R. J. Reed, 1988: A generalization of Petterssen's frontogenesis function and its relation to the forcing of vertical motion. *Mon. Wea. Rev.,* **116,** 762–780.

King, S., 2000: *On Writing: A Memoir of the Craft.* Pocket Books, 297 pp.

Kingsmill, D. E., and N. A. Crook, 2003: An observational study of atmospheric bore formation from colliding density currents. *Mon. Wea. Rev.,* **131,** 2985–3002.

Kinsman, B., 1957: Proper and improper use of statistics in geophysics. *Tellus,* **9,** 408–418.

Knox, J. A., and P. J. Croft, 1997: Storytelling in the meteorology classroom. *Bull. Amer. Meteor. Soc.,* **78,** 897–906.

Kohonen, T., 1990: The self-organizing map. *Proc. IEEE,* **78,** 1464–1480.

Koop, T., and U. Pöschl, 2006: An open, two-stage peer-review journal. *Nature,* doi:10.1038/nature04988.

Kosslyn, S. M., 2007: *Clear and to the Point: 8 Psychological Principles for Compelling PowerPoint Presentations.* Oxford University Press, 222 pp.

Kovach, B., and T. Rosenstiel, 2001: *The Elements of Journalism: What Newspeople Should Know and the Public Should Expect.* Three Rivers Press, 208 pp.

Krichak, S. O., and P. Alpert, 2002: A fractional approach to the factor separation method. *J. Atmos. Sci.,* **59,** 2243–2252.

Kuhn, T. S., 1970: *The Structure of Scientific Revolutions*. 2nd ed. Univ. of Chicago Press, 210 pp.

Kundzewicz, Z. W., and D. Koutsoyiannis, 2005: Editorial—The peer-review system: Prospects and challenges. *Hydrol. Sci.,* **50,** 577–590.

Kuo, Y.-H., S. Low-Nam, and R. J. Reed, 1991: Effects of surface energy fluxes during the early development and rapid intensification stages of seven explosive cyclones in the western Atlantic. *Mon. Wea. Rev.,* **119,** 457–476.

Laprise, R., 1992: The resolution of global spectral models. *Bull. Amer. Meteor. Soc.,* **73,** 1453–1454.

Lawrence, P. A., 2001: Science or alchemy? *Nat. Rev. Genetics,* **2,** 139–142.

Lawrence, S., 2001: Online or invisible? *Nature,* **411,** 521.

Leenaars, A. A., W. G. Bringmann, and W. D. G. Balance, 2006: The effects of positive vs. negative wording on subjects' validity ratings of "true" and "false" feedback statements. *J. Clin. Psychol.,* **34,** 369–370.

Lesser, L. I., C. B. Ebbeling, M. Goozner, D. Wypij, and D. S. Ludwig, 2007: Relationship between funding source and conclusion among nutrition-related scientific articles. *PLoS Med.,* **4** (1), e5, doi:10.1371/journal.pmed.0040005.

Lewis, J. M., 1996: Joseph G. Galway. *Wea. Forecasting,* **11,** 263–268.

Lewis, J. M., 2005: Roots of ensemble forecasting. *Mon. Wea. Rev.,* **133,** 1865–1885.

Li, Z., and Coauthors, 2007: The rapid growth of publications by atmospheric and oceanic scientists of Chinese origin. *Bull. Amer. Meteor. Soc.,* **88,** 846–848.

Light, A., and P. J. Bartlein, 2004: The end of the rainbow? Color schemes for improved data graphics. *Eos, Trans. Amer. Geophys. Union,* **85,** pp. 385, 391.

Lipton, W. J., 1998: *The Science Editor's Soapbox.* 93 pp. [Available from Science Soapbox, P.O. Box 16103, Fresno, CA 93755-6103.]

MacKeen, P. L., H. E. Brooks, and K. L. Elmore, 1999: Radar reflectivity–derived thunderstorm parameters applied to storm longevity forecasting. *Wea. Forecasting,* **14,** 289–295.

Maddox, J., 1990: Does the literature deserve the name? *Nature,* **348,** 191.

Markowski, P. M., and J. R. Stonitsch, 2007: Reply. *Mon. Wea. Rev.,* **135,** 4240–4246.

Martinson, B. C., M. S. Anderson, and R. de Vries, 2005: Scientists behaving badly. *Nature,* **435,** 737–738.

Martius, O., C. Schwierz, and H. C. Davies, 2006: A refined Hovmöller diagram. *Tellus,* **58A,** 221–226.

Martner, B. E., P. J. Neiman, and A. B. White, 2007: Collocated radar and radiosonde observations of a double-brightband melting layer in northern California. *Mon. Wea. Rev.,* **135,** 2016–2024.

Mass, C. F., 2003: IFPS and the future of the National Weather Service. *Wea. Forecasting,* **18,** 75–79.

McIntyre, M. E., 1997: Lucidity and science I: Writing skills and the pattern perception hypothesis. *Interdiscip. Sci. Rev.,* **22,** 199–216.

McManus, P., 2000: *The Deer on a Bicycle: Excursions into the Writing of Humor.* Eastern Washington University Press, 188 pp.

McNulty, R. P., 1978: On upper tropospheric kinematics and severe weather occurrence. *Mon. Wea. Rev.,* **106,** 662–672.

Miller, J. E., 1948: On the concept of frontogenesis. *J. Meteor.,* **5,** 169–171.

Montgomery, S. L., 2003: *The Chicago Guide to Communicating Science.* University of Chicago Press, 228 pp.

Morss, R. E., O. V. Wilhemi, M. W. Downton, and E. Gruntfest, 2005: Flood risk, uncertainty, and scientific information for decision making: Lessons from an interdisciplinary project. *Bull. Amer. Meteor. Soc.,* **86,** 1593–1601.

Moser, S. C., and L. Dilling, Eds., 2007: *Creating a Climate for Change: Communicating Climate Change and Facilitating Social Change.* Cambridge University Press, 527 pp.

Murphy, A. H., 1991: Forecast verification: Its complexity and dimensionality. *Mon. Wea. Rev.,* **119,** 1590–1601.

Nasar, S., and J. Cohen, Eds., 2008: *The Best American Science Writing 2008.* Harper Perennial, 336 pp.

Nature, 2006: Nature's peer review trial. *Nature.* [Available online at www.nature.com/nature/peerreview/debate/nature05535.html.]

Nelkin, D., 1995: *Selling Science: How the Press Covers Science and Technology.* W. H. Freeman, 217 pp.

Novak, D. R., B. A. Colle, and S. E. Yuter, 2008: High-resolution observations and model simulations of the life cycle of an intense mesoscale snowband over the northeastern United States. *Mon. Wea. Rev.,* **136,** 1433–1456.

Oceanography Society, The, 2005: *Scientifically Speaking.* The Oceanography Society, 24 pp. [Available online at www.tos.org/resources/publications/sci_speaking.html.]

Oppenheimer, D. M., 2006: Consequences of erudite vernacular utilized irrespective of necessity: Problems with using long words needlessly. *Appl. Cogn. Psychol.,* **20,** 139–150.

Orville, H. D., 1999: On scientific accountability and professionalism. *Bull. Amer. Meteor. Soc.,* **80,** 1434.

Orwell, G., 1945: Politics and the English language. [Available online at www.k-1.com/Orwell/index.cgi/work/essays/language.html.]

Oxman, A. D., I. Chalmers, and A. Liberati, 2004: A field guide to experts. *Brit. Med. J.,* **329,** 1460–1463.

Pagel, W. J., F. E. Kendall, and H. R. Gibbs, 2002: Self-identified publishing needs of nonnative English-speaking faculty and fellows at an academic medical institution. *Sci. Ed.,* **25,** 111–114.

Perelman, L. C., J. Paradis, and E. Barrett, 1998: *The Mayfield Handbook of Technical & Scientific Writing.* Mayfield Publishing, 508 pp. [Available online at www.mhhe.com/mayfieldpub/tsw/home.htm.]

Petterssen, S., 1936: Contribution to the theory of frontogenesis. *Geofys. Publ.,* **11** (6), 1–27.

Pfeifer, M., and W. A. Gallus, 2007: Intercomparison of simulations using 4 WRF microphysical schemes with dual-polarization data for a German squall line. Preprints, *33rd Conf. Radar Meteorology,* Cairns, Queensland, Australia, Amer. Meteor. Soc., P6B.9. [Available online at ams.confex.com/ams/33Radar/techprogram/paper_123555.htm.]

Pielke, R. A., Sr., 1991: A recommended specific definition of "resolution." *Bull. Amer. Meteor. Soc.,* **72,** 1914.

Pielke, R. A., Sr., 2001: Further comments on "The differentiation between grid spacing and resolution and their application to numerical modeling." *Bull. Amer. Meteor. Soc.,* **82,** 699.

Plotkin, H., 2004: How to get your paper rejected. *Brit. Med. J.,* **329,** 1469.

Podsakoff, P. M., S. B. MacKenzie, J.-Y. Lee, and N. P. Podsakoff, 2003: Common method biases in behavioral research: A critical review of the literature and recommended remedies. *J. Appl. Pyschol.,* **88,** 879–903.

Pöschl, U., 2004: Interactive journal concept for improved scientific publishing and quality assurance. *Learned Publ.,* **17,** 105–113.

Provenzale, J. M., and R. J. Stanley, 2005: A systematic guide to reviewing a manuscript. *Amer. J. Roentgenol.,* **185,** 848–854.

Rasmussen, R. M., J. Vivekanandan, J. Cole, B. Myers, and C. Masters, 1999: The estimation of snowfall rate using visibility. *J. Appl. Meteor.,* **38,** 1542–1563.

Reiter, E. R., 1963: *Jet Stream Meteorology.* University of Chicago Press, 515 pp.

Reynolds, G., 2008: *Presentation Zen: Simple Ideas on Presentation Design and Delivery.* New Riders, 229 pp.

Richter, H., and L. F. Bosart, 2002: The suppression of deep moist convection near the southern Great Plains dryline. *Mon. Wea. Rev.,* **130,** 1665–1691.

Ridker, P. M., and J. Torres, 2006: Reported outcomes in major cardiovascular clinical trials funded by for-profit and not-for-profit organizations: 2000–2005. *J. Amer. Med. Assoc.,* **295,** 2270–2276.

Rinehart, R. E., 2004: *Radar for Meteorologists.* 3rd ed. Rinehart Publications, 428 pp.

Roebber, P. J., S. L. Bruening, D. M. Schultz, and J. V. Cortinas, Jr., 2003: Improving snowfall forecasting by diagnosing snow density. *Wea. Forecasting,* **18,** 264–287.

Roig, M., 2006: Avoiding plagiarism, self-plagiarism, and other questionable writing practices: A guide to ethical writing. [Available online at facpub.stjohns. edu/~roigm/plagiarism/.]

Rosales, R. G., 2006: *The Elements of Online Journalism.* iUniverse, 66 pp.

Rose, S. F., P. V. Hobbs, J. D. Locatelli, and M. T. Stoelinga, 2004: A 10-yr climatology relating the locations of reported tornadoes to the quadrants of upper-level jet streaks. *Wea. Forecasting,* **19,** 301–309.

Rosenthal, E. L., J. L. Masdon, C. Buckman, and M. Hawn, 2003: Duplicate publications in the otolaryngology literature. *Laryngoscope,* **113,** 772–774.

Rosner, J. L., 1990: Reflections of science as a product. *Nature,* **345,** 108.

Rossner, M., and K. M. Yamada, 2004: What's in a picture? The temptation of image manipulation. *J. Cell Biology,* **166,** 11–15.

Roundy, N., and D. Mair, 1982: The composing process of technical writers: A preliminary study. *J. Adv. Compos.,* **3** (1–2), 89–101.

Sand-Jensen, K., 2007: How to write consistently boring scientific literature. *Oikos,* **116,** 723–727.

Sanders, F., 1999: A proposed method of surface map analysis. *Mon. Wea. Rev.,* **127,** 945–955.

Sanders, F., and C. A. Doswell III, 1995: A case for detailed surface analysis. *Bull. Amer. Meteor. Soc.,* **76,** 505–521.

Schall, J., 2006: *Style for Students.* Thomson, 260 pp. [Available online at www.e-education. psu.edu/styleforstudents.]

Scherrer, S. C., C. Appenzeller, and M. A. Liniger, 2005: Temperature trends in Switzerland and Europe: Implications for climate normals. *Int. J. Climatology,* **26,** 565–580.

Schmidt, R. H., 1987: A worksheet for authorship of scientific articles. *Bull. Ecol. Soc. Amer.,* **68,** 8–10.

Schultz, D. M., 2004: Historical research in the atmospheric sciences: The value of literature reviews, libraries, and librarians. *Bull. Amer. Meteor. Soc.,* **85,** 995–999.

Schultz, D. M., 2007: Comments on "Unusually long duration, multiple-Doppler radar observations of a front in a convective boundary layer." *Mon. Wea. Rev.,* **135,** 4237–4239.

Schultz, D. M., 2008: The past, present, and future of *Monthly Weather Review. Mon. Wea. Rev.,* **136,** 3–6.

Schultz, D. M., 2009: Are three heads better than two? How the number of reviewers and editor behavior affect the rejection rate. *Scientometrics,* doi: 10.1007/ s11192-009-0084-0.

Schultz, D. M., 2010: Rejection rates for journals publishing atmospheric science. *Bull. Amer. Meteor. Soc.,* in press.

Schultz, D. M., and C. F. Mass, 1993: The occlusion process in a midlatitude cyclone over land. *Mon. Wea. Rev.,* **121,** 918–940.

Schultz, D. M., and P. N. Schumacher, 1999: The use and misuse of conditional symmetric instability. *Mon. Wea. Rev.,* **127,** 2709–2732; Corrigendum, **128,** 1573.

Schultz, D. M., and W. J. Steenburgh, 1999: The formation of a forward-tilting cold front with multiple cloud bands during Superstorm 1993. *Mon. Wea. Rev.,* **127,** 1108–1124.

Schultz, D. M., and R. J. Trapp, 2003: Nonclassical cold-frontal structure caused by dry subcloud air in northern Utah during the Intermountain Precipitation Experiment (IPEX). *Mon. Wea. Rev.,* **131,** 2222–2246.

Schultz, D. M., and J. A. Knox, 2007: Banded convection caused by frontogenesis in a conditionally, symmetrically, and inertially unstable environment. *Mon. Wea. Rev.,* **135,** 2095–2110.

Schultz, D. M., D. Keyser, and L. F. Bosart, 1998: The effect of large-scale flow on low-level frontal structure and evolution in midlatitude cyclones. *Mon. Wea. Rev.,* **126,** 1767–1791.

Schultz, D. M., P. N. Schumacher, and C. A. Doswell III, 2000: The intricacies of instabilities. *Mon. Wea. Rev.,* **128,** 4143–4148.

Schultz, D. M., J. V. Cortinas Jr., and C. A. Doswell III, 2002: Comments on "An operational ingredients-based methodology for forecasting midlatitude winter season precipitation." *Wea. Forecasting,* **17,** 160–167.

Schultz, D. M., D. S. Arndt, D. J. Stensrud, and J. W. Hanna, 2004: Snowbands during the cold-air outbreak of 23 January 2003. *Mon. Wea. Rev.,* **132,** 827–842.

Schultz, D. M., K. Seitter, L. Bosart, C. Gorski, and C. Iovinella, 2007a: Factors affecting the increasing costs of AMS conferences. *Bull. Amer. Meteor. Soc.,* **88,** 408–417.

Schultz, D. M., S. Mikkonen, A. Laaksonen, and M. B. Richman, 2007b: Weekly precipitation cycles? Lack of evidence from United States surface stations. *Geophys. Res. Lett.,* **34,** L22815, doi:10.1029/2007GL031889.

Scorer, R. S., 2004: The meaningfulness of mathematical theories of atmospheric dispersion. *Meteor. Appl.,* **11,** 363–367.

Sears-Collins, A. L., D. M. Schultz, and R. H. Johns, 2006: Spatial and temporal variability of nonfreezing drizzle in the United States and Canada. *J. Climate,* **19,** 3629–3639; Corrigendum, **21,** 1447–1448.

Seglen, P. O., 1997: Why the impact factor of journals should not be used for evaluating research. *Brit. Med. J.,* **314,** 497–502.

Shapiro, A., 2005: Drag-induced transfer of horizontal momentum between air and raindrops. *J. Atmos. Sci.,* **62,** 2205–2219.

Shaw, W. N., 1911: *Forecasting Weather.* Van Nostrand, 380 pp.

Shepherd, G. B., 1994: *Rejected: Leading Economists Ponder the Publication Process.* Thomas Horton & Daughters, 150 pp.

Shermer, M., 2002: *Why People Believe Weird Things: Pseudoscience, Superstition, and Other Confusions of Our Time.* Holt, 349 pp.

Sherwood, S. C., 2000: On moist instability. *Mon. Wea. Rev.,* **128,** 4139–4142.

Sigma Xi, 1986: *Honor in Science.* 2nd ed. Sigma Xi, 41 pp. [Available from Publications Office, Sigma Xi, The Scientific Research Society, P.O. Box 13975, Research Triangle Park, NC 27709.]

Sigma Xi, 1999: *The Responsible Researcher: Paths and Pitfalls.* Sigma Xi, 64 pp. [Available from Publications Office, Sigma Xi, The Scientific Research Society, P.O. Box 13975, Research Triangle Park, NC 27709.]

Simkin, M. V., and V. P. Roychowdhury, 2003: Read before you cite! *Complex Syst.,* **14,** 269–274.

Skamarock, W. C., 2004: Evaluating mesoscale NWP models using kinetic energy spectra. *Mon. Wea. Rev.,* **132,** 3019–3032.

Smith, A. J., 1990: The task of the referee. *Computer,* **23** (4), 65–71.

Smith, R. K., and M. J. Reeder, 1988: On the movement and low-level structure of cold fronts. *Mon. Wea. Rev.,* **116,** 1927–1944.

Snellman, L., 1982: Impact of AFOS on operational forecasting. Preprints, *Ninth Conf. on Weather Forecasting and Analysis,* Seattle, WA, Amer. Meteor. Soc., 13–16.

Souther, J. W., 1985: What to report. *IEEE Trans. Prof. Commun.,* **28** (3), 5–8.

Spekat, A., and F. Kreienkamp, 2007: Somewhere over the rainbow—Advantages and pitfalls of colourful visualizations in geosciences. *Adv. Sci. Res.,* **1,** 15–21.

Stanitski, D. M., and D. J. Charlevoix, 2008: Who are the student members of the AMS? *Bull. Amer. Meteor. Soc.,* **89,** 892–895.

Stein, U., and P. Alpert, 1993: Factor separation in numerical simulations. *J. Atmos. Sci.,* **50,** 2107–2115.

Stensrud, D. J., 1996: Importance of low-level jets to climate: A review. *J. Climate,* **9,** 1698–1711.

Stensrud, D. J., and H. E. Brooks, 2005: The future of peer review? *Wea. Forecasting*, **20,** 825–826.

Stoelinga, M. T., J. D. Locatelli, and P. V. Hobbs, 2002: Warm occlusions, cold occlusions, and forward-tilting cold fronts. *Bull. Amer. Meteor. Soc.*, **83,** 709–721.

Stohl, A., 2008: The travel-related carbon dioxide emissions of atmospheric researchers. *Atmos. Chem. Phys.*, **8,** 6499–6504.

Strunk, W., Jr., and E. B. White, 2000: *The Elements of Style*. 4th ed. Allyn and Bacon, 105 pp.

Student, 1908: The probable error of a mean. *Biometrika*, **6,** 1–25.

Sun, X.-L., and J. Zhou, 2002: English versions of Chinese authors' names in biomedical journals: Observations and recommendations. *Sci. Ed.*, **25,** 3–4.

Swan, A., 2007: Open access and the progress of science. *Amer. Sci.*, **95,** 198–200. [Available online at eprints.ecs.soton.ac.uk/13860.]

Taylor & Francis, 2006: Statement of retraction. *Int. J. Remote Sens.*, **27,** 3749–3750.

Thrower, P. A., 2007: Writing a scientific paper: I. Titles and abstracts. *Carbon*, **45,** 2143–2144.

Todd, P. A., and R. J. Ladle, 2008: Hidden dangers of a "citation culture." *Ethics Sci. Environ. Polit.*, **8,** 13–16.

Tscharntke, T., M. E. Hochberg, T. A. Rand, V. H. Resh, and J. Krauss, 2007: Author sequence and credit for contributions in multiauthored publications. *PLoS Biol.*, **5** (1), e18.

Tufte, E. R., 1990: *Envisioning Information*. Graphics Press, 126 pp.

Tufte, E. R., 1997: *Visual Explanations*. Graphics Press, 157 pp.

Tufte, E. R., 2001: *The Visual Display of Quantitative Information*. 2d ed. Graphics Press, 197 pp.

Tukey, J. W., 1977: *Exploratory Data Analysis*. Addison-Wesley, 688 pp.

Tuovinen, J., A.-J. Punkka, J. Rauhala, H. Hohti, and D. M. Schultz, 2009: Climatology of severe hail in Finland: 1930–2006. *Mon. Wea. Rev.*, **137,** 2238–2249.

Tweney, R. D. and D. Swart, 1977: Experimental control of reaction times to negative and expletive sentences. *Amer. J. Psychol.*, **90,** 299–308.

Twomey, S., 1974: Pollution and the planetary albedo. *Atmos. Environ.*, **8,** 1251–1256.

U.S. Air Force, 2004: *The Tongue and Quill*. Air Force Handbook AFH 33–337, 376 pp. [Available online at www.e-publishing.af.mil/shared/media/epubs/afh33-337.pdf.]

The University of Chicago Press, 1993: *The Chicago Manual of Style*. 14th ed. The University of Chicago Press, 933 pp.

The University of Chicago Press, 2003: *The Chicago Manual of Style*. 15th ed. The University of Chicago Press, 984 pp.

Valiela, I., 2001: *Doing Science: Design, Analysis, and Communication of Scientific Research*. Oxford University Press, 294 pp.

Vonnegut, B., 1994: The atmospheric electricity paradigm. *Bull. Amer. Meteor. Soc.*, **75,** 53–61.

Wakimoto, R. M., and B. E. Martner, 1992: Observations of a Colorado tornado. Part II: Combined photogrammetric and Doppler radar analysis. *Mon. Wea. Rev.*, **120,** 522–543.

Walters, M. K., 2000: Comments on "The differentiation between grid spacing and resolution and their application to numerical modeling." *Bull. Amer. Meteor. Soc.,* **81,** 2475–2477.

Wang, P.-Y., J. E. Martin, J. D. Locatelli, and P. V. Hobbs, 1995: Structure and evolution of winter cyclones in the central United States and their effects on the distribution of precipitation. Part II: Arctic fronts. *Mon. Wea. Rev.,* **123,** 1328–1344.

Weller, A. C., 2001: *Editorial Peer Review: Its Strengths and Weaknesses.* ASIST Monograph Series, Information Today, Inc., 342 pp.

Wells, W. G., Jr., 1992: *Working With Congress: A Practical Guide for Scientists and Engineers.* 2nd ed. American Association for the Advancement of Science, 148 pp.

Weston, A., 2009: *A Rulebook for Arguments.* 4th ed. Hackett Publishing, 88 pp.

Wilkinson, A. M., 1991: *The Scientists Handbook for Writing Papers and Dissertations.* Prentice Hall, 522 pp.

Wilks, D. S., 2006: *Statistical Methods in the Atmospheric Sciences.* 2nd ed. Academic Press, 627 pp.

Williams, D. R., 2004: *Sin Boldly! Dr. Dave's Guide to Writing the College Paper.* 2nd ed. Basic Books, 226 pp.

Williams, J. M., 2006: *Style: Ten Lessons in Clarity and Grace.* 9th ed. Longman, 304 pp.

Wilson, J. R., 2002: Responsible authorship and peer review. *Sci. Eng. Ethics,* **8,** 155–174.

Wiseman, R., 2008: *Quirkology: The Curious Science of Everyday Lives.* Pan MacMillan, 299 pp.

Wu, G., H. Wang, and D.-L. Zhang, 2004: Editorial statement—Action and policy. *Adv. Atmos. Sci.,* **21,** 382.

Xu, K.-M., and K. A. Emanuel, 1989: Is the tropical atmosphere conditionally unstable? *Mon. Wea. Rev.,* **117,** 1471–1479.

Yilmaz, I., 2007: Plagiarism? No, we're just borrowing better English. *Nature,* **449,** 658.

INDEX

Clausius-Clapeyron relation, 352
clip art, 286
Cohen, J., 389
cold front aloft, 96–97
cold-type occlusions, existence of,
    352
collaboration versus coordination,
    352
Colomb, G. G., 386
colony collapse disorder, 329
commas, 345–346
comment-reply exchanges, 9
communication
    with the public and media,
        329–334
    in the workplace, 323–328
comparisons, 83–85
"conciseness is interpreted as
    intelligence", 172
condensation, 352
conference abstracts, 259–264
conflicts of interest, 186–187
Congratulations!, 203
contextualizing background
    information, 34
continuing education, 335–341
convective initiation, 352
convective temperature, 352–353
Cook, C. K., 383
copy editing, 8–9
copyright, 6
copyright infringement, 189
Corfidi, S., 355
"correlation", 121, 353
corresponding author, 3
cover page, 31
CPR (concision, precision, and
    revision), 158
Criswell, 160
cryptomnesia, 188
culture, low-context versus high-
    context, 192
curricula vita, 324–325
Curtis, R., 204
Czavinszky, Peter, 245

D

Damron, R. L., 388
data and methods, 40–41, 62, 353

dates and times, 353
Day, R. A., 23, 32, 383, 388
Death by PowerPoint, 273
devil's advocate, 61
diagrams. *See* figures
difluence/divergence, 353–354
divergence/convergence, 354
Doswell, Charles, 39, 210, 267, 302,
    339, 351, 353–354, 359–360,
    362, 363–364, 377, 381, 382,
    388
double-blind peer review, 239
duplicate publication, 189, 375
dynamics, 354

E

Ebel, H. F., 384
editing, 157–174
    condensing text through précis,
        160–161
    determining correct length of
        manuscript, 172–173
    example of editing process,
        161–167
    feedback, 170–171
    final edits, 173–174
    need for concision, 171–172
    parallel, 241
    revisions, 157–159
    serial, 241
editors, 5–8, 225–228
Einstein, A., 13, 61
*Electronic Journal of Severe Storms
    Meteorology*, 16, 239, 377, 386
Elements of Style (Strunk and White),
    172
Eloquent Science, *xxi–xxiii*, 11, 53,
    57, 390
em dashes, 348–349
Emanuel, Kerry, *xix–xx*, 60, 266
en dashes, 348
English as a Second Language
    authors, 191–202, 376
    collaborating with other authors,
        199–202
    common weaknesses in
        manuscripts of, 193–194
    and cultural differences, 192
    learning from examples, 195–197

seeking help, 198–199
    translating native language versus
        writing in English, 197
Epictetus, 102
equations, 141–142, 354
Errico, R. M., *xxv*, 368, 386
ESL authors. *See* English as a Second
    Language authors
ethics, scientific, 183–190

F

fabrication of data, 184–185
fabstracts, 259
Fairbairn, G., 385
Fairbairn, S., 385
"fall in love with your own text", 58
false alarm rates versus false alarm
    ratio, 354–355
falsification of data, 184–185
Faustian bargain, 198, 376
Feibelman, P. J., 266, 389
Fernau, M. E., 387
Fert, Albert, 249
Feynman, R., 42
figures, 103–141
    aesthetics, 110–112
    annotation, 113–115
    bar charts, 122–126
    captions for, 137
    color, 115–118
    conceptual models, 133–134
    consistency, 112–113
    design, 108–110
    direct versus indirect citation,
        139–140
    discussing in the text, 137–138
    grayscaling, 115–118
    horizontal maps, 127–130
    Hovmöller diagrams, 131
    instrumentation figures, 133
    line graphs, 118–119
    numbering, 140–141
    from other sources, 134
    oversimplified comparisons,
        138–139
    photographs, 128
    pie charts, 132
    placing in the manuscript, 141
    scatterplots, 119–121

random, 361–362
Rauber, R. M., 387
readership, actual versus potential, 15–16
references, 46, 153–154, 155. *See also* citations
rejection, dealing with, 248–249
rejection rates of journals, 16, 367–368
remote-control clicker, 289, 295
reradiation/reemission, 361–362
Research Experience for Undergraduates, xii
research, why they call it, 40
resolution, 362
resounding banality, 35
resumés, 324–325
reviewers, 5–8, 229–241
    anonymity, 239
    choosing to review, 230–231
    decision to revise or reject, 234–235
    guidelines for reviewing, 231–234
    obligations of, 231
    providing comments to others, 239–241
    writing the review, 235–238
reviews, responding to, 243–252
Reynolds, G., 301, 379, 383
Roebber, P., 52, 319
Roig, Miguel, 388
"role", 94, 371
Romanian Journal of Meteorology, 17
Rosales, R.G., 389
Rotunno, R., 67
royal "we", 77

S

salami-slicing (and self-plagiarism), 189
Saltikoff, E., 260
Samelson, Roger, 245
Samsury, Chris, 326–327
Sand-Jensen, K., 390
Sanders, Fred, 12, 225, 228
sandwich effect, 233, 237, 238
Schall, Joe, 150, 384, 388, 389
Science, 16, 17, 18, 24
scientific meetings, 253–258

scientific papers
    abstract, 32–33
    acknowledgments, 45–46
    alternative structures to the manuscript, 46–47
    appendices, 46
    conclusion, 43–45
    cover page, 31–32
    data and methods, 40–41
    decision to publish, 11–19
    differences between literary and scientific writing, 59–60
    discussion, 42–43
    introduction, 33–37
    keywords, 33
    literature synthesis, 37–40
    making writing more accessible, 60–61
    and nonlinear reading, 30–31
    parts of, 30
    references, 46
    results, 41–42
    sections and subsections, 73
    structure of, 29–47
    submission of, 3–5
    titles of, 21–27
Scorer, R. S., 363
"see", 153
self-discovery poster, 306–308
self-plagiarism, 188–189
sentences, constructing effective, 75–86
    active versus passive voice, 76–80
    comparisons, 83–85
    misplaced modifiers, 85–86
    parallel structure, 82
    rhythm and aesthetics, 86
    subject-verb distance, 80–81
    verb tense, 81–82
severe storms, 362
severe weather, definition of, 362
Shermer, M., 391
short-wave, 362
significance/significant, 362
significant severe weather, definition of, 362
Sjodin, T., 268
slides, 270, 273–279, 286–288, 300–301

slideware, 273, 274, 278–279, 289
Soccio, Mary Grace, xi, *xvii–xviii*
Someone-else et al., 178
sources, citing, 148
Spiderman, 111
Springsteen, Bruce, 340
"state", 362–363
statistical association, 363
Steenburgh, Jim, 167, 229
Stensrud, D. J., 359
Stoelinga, Mark, 96–97
Stohl, A., 391
stream-of-consciousness writing, 53, 55
Strunk, W., Jr., 74, 172, 383, 388
subject-verb distance, 80–81
submission of scientific papers, 3–5
Sun, X.-L., 392
Swan, J. A., 79, 383
syntax of citations, 151–153

T

t test, 363
tables, 134–141
    captions for, 137
    direct versus indirect citation, 139–140
    discussing in the text, 137–138
    numbering, 140–141
    placing in the manuscript, 141
target journal, 3–4, 15–19
Tarp, Keli Pirtle, 334
technical editing, 8–9, 201
temperatures, cold/warm, 363
Template for Reading Critically, 232–233, 237
terminology, scientific, 96
text recycling, 189
"thank you", 297, 301
theory, 363
Thuburn, John, 33
thunderstorm, 363
title
    characteristics of effective title, 21–22
    colonic, 24–25
    examples of, 26–27
    multipart papers, 25–26
    structuring, 22–25

topic sentence, 65
transitional devices, 70–71
trigger, 363–364
TRMM rainfall, 364
Twain, Mark, 12
Two Cultures, *xx*

U

"ummmm", 298
U.S. Air Force, 37, 89, 370, 384, 389
"using", 85
UTC, 364

V

Valiela, I., 139, 386
Vancouver reference system, 144–146
verb tense, 81–82
vertical motion, 364
vocabulary. *See* words, using effectively
Vonnegut, Bernard, 340, 381
Vonnegut, Kurt, 340, 381

vorticity, definition versus equation, 364
vorticity generation by shear, 364–365

W

Wakimoto, Roger, 128
Ward, Bud, 389
warm, puppies not temperatures, 99
Weston, A., 386
White, E. B., 74, 172, 383, 388
Whiteman, C. David, 226
"why", 365
Wilkinson, A. M., 65, 385
Williams, D. R., 385, 389
Williams, J. M., 384, 386
Wilson, J. R., 387
Wiscombe, W., 310, 311, 314
Wizard of Oz, 225
"Woof", 75
word bank, 195

words, using effectively, 87–102
  bias, eliminating, 100–102
  concision, 88–90
  misinterpretations, minimizing, 102
  precision, 90–98
  proper form, 98–99
writer's block, 50–52
writing environment, preparing, 52–53
writing, motivation for, 49–54
writing skills, teaching, 338

Y

"yes men and women", 170
You et al., 178
"you know", 298

Z

Zhang, F., 194
Zhou, J., 392
Zrnić, Dusan, 280